Springer INdAM Series

Volume 8

For further volumes:
http://www.springer.com/series/10283

Vincenzo Ancona • Elisabetta Strickland
Editors

Trends in Contemporary Mathematics

Springer

Editors
Vincenzo Ancona
Istituto Nazionale di Alta Matematica
"Francesco Severi"
Roma
Italy

Elisabetta Strickland
Dipartimento di Matematica
Università degli Studi di Roma
"Tor Vergata"
Roma
Italy

ISSN 2281-518X ISSN 2281-5198 (electronic)
Springer INdAM Series
ISBN 978-3-319-05253-3 ISBN 978-3-319-05254-0 (eBook)
DOI 10.1007/978-3-319-05254-0
Springer Cham Heidelberg New York Dordrecht London

Library of Congress Control Number: 2014948184

Printed on acid-free paper

Springer is part of Springer Science+Business Media (www.springer.com)

Preface

INdAM, the Istituto Nazionale di Alta Matematica, has been a leading Italian mathematics research institute ever since it was founded in 1939. Throughout its existence, one of its principal aims has been to invite leading scientists in order to present their research and to deliver high-level training, while at the same time interacting with the mathematical communities scattered around the country.

An important initiative in accomplishing this goal is the INdAM Day, an event conceived a decade ago by Corrado De Concini, then President of INdAM, with the intention of providing an insight into the state of the art in contemporary mathematics by means of four high-level expository lectures. Since the first INdAM Day was held on 18 June 2004 in Rome, each year speakers have been chosen by INdAM from among leading mathematicians around the world, and various Departments of Mathematics around Italy have hosted the initiative: Naples in 2005, Milan in 2006, Pisa in 2007, Padua in 2008, Turin in 2009, Catania in 2010, L'Aquila in 2011, Genoa in 2012, and Palermo in 2013.

To date, more than 40 mathematicians of international renown have delivered talks covering a wide spectrum of current trends in mathematics. These talks have not only been of obvious scientific interest but have also managed to prove the cultural relevance of mathematics. None of us on the INdAM staff would ever pretend to be capable of emulating Hilbert more than a century ago. As we could not possibly match his breadth of vision, we have simply reached a compromise in focusing on certain topics. These have not always been in areas in which we have extensive personal knowledge, but for whatever reason our selection seems to have repeatedly captured the attention of people over the years. We could never predict which areas of mathematics are likely to be fashionable, but our speakers have certainly succeeded in making predictions about mathematics as a whole.

We have also had some strokes of luck, as when Claire Voisin was awarded the Clay Research Award shortly before delivering her talk in Padua in 2008, or when Cédric Villani was awarded the Fields Medal at the ICM in Hyderabad just 2 months after giving his talk in Catania in 2010.

This volume presents a selection of these talks in order to leave a visible trace of the original efforts of INdAM.

Rome, Italy, 2014

Vincenzo Ancona
Elisabetta Strickland

Acknowledgements

The Editors of this volume would like to thank Dott.ssa Elisabetta Esposito of INdAM, who contributed patiently to the birth of this collection of papers, a task which required special devotion in order to achieve the goals typical of a research institute.

Contents

Chapter 1
Interpolation and Comparison Methods in the Mean Field Spin Glass Model

Francesco Guerra

Abstract We give a short overview of the recent rigorous mathematical methods developed for the study of complex disordered systems, in particular spin glasses in the mean field Sherrington-Kirkpatrick formulation. We show that interpolation methods, and related comparison arguments, are very powerful tools in order to study these models. We consider the problem of the infinite volume limit for the free energy, Then we introduce the Parisi solution for the spin glass, based on the spontaneous breaking of replica symmetry, and characterized by a functional order parameter entering in a variational principle. We show how the validity of the Parisi representation can be rigorously established. Finally, we point out some perspective for future developments.

1.1 Introduction

In a famous paper on Physical Review Letters, more than 30 years ago, David Sherrington and Scott Kirkpatrick introduced a celebrated mean field model for spin glasses [1, 2], then considered to be a "solvable model".

The impact of this model on the theoretical physics research has been impressive. During the three decades after its introduction, hundreds and hundreds of papers have been devoted to the study of its properties, even through numerical methods.

Expanded version of an invited lecture delivered at the INdAM Day, Rome, June 18, 2004.

F. Guerra (✉)
Department of Physics, University of Rome, and INFN, Section of Rome
Piazzale Aldo Moro 5, 00185 Roma, Italy
e-mail: francesco.guerra@roma1.infn.it

V. Ancona and E. Strickland (eds.), *Trends in Contemporary Mathematics*,
Springer INdAM Series 8, DOI 10.1007/978-3-319-05254-0_1,

The relevance of the model surely comes from the fact that it is able to represent successfully, at least at the level of the mean field approximation, some important features of the physical spin glass systems, of great interest for their peculiar properties.

Some dilute magnetic alloys called spin glasses (see [3] and [4] for extensive reviews) are extremely interesting systems from a physical point of view. Their peculiar feature is to exhibit a new magnetic phase, where magnetic moments are frozen into disordered equilibrium orientations, without any long-range order. Moreover, these materials have some very slowly relaxing modes, with consequent memory effects.

The Sherrington-Kirkpatrick (SK) model is a simplified mean field model, intended to capture some basic properties of spin glasses.

There is also an additional very important reason for the relevance of this model, and related ones. In fact, recently it has become progressively clear that disordered systems of the Sherrington-Kirkpatrick type, and their generalizations, seem to play a very important role for theoretical and practical applications to hard optimization problems, as it is shown for example by Marc Mézard, Giorgio Parisi and Riccardo Zecchina in [5].

It is interesting to remark that the original paper was entitled "Solvable Model of a Spin-Glass", while a previous draft, according to what reported by David Sherrington, contained even the stronger denomination "Exactly Solvable". However, it turned out that the very natural solution devised by the authors is valid only at high temperatures, or for large external magnetic fields. At low temperatures, the proposed solution exhibits a nonphysical drawback given by a negative entropy, as properly recognized by the authors in their very first paper.

It took a few years to find an acceptable solution. This was done by Giorgio Parisi in a series of papers, by marking a radical departure from the previous methods. In fact, a very deep method of "spontaneous replica symmetry breaking" was developed. As a consequence the physical content of the theory was encoded in a functional order parameter of new type, and a remarkable structure began to show up for the pure states of the theory, characterized by a kind of hierarchical, ultrametric organization. These very interesting developments, due to Giorgio Parisi, and his coworkers, are explained in a challenging way in the classical book [6]. Part of this structure will be recalled in the following.

It is important to remark that the Parisi solution is presented in the form of an ingenious and clever *Ansatz*. Until a few years ago it was not known whether this *Ansatz* would give the true solution for the model, in the so-called thermodynamic limit, when the size of the system becomes infinite, or it would be only a very good approximation to the true solution.

The general structures offered by the Parisi solution, and their possible generalizations for similar models, exhibit an extremely rich and interesting mathematical content. In a very significant way, Michel Talagrand inserted a strongly suggestive sentence in the title to his book [7]: "Spin glasses: a challenge for mathematicians".

As a matter of fact, the problem of giving a proper mathematical understanding of the spin glass structure is extremely difficult. In this talk, we would like to recall the main features of a very powerful method, yet extremely simple in its very essence, based on comparison and interpolation arguments on families of Gaussian random variables.

The method found its first simple application in [8], where it was shown that the Sherrington-Kirkpatrick replica symmetric approximate solution is a rigorous lower bound for the quenched free energy of the system, uniformly in the size, for any value of the temperature and the external magnetic field. Then, it was possible to reach a long awaited result [9]: the convergence of the free energy density in the thermodynamic limit.

Moreover, still by a generalized interpolation on families of Gaussian random variables, the first mentioned result, on the replica symmetric solution, was extended to give a rigorous proof that the expression given by the Parisi *Ansatz* is also a lower bound for the quenched free energy of the system, uniformly in the size [10]. The method gives not only the bound, but also the explicit form of the correction terms in the form of a sum rule. In a subsequent very important result, Michel Talagrand has been able to dominate these correction terms, showing that they vanish in the thermodynamic limit. This extraordinary achievement was firstly announced in a short note [11], containing only a synthetic sketch of the proof, and then presented with all details in a long paper in Annals of Mathematics [12].

The interpolation method is also at the basis of the far-reaching generalized variational principle proven by Michael Aizenman, Robert Sims and Shannon Starr in [13].

In this lecture, we will concentrate mostly on the main questions connected with the free energy. In particular, we will consider the subadditivity of the quenched free energy with respect to the system size, the existence of the infinite-volume limit, the broken replica symmetry sum rules and bounds, and the Parisi variational principle. Our treatment will be as simple as possible, by relying on the basic structural properties, and by describing methods of presumably very long lasting power.

The organization of the paper is as follows. In Sect. 1.2 we explain the basic features of the mean field spin glass models, by introducing all necessary definitions. In next Sect. 1.3 we give a simple application of the interpolation method to the mean-field spin glass model, by showing the sub-additivity of the quenched free energy with respect to the system size, and the existence of the infinite-volume limit [9].

Section 1.4 is devoted to a description of the main features of the Parisi representation for the free energy and to its rigorous establishment.

Section 1.5 is devoted to some results, which have been obtained after the talk given at INdAM, and to perspectives for further developments.

In conclusion, the author would like to thank the organizers of the first 2004 INdAM Day in Rome, in particular Corrado De Concini, for the kind invitation and exquisite hospitality.

1.2 Basic Definitions for the Mean Field Spin Glass Model

The generic configuration of the mean field spin glass model is defined through Ising spin variables $\sigma_i = \pm 1$, attached to each site $i = 1, 2, \ldots, N$.

But now there is also an external quenched disorder given by the $N(N-1)/2$ independent and identical distributed random variables J_{ij}, defined for each couple of sites. For the sake of simplicity, we assume each J_{ij} to be a centered unit Gaussian with averages $E(J_{ij}) = 0$, $E(J_{ij}^2) = 1$. By quenched disorder we mean that the J have a kind of stochastic external influence on the system, without participating to the thermal equilibrium.

Now the Hamiltonian of the model is given by the mean field expression

$$H_N(\sigma, J) = -\frac{1}{\sqrt{N}} \sum_{(i,j)} J_{ij}\sigma_i\sigma_j. \tag{1.1}$$

Here, the sum runs over all couples of sites. Notice that the term $\sqrt{N}$ is necessary in order to ensure a good thermodynamic behavior to the free energy, extensive in the system size N. For the sake of simplicity, we have considered only the case of zero external field. But the general case, with a magnetic external field, can be treated without any essential additional complication.

For a given inverse temperature β, let us now introduce the disorder-dependent partition function $Z_N(\beta, J)$ and the quenched average of the free energy per site $f_N(\beta)$, according to the definitions

$$Z_N(\beta, J) = \sum_{\sigma_1 \ldots \sigma_N} \exp(-\beta H_N(\sigma, J)), \tag{1.2}$$

$$-\beta f_N(\beta) = N^{-1} E \log Z_N(\beta, J). \tag{1.3}$$

Notice that in (1.3) the average E with respect to the external noise is made *after* the log is taken. This procedure is called quenched averaging. It represents the physical idea that the external noise does not participate in the thermal equilibrium. Only the σ_i variables are thermalized.

For the sake of simplicity, it is also convenient to write the partition function in the following equivalent form. First of all let us introduce a family of centered Gaussian random variables $\mathcal{K}(\sigma)$, indexed by the configurations σ, and characterized by the covariances

$$E\big(\mathcal{K}(\sigma)\mathcal{K}(\sigma')\big) = q^2(\sigma, \sigma'), \tag{1.4}$$

where $q(\sigma, \sigma')$ are the overlaps between two generic configurations, defined by

$$q(\sigma, \sigma') = N^{-1} \sum_i \sigma_i \sigma'_i, \tag{1.5}$$

with the obvious bounds $-1 \leq q(\sigma,\sigma') \leq 1$, and the normalization $q(\sigma,\sigma) = 1$. Then, starting from the definition (1.1), it is immediately seen that the partition function in (1.2) can be also written, by neglecting unessential constant terms, in the form

$$Z_N(\beta,\mathcal{K}) = \sum_{\sigma_1\ldots\sigma_N} \exp(\beta\sqrt{\frac{N}{2}}\mathcal{K}(\sigma)), \tag{1.6}$$

which will be the starting point of our treatment. Here the dependence of the partition function on the random variables $\mathcal{K}$ has been stressed in the notation.

According to the general well established strategy of statistical mechanics [14], firstly we consider the problem of the infinite volume limit.

1.3 The Thermodynamic Limit for the Free Energy

The proof of the convergence of the free energy per site in the thermodynamic limit was a result long awaited since decades. In [9] it was possible to give an unexpected very simple proof. Let us show the argument. Consider a system of size N and two smaller systems of sizes N_1 and N_2 respectively, with $N = N_1 + N_2$. Let us now compare

$$E \log Z_N(\beta,\mathcal{K}) = E \log \sum_{\sigma_1\ldots\sigma_N} \exp(\beta\sqrt{\frac{N}{2}}\mathcal{K}(\sigma)), \tag{1.7}$$

with

$$E \log \sum_{\sigma_1\ldots\sigma_N} \exp(\beta\sqrt{\frac{N_1}{2}}\mathcal{K}_1(\sigma^{(1)})) \exp(\beta\sqrt{\frac{N_2}{2}}\mathcal{K}_2(\sigma^{(2)})) =$$
$$E \log Z_{N_1}(\beta,\mathcal{K}_1) + E \log Z_{N_2}(\beta,\mathcal{K}_2), \tag{1.8}$$

where $\sigma^{(1)}$ are the $(\sigma_i,\ i = 1,\ldots,N_1)$, and $\sigma^{(2)}$ are the $(\sigma_i,\ i = N_1+1,\ldots,N)$. Covariances for $\mathcal{K}_1$ and $\mathcal{K}_2$ are expressed as in (1.4), but now the overlaps are replaced with the partial overlaps of the first and second block, q_1 and q_2 respectively, defined as

$$q_1(\sigma,\sigma') = N_1^{-1} \sum_{i=1}^{N_1} \sigma_i \sigma_i', \tag{1.9}$$

and analogously for the q_2 of the second block.

The key idea now is to build an interpolation scheme, between the large system and the two small systems. This is easily achieved by introducing the interpolation parameter $0 \leq t \leq 1$, and the interpolating auxiliary function $\phi(t)$, defined as

$$\phi(t) = E \log \sum_{\sigma_1 \ldots \sigma_N} \exp(\sqrt{t}\beta\sqrt{\frac{N}{2}}\mathcal{K} + \sqrt{1-t}\beta\sqrt{\frac{N_1}{2}}\mathcal{K}_1 + \sqrt{1-t}\beta\sqrt{\frac{N_2}{2}}\mathcal{K}_2). \tag{1.10}$$

Here, we have realized the families of random variables $\mathcal{K}, \mathcal{K}_1, \mathcal{K}_2$ as independent on the same probability space. The interpolation through the $\sqrt{t}$ and $\sqrt{1-t}$ assures a linear interpolation between the respective covariances. Obviously, we have

$$\phi(1) = E \log Z_N(\beta, \mathcal{K}),$$

while

$$\phi(0) = E \log Z_{N_1}(\beta, \mathcal{K}_1) + E \log Z_{N_2}(\beta, \mathcal{K}_2).$$

Now it is easy to calculate directly the t derivative of ϕ (see for example [15]), with the result

$$\frac{d}{dt}\phi(t) = \frac{\beta^2}{4}\frac{N_1 N_2}{N}\langle(q_1 - q_2)^2\rangle_t, \tag{1.11}$$

where $\langle\rangle_t$ is a quite complicated, but explicitly given, t dependent probability measure on the random variables (q_1, q_2) [15]. In this derivation we have exploited the simple connection between the global overlap and the block overlaps

$$Nq = N_1 q_1 + N_2 q_2. \tag{1.12}$$

Since in any case the square in (1.11) is positive, by integrating on t and by exploiting the recognized boundary values at $t = 0$ and $t = 1$, we reach the superadditivity property

$$E \log Z_N(\beta, \mathcal{K}) \geq E \log Z_{N_1}(\beta, \mathcal{K}_1) + E \log Z_{N_2}(\beta, \mathcal{K}_2), \tag{1.13}$$

firstly established in [9]. Of course, the corresponding free energies show a subadditive property, because of the minus sign involved in their definition.

From the superaddivity property, through standard methods [14], the existence of the limit follows in the form

$$\lim_{N\to\infty} N^{-1} E \log Z_N(\beta, \mathcal{K}) = \sup_N N^{-1} E \log Z_N(\beta, h, \mathcal{K}) \equiv -\beta f(\beta). \tag{1.14}$$

1.4 Comparison with the Parisi Representation for the Free Energy

We refer to the original paper [16], and to the extensive review given in [6], for the general motivations, and the derivation of the broken replica symmetry *Ansatz*, in the frame of the ingenious replica trick. Here we limit ourselves to a synthetic description of its general structure, independently from the replica trick. The deep motivation for the introduction of the Parisi trial functional is sketched in [17], in the frame of the cavity method (see also [18]).

First of all, let us introduce the convex space $\mathcal{X}$ of the functional order parameters x, as nondecreasing functions of the auxiliary variable q, both x and q taking values on the interval $[0, 1]$, i.e.

$$\mathcal{X} \ni x : [0,1] \ni q \to x(q) \in [0,1]. \tag{1.15}$$

Notice that we call x the function, and $x(q)$ its values. We introduce a metric on $\mathcal{X}$ through the $L^1([0,1], dq)$ norm, where dq is the Lebesgue measure.

For our purposes, we will consider the case of piecewise constant functional order parameters, characterized by an integer K, and two sequences $q_0, q_1, \dots, q_K$, $m_1, m_2, \dots, m_K$ of numbers satisfying

$$0 = q_0 \leq q_1 \leq \dots \leq q_{K-1} \leq q_K = 1,\ \ 0 \leq m_1 \leq m_2 \leq \dots \leq m_K \leq 1, \tag{1.16}$$

such that

$$x(q) = m_1 \text{ for } 0 = q_0 \leq q < q_1,\ \ x(q) = m_2 \text{ for } q_1 \leq q < q_2, \\ \dots, x(q) = m_K \text{ for } q_{K-1} \leq q \leq q_K. \tag{1.17}$$

In the following, we will find it convenient to define also $m_0 \equiv 0$, and $m_{K+1} \equiv 1$. The replica symmetric case of Sherrington and Kirkpatrick corresponds to

$$K = 2,\ q_1 = \bar{q},\ m_1 = 0,\ m_2 = 1. \tag{1.18}$$

Let us now introduce the function f, with values $f(q, y; x, \beta)$, of the variables $q \in [0,1]$, $y \in R$, depending also on the functional order parameter x, and on the inverse temperature β, defined as the solution of the nonlinear antiparabolic equation

$$(\partial_q f)(q,y) + \frac{1}{2}(\partial_y^2 f)(q,y) + \frac{1}{2}x(q)(\partial_y f)^2(q,y) = 0, \tag{1.19}$$

with final condition

$$f(1,y) = \log\cosh(\beta y). \tag{1.20}$$

Here, we have stressed only the dependence of f on q and y.

It is very simple to integrate Eq. (1.19) when x is piecewise constant. In fact, consider $x(q) = m_a$, for $q_{a-1} \leq q \leq q_a$, firstly with $m_a > 0$. Then, it is immediately seen that the correct solution of Eq. (1.19) in this interval, with the right final boundary condition at $q = q_a$, is given by

$$f(q, y) = \frac{1}{m_a} \log \int \exp\big(m_a f(q_a, y + z\sqrt{q_a - q})\big)\, d\mu(z), \tag{1.21}$$

where $d\mu(z)$ is the centered unit Gaussian measure on the real line. On the other hand, if $m_a = 0$, then (1.19) loses the nonlinear part and the solution is given by

$$f(q, y) = \int f(q_a, y + z\sqrt{q_a - q})\, d\mu(z), \tag{1.22}$$

which can be seen also to follow from (1.21) in the limit $m_a \to 0$. Starting from the last interval K, and using (1.21) iteratively on each interval, we easily get the solution of (1.19) and (1.20), in the case of piecewise constant order parameter x, as in (1.17), through a chain of Gaussian integrations.

Now we introduce the following important definitions. The trial auxiliary function, associated to a given mean field spin glass system, as described in Sect. 1.3, depending on the functional order parameter x, is defined as

$$\log 2 + f(0, 0; x, \beta) - \frac{\beta^2}{2} \int_0^1 q\, x(q)\, dq. \tag{1.23}$$

Notice that in this expression the function f appears evaluated at $q = 0$, and $y = 0$.

The Parisi spontaneously broken replica symmetry expression for the free energy is given by the definition

$$-\beta f_P(\beta) \equiv \inf_x \Big(\log 2 + f(0, 0; x, \beta) - \frac{\beta^2}{2} \int_0^1 q\, x(q)\, dq\Big), \tag{1.24}$$

where the infimum is taken with respect to all functional order parameters x.

Notice that the infimum appears here, as compared to the supremum that would appear in a variational principle of the usual entropy type in statistical mechanics. Therefore, Parisi variational principle is really a new structure in statistical mechanics, that deserves careful study in itself.

In [10], by exploiting a suitable interpolation scheme, we have established a rigorous connection between the partition function of the mean field spin glass and the Parisi *Ansatz*. We skip all details and state only the final result, in the form of the sum rule

$$\log 2 + f(0, 0; x, \beta) - \frac{\beta^2}{2} \int_0^1 q\, x(q)\, dq =$$

$$N^{-1} E \log Z_N(\beta, \mathcal{K}) + \frac{\beta^2}{4} \langle (q_{12} - q_a)^2 \rangle, \tag{1.25}$$

where $\langle\rangle$ is an explicitly given but quite complicated measure average over the variables σ, σ', appearing in the two replica overlap q_{12}, and the variable $q_{.}$, taking the values q_a. The sum rule holds for any value of the order parameter x. One of the miracles occurring in the proof of this sum rule is that the second term appearing in the Parisi trial functional here comes for free from the completion of the square in the third term of the sum rule.

In any case, the third term, being the average of a square, is positive. Therefore we have the following important result.

Theorem 1.1. *For all values of the inverse temperature β, and for any functional order parameter x, the following bound holds*

$$N^{-1} E \log Z_N(\beta, \mathcal{K}) \leq \log 2 + f(0,0;x,\beta) - \frac{\beta^2}{2}\int_0^1 q\,x(q)\,dq,$$

uniformly in N. Consequently, we have also

$$N^{-1} E \log Z_N(\beta, \mathcal{K}) \leq \inf_x \Big(\log 2 + f(0,h;x,\beta) - \frac{\beta^2}{2}\int_0^1 q\,x(q)\,dq\Big),$$

uniformly in N.

This result can be understood also in the frame of the generalized variational principle established by Aizenman-Sims-Starr [13], as shown for example in [15], by exploiting the general structure of the Derrida-Ruelle-Parisi probabibility cascades.

Up to this point we have seen how to obtain upper bounds. The problem arises whether we can also get lower bounds, so as to shrink the thermodynamic limit to the value given by the $\inf_x$ in Theorem 1.1. After a short announcement in [11], Michel Talagrand wrote an extended paper [12], where the complete proof of the control of the lower bound is firmly established. We refer to the original paper for the complete details of this remarkable achievement. About the methods, here we only recall that the sum rule in [10], explained above, gives also the corrections to the bounds appearing in Theorem 1.1, albeit in a quite complicated form. Talagrand has been able to establish that these corrections do in fact vanish in the thermodynamic limit. In order to be able to reach this important result it is necessary to prove an extension of the broken replica symmetry bounds of Theorem 1.1 to the case where two replicas of the system are coupled together. This task has not been reached yet in its full generality, but the treatment given by Talagrand is sufficient to prove the vanishing of the correction terms in the infinite volume limit.

In conclusion, we can establish the following conclusive result about the expression of the free energy in the mean field spin glass.

Theorem 1.2. *For the mean field spin glass model we have*

$$\lim_{N\to\infty} N^{-1} E \log Z_N(\beta,\mathcal{K}) = \sup_N N^{-1} E \log Z_N(\beta,\mathcal{K}) \tag{1.26}$$

$$= \inf_x \Big(\log 2 + f(0,0;x,\beta) - \frac{\beta^2}{2} \int_0^1 q\, x(q)\, dq\Big). \tag{1.27}$$

1.5 Further Developments and Outlook

As we have seen, in these last few years there has been an impressive progress in the understanding of the mathematical structure of spin glass models, mainly due to the systematic exploitation of interpolation methods. However many important problems are still open. The most important one is the full understanding of the hierarchical ultrametric organization of the overlap distributions, as appears in Parisi theory, and the decomposition in pure states of the glassy phase, at low temperatures. An important step in this direction have been obtained through the establishment of the so called Ghirlanda-Guerra identities [19]. Based on these, Dmitry Panchenko [20] has been able to prove ultra-metricity of the overlap distribution, a very remarkable achievement of the last years.

Moreover, interpolation and comparison methods have been extended to other important disordered models, such as for example neural networks, bipartite models, multi-species models. Here the difficulty is that the positivity arguments, so essential in the application of the interpolation methods, do not seem to emerge naturally inside the structure of the theory. For recent results see [21–27].

Even for a class of simple mean field diluted ferromagnetic systems, the treatment of the infinite volume limit has not been reached yet, due to the lack of positivity arguments. Only the $\beta \to \infty$ limit is well understood [28].

For extensions to diluted spin glass models we refer for example to [29–31].

Finally, the problem of connecting properties of the short-range model with those arising in the mean field case is still almost completely open. For partial results, and different points of view, see [32–37].

Finally, we mention a pedagogically very useful complete review appeared [38], about the application of the interpolation methods, and the other methods of spin glass theory, to the simple case of the ferromagnetic mean field model.

Acknowledgements We gratefully acknowledge useful conversations with Michael Aizenman, Adriano Barra, Enzo Marinari, Dmitry Panchenko, Giorgio Parisi, and Michel Talagrand.

This work was supported in part by MIUR (Italian Ministry of Instruction, University and Research), and by INFN (Italian National Institute for Nuclear Physics).

References

1. D. Sherrington, S. Kirkpatrick, Solvable model of a spin-glass. Phys. Rev. Lett. **35**, 1792–1796 (1975)
2. S. Kirkpatrick, D. Sherrington, Infinite-ranged models of spin-glasses. Phys. Rev. **B17**, 4384–4403 (1978)
3. P. Young (ed.), *Spin Glasses and Random Fields* (World Scientific, Singapore, 1987)
4. D.L. Stein, Disordered systems: mostly spin glasses, in *Lectures in the Sciences of Complexity*, ed. by D.L. Stein. (Addison-Wesley, New York, 1989)
5. M. Mézard, G. Parisi, R. Zecchina, Analytic and algorithmic solution of random satisfiability problems. Science **297**, 812 (2002)
6. M. Mézard, G. Parisi, M.A. Virasoro, *Spin Glass Theory and Beyond* (World Scientific, Singapore, 1987)
7. M. Talagrand, *Spin Glasses: A Challenge for Mathematicians. Mean Field Models and Cavity Method* (Springer, Berlin, 2003)
8. F. Guerra, Sum rules for the free energy in the mean field spin glass model. Fields Inst. Commun. **30**, 161 (2001)
9. F. Guerra, F.L. Toninelli, The thermodynamic limit in mean field spin glass models. Commun. Math. Phys. **230**, 71–79 (2002)
10. F. Guerra, Broken replica symmetry bounds in the mean field spin glass model. Commun. Math. Phys. **233**, 1–12 (2003)
11. M. Talagrand, The generalized Parisi formula. Comptes Rendus de l'Académie des Sciences, Paris **337**, 111–114 (2003)
12. M. Talagrand, The Parisi formula. Ann. Math. **163**, 221–263 (2006)
13. M. Aizenman, R. Sims, S. Starr, Extended variational principle for the Sherrington-Kirkpatrick spin-glass model. Phys. Rev. **B68**, 214403 (2003)
14. D. Ruelle, *Statistical Mechanics. Rigorous Results* (W.A. Benjamin Inc., New York, 1969)
15. F. Guerra, An introduction to mean field spin glass theory: methods and results, in *Mathematical Statistical Physics*, ed. by A. Bovier et al. (Elsevier, Oxford/Amsterdam, 2006), pp. 243–271
16. G. Parisi, A sequence of approximate solutions to the S-K model for spin glasses. J. Phys. **A13**, L-115 (1980)
17. F. Guerra, Fluctuations and thermodynamic variables in mean field spin glass models, in *Stochastic Processes, Physics and Geometry, II*, ed. by S. Albeverio, U. Cattaneo, D. Merlini (World Scientific, Singapore, 1995)
18. F. Guerra, About the cavity fields in mean field spin glass models, invited lecture at the international congress of mathematical physics, Lisboa, 2003, available on http://arxiv.org/abs/cond-mat/0307673
19. F. Guerra, S. Ghirlanda, General properties of overlap probability distributions in disordered spin systems. Towards Parisi ultrametricity. J. Phys. A-Math. Gen. **31**, 9149–9155 (1998)
20. D. Panchenko, *The Sherrington-Kirkpatrick Model* (Springer, New York, 2013)
21. A. Barra, F. Guerra, About the ergodic regime in the analogical Hopfield neural networks: moments of the partition function. J. Math. Phys. **50**, 125217 (2008)
22. A. Barra, F. Guerra, *Constraints for the order parameters in analogical neural networks*, Percorsi d'Ateneo, S. Vitolo Ed., Salerno (2008)
23. A. Barra, G. Genovese, F. Guerra, The replica symmetric approximation of the analogical neural network. J. Stat. Phys. **140**, 784–796 (2010)
24. A. Barra, G. Genovese, F. Guerra, Equilibrium statistical mechanics of bipartite spin systems. J. Phys. A: Math. Theor. **44**, 245002 (2011)
25. A. Barra, G. Genovese, F. Guerra, D. Tantari, How glassy are neural networks? J. Stat. Mech. **2012**, P07009 (2012)
26. A. Barra, P. Contucci, E. Mingione, D. Tantari, *Multi-species mean-field spin-glasses. Rigorous results*, arXiv:1307.5154

27. D. Panchenko, The free energy in a multi-species Sherrington-Kirkpatrick model. arXiv:1310.6679
28. L. De Sanctis, F. Guerra, Mean field dilute ferromagnet: high temperature and zero temperature behavior. J. Stat. Phys. **132**, 759–785 (2008)
29. S. Franz, M. Leone, Replica bounds for optimization problems and diluted spin systems. J. Stat. Phys. **111**, 535–564 (2003)
30. F. Guerra, F.L. Toninelli, The high temperature region of the Viana-Bray diluted spin glass model. J. Stat. Phys. **115**, 531–555 (2004)
31. D. Panchenko, M. Talagrand, Bounds for diluted mean-field spin glass models. Probab. Theory Relat. Fields **130**, 319–336 (2004)
32. F. Guerra, F.L. Toninelli, Some comments on the connection between disordered long range spin glass models and their mean field version. J. Phys. A: Math. Gen. **36**, 10987–10995 (2003)
33. S. Franz, F.L. Toninelli, The Kac limit for finite-range spin glasses. Phys. Rev. Lett. **92**, 030602 (2004)
34. S. Franz, F.L. Toninelli, Finite-range spin glasses in the Kac limit: free energy and local observables. J. Phys. A: Math. Gen. **37**, 7433 (2004)
35. E. Marinari, G. Parisi, J.J. Ruiz-Lorenzo, Numerical simulations of spin glass systems, in ed. by P. Young *Spin Glasses and Random Fields* (World Scientific, Singapore, 1987), pp. 59–98
36. E. Marinari, G. Parisi, F. Ricci-Tersenghi, J.J. Ruiz-Lorenzo, F. Zuliani, Replica symmetry breaking in short range spin glasses: a review of the theoretical foundations and of the numerical evidence. J. Stat. Phys. **98**, 973–1074 (2000)
37. C.M. Newman, D.L. Stein, Simplicity of state and overlap structure in finite-volume realistic spin glasses. Phys. Rev. E **57**, 1356–1366 (1998)
38. A. Barra, The mean field ising model trough interpolating techniques. J. Stat. Phys. **132**, 787–809 (2008)

Chapter 2
Integrability of Dirac Reduced Bi-Hamiltonian Equations

Alberto De Sole, Victor G. Kac, and Daniele Valeri

Abstract First, we give a brief review of the theory of the Lenard-Magri scheme for a non-local bi-Poisson structure and of the theory of Dirac reduction. These theories are used in the remainder of the paper to prove integrability of three hierarchies of bi-Hamiltonian PDE's, obtained by Dirac reduction from some generalized Drinfeld-Sokolov hierarchies.

2.1 Introduction

It has been demonstrated in a series of papers [1–5] that the framework of Poisson vertex algebras is extremely useful for the theory of Hamiltonian PDE's. For example, the theories of non-local Poisson structures [2], and of the infinite dimensional Dirac reduction [5], have been developed in this framework. Moreover, this languages turned out to be very convenient not only for the development of the general theory, but also for the study of concrete bi-Hamiltonian systems, like the generalized Drinfeld-Sokolov (DS) hierarchies, considered in [3, 4]. In these two papers we studied in more detail three integrable bi-Hamiltonian hierarchies: the homogeneous DS hierarchy, associated to a simple Lie algebra $\mathfrak{g}$, studied already in [6], and the generalized DS hierarchies attached to a minimal and to a short nilpotent

A. De Sole (✉)
Dipartimento di Matematica, Università degli Studi di Roma "La Sapienza", Piazzale Aldo Moro 2, 00185 Rome, Italy
e-mail: desole@mat.uniroma1.it

V.G. Kac
Department of Mathematics, MIT, 77 Massachusetts Avenue, Cambridge, MA 02139, USA
e-mail: kac@math.mit.edu

D. Valeri
SISSA, via Bonomea 265, 34136 Trieste, Italy
e-mail: dvaleri@sissa.it

V. Ancona and E. Strickland (eds.), *Trends in Contemporary Mathematics*, Springer INdAM Series 8, DOI 10.1007/978-3-319-05254-0_2,

element of $\mathfrak{g}$. We also considered the Dirac reductions of the last two hierarchies by elements of conformal weight 1. In the case of a "short" hierarchy we thus obtain Svinolupov's integrable hierarchy [7], constructing thereby (non-local) bi-Poisson structures for them. However, it is not at all clear (and probably false in general) that the equations obtained by Dirac reduction from integrable bi-Hamiltonian equations remain bi-Hamiltonian integrable. We were able to prove this in [5] for the reduced "minimal" hierarchy only in the first non-trivial case of $\mathfrak{g} = \mathfrak{sl}_3$.

In the present paper, using the theory of singular degree of a rational matrix pseudodifferential operator [8], we prove integrability of the reduced "minimal" and "short" hierarchies for arbitrary $\mathfrak{g}$. Furthermore, considering Dirac reduction of the homogeneous DS hierarchy, associated to a fixed regular element s in a Cartan subalgebra $\mathfrak{h}$ of $\mathfrak{g}$, we prove integrability of the following bi-Hamiltonian PDE, for all $a \in \mathfrak{h}$:

$$\frac{de_\alpha}{dt} = \frac{\alpha(a)}{\alpha(s)} e'_\alpha + \sum_{\beta \in \Delta \setminus \{-\alpha\}} \frac{\beta(a)}{\beta(s)} e_{-\beta}[e_\beta, e_\alpha] \ , \quad \alpha \in \Delta \, , \tag{2.1}$$

where Δ is the root system of $\mathfrak{g}$ and $\{e_\alpha\}_{\alpha\in\Delta}$ are root vectors such that $(e_\alpha|e_{-\alpha}) = 1$ with respect to an invariant non-degenerate bilinear form $(\cdot\,|\,\cdot)$ on $\mathfrak{g}$. Equation (2.1) is bi-Hamiltonian with respect to the following two compatible Poisson structures ($\alpha, \beta \in \Delta$):

$$(H_0)_{\alpha,\beta}(\partial) = \delta_{\alpha,-\beta}\beta(s) \, , \tag{2.2}$$

and

$$\begin{aligned} (H_1)_{\alpha,\beta}(\partial) &= [e_\beta, e_\alpha] - (\alpha|\beta) e_\alpha \partial^{-1} \circ e_\beta \quad \text{for } \beta \neq -\alpha \, , \\ (H_1)_{\alpha,-\alpha}(\partial) &= \partial + (\alpha|\alpha) e_\alpha \partial^{-1} \circ e_{-\alpha} \, . \end{aligned} \tag{2.3}$$

The corresponding first two conserved Hamiltonian densities are

$$h_0 = a \ , \quad h_1 = \frac{1}{2} \sum_{\alpha\in\Delta} \frac{\alpha(a)}{\alpha(s)} e_\alpha e_{-\alpha} \, . \tag{2.4}$$

The proof of integrability in all cases is based on the Lenard-Magri scheme of integrability for non-local bi-Poisson structures, developed in [2].

2.2 Non-local Poisson Structures and Hamiltonian Equations

2.2.1 *Evolutionary Vector Fields, Frechet Derivatives and Variational Derivatives*

Let $\mathcal{V}$ be the algebra of differential polynomials in ℓ variables: $\mathcal{V} = \mathbb{F}[u_i^{(n)} \mid i \in I, n \in \mathbb{Z}_+]$, where $I = \{1, \dots, \ell\}$, over a field $\mathbb{F}$ of characteristic zero. (In fact, most of the results hold in the generality of algebras of differential functions, as defined in [2].) It is a differential algebra with derivation defined by $\partial(u_i^{(n)}) = u_i^{(n+1)}$. We also let $\mathcal{K}$ be the field of fractions of $\mathcal{V}$ (it is still a differential algebra). We also denote by $\tilde{\mathcal{K}}$ the *linear closure* of $\mathcal{K}$, which is the smallest differential field extension of $\mathcal{K}$ containing solutions to any linear differential equation with coefficients in $\tilde{\mathcal{K}}$, and whose subfield of constants is $\overline{\mathbb{F}}$, the algebraic closure of $\mathbb{F}$, see e.g. [9].

For $P \in \mathcal{V}^\ell$ we have the associated *evolutionary vector field*

$$X_P = \sum_{i\in I, n\in\mathbb{Z}_+} (\partial^n P_i)\frac{\partial}{\partial u_i^{(n)}} \in \mathrm{Der}(\mathcal{V})\,.$$

This makes $\mathcal{V}^\ell$ into a Lie algebra, with Lie bracket $[X_P, X_Q] = X_{[P,Q]}$, given by

$$[P, Q] = X_P(Q) - X_Q(P) = D_Q(\partial)P - D_P(\partial)Q\,,$$

where $D_P(\partial)$ and $D_Q(\partial)$ denote the Frechet derivatives of $P, Q \in \mathcal{V}^\ell$.

In general, for $\theta = \left(\theta_\alpha\right)_{\alpha=1}^m \in \mathcal{V}^m$, the *Frechet derivative* $D_\theta(\partial) \in \mathrm{Mat}_{m\times\ell}\,\mathcal{V}[\partial]$ is defined by

$$D_\theta(\partial)_{\alpha i} = \sum_{n\in\mathbb{Z}_+} \frac{\partial\theta_\alpha}{\partial u_i^{(n)}}\partial^n\,, \quad \alpha = 1,\dots,m\,,\ i = 1,\dots,\ell\,. \tag{2.5}$$

Its adjoint $D_\theta^*(\partial) \in \mathrm{Mat}_{\ell\times m}\,\mathcal{V}[\partial]$ is then given by

$$D_\theta^*(\partial)_{i\alpha} = \sum_{n\in\mathbb{Z}_+} (-\partial)^n \frac{\partial\theta_\alpha}{\partial u_i^{(n)}}\,, \quad \alpha = 1,\dots,m\,,\ i = 1,\dots,\ell\,.$$

For $f \in \mathcal{V}$ its *variational derivative* is $\frac{\delta f}{\delta u} = \left(\frac{\delta f}{\delta u_i}\right)_{i\in I} \in \mathcal{V}^{\oplus\ell}$, where

$$\frac{\delta f}{\delta u_i} = \sum_{n\in\mathbb{Z}_+} (-\partial)^n \frac{\partial f}{\partial u_i^{(n)}}\,.$$

Given an element $\xi \in \mathcal{V}^{\oplus\ell}$, the equation $\xi = \frac{\delta h}{\delta u}$ can be solved for $h \in \mathcal{V}$ if and only if $D_\xi(\partial)$ is a self-adjoint operator: $D_\xi(\partial) = D_\xi^*(\partial)$ (see e.g. [1]).

2.2.2 Rational Matrix Pseudodifferential Operators

Consider the skewfield $\mathcal{K}((\partial^{-1}))$ of pseudodifferential operators with coefficients in $\mathcal{K}$, and the subalgebra $\mathcal{V}[\partial]$ of differential operators on $\mathcal{V}$.

The algebra $\mathcal{V}(\partial)$ of *rational* pseudodifferential operators consists of pseudodifferential operators $L(\partial) \in \mathcal{V}((\partial^{-1}))$ which admit a fractional decomposition $L(\partial) = A(\partial)B(\partial)^{-1}$, for some $A(\partial), B(\partial) \in \mathcal{V}[\partial]$, $B(\partial) \neq 0$. The algebra of *rational matrix pseudodifferential operators* is, by definition, $\mathrm{Mat}_{\ell\times\ell}\,\mathcal{V}(\partial)$ [9,10].

A matrix differential operator $B(\partial) \in \mathrm{Mat}_{\ell\times\ell}\,\mathcal{V}[\partial]$ is called *non-degenerate* if it is invertible in $\mathrm{Mat}_{\ell\times\ell}\,\mathcal{K}((\partial^{-1}))$. Any matrix $H(\partial) \in \mathrm{Mat}_{\ell\times\ell}\,\mathcal{V}(\partial)$ can be written as a ratio of two matrix differential operators: $H(\partial) = A(\partial)B^{-1}(\partial)$, with $A(\partial), B(\partial) \in \mathrm{Mat}_{\ell\times\ell}\,\mathcal{V}[\partial]$, and $B(\partial)$ non-degenerate.

2.2.3 Singular Degree of a Rational Matrix Pseudodifferential Operator

The *Dieudonné determinant* of $A \in \mathrm{Mat}_{\ell\times\ell}\,\mathcal{K}((\partial^{-1}))$ is defined as follows. If A is degenerate, then $\det(A) = 0$. Otherwise, $\det(A)$ is a pair

$$\det(A) = (\det_1(A), \deg(A)) \in \mathcal{K} \times \mathbb{Z}\,,$$

where $\det_1(A)$ and $\deg(A)$ are defined by the following conditions:

(i) $\det_1(AB) = \det_1(A)\det_1(B)$ for all non-degenerate $A, B \in \mathrm{Mat}_{\ell\times\ell}\,\mathcal{K}((\partial^{-1}))$;
(ii) $\deg(AB) = \deg(A)+\deg(B)$ for all non-degenerate $A, B \in \mathrm{Mat}_{\ell\times\ell}\,\mathcal{K}((\partial^{-1}))$;
(iii) if A is upper triangular, with diagonal entries $A_i = a_i\partial^{d_i}$+lower terms, $i = 1,\dots,\ell$, with $a_i \neq 0$, then

$$\det_1(H) = \prod_{i=1}^{\ell} a_i\ , \quad \deg(A) = \sum_{i=1}^{\ell} d_i\,.$$

For a non-degenerate $A \in \mathrm{Mat}_{\ell\times\ell}\,\mathcal{K}((\partial^{-1}))$, the integer $\deg(A)$ is called the *degree* of A. (It is a non-negative integer if A is a matrix differential operator.)

Let $H \in \mathrm{Mat}_{\ell\times\ell}\,\mathcal{V}(\partial)$ be a rational matrix pseudodifferential operator. The *singular degree* of H, denoted $\mathrm{sdeg}(H)$ [8], is, by definition, the minimal possible value of $\deg(B)$ among all fractional decomposition $H = AB^{-1}$, with $A, B \in \mathrm{Mat}_{\ell\times\ell}\,\mathcal{V}[\partial]$, and $B(\partial)$ non-degenerate.

Suppose that we have a rational expression for $H \in \mathrm{Mat}_{\ell\times\ell}\,\mathcal{V}(\partial)$ of the form

$$H = \sum_{\alpha\in\mathcal{A}} A_1^{\alpha}(B_1^{\alpha})^{-1}\dots A_n^{\alpha}(B_n^{\alpha})^{-1}\,, \tag{2.6}$$

with $A_i^{\alpha}, B_i^{\alpha} \in \mathrm{Mat}_{\ell\times\ell}\,\mathcal{K}[\partial]$ and B_i^{α} non-degenerate, for all $i \in \mathcal{I} = \{1,\dots,n\}, \alpha \in \mathcal{A}$ (a finite index set). It is not hard to show that $\mathrm{sdeg}(H) \leq \sum_{\alpha\in\mathcal{A}}\sum_{i=1}^{n}\deg(B_i^{\alpha})$, [8]. We say that the rational expression (2.6) is *minimal* if equality holds.

Theorem 2.1 ([8, Cor.4.11]). *The rational expression* (2.6) *is minimal if and only if both the following systems of differential equations in the variables* $\{F_i^{\alpha}\}_{\alpha\in\mathcal{A},i\in\{1,\dots,n\}}$

$$\begin{cases} B_n^\alpha F_n^\alpha = 0\,, \ \alpha \in \mathcal{A} \\ A_i^\alpha F_i^\alpha = B_{i-1}^\alpha F_{i-1}^\alpha\,, \ 2 \le i \le n, \alpha \in \mathcal{A} \\ \sum_{\alpha \in \mathcal{A}} A_1^\alpha F_1^\alpha = 0 \end{cases} \tag{2.7}$$

and

$$\begin{cases} B_1^{\alpha *} F_1^\alpha = 0\,, \ \alpha \in \mathcal{A} \\ A_i^{\alpha *} F_{i-1}^\alpha = B_i^{\alpha *} F_i^\alpha\,, \ 2 \le i \le n, \alpha \in \mathcal{A} \\ \sum_{\alpha \in \mathcal{A}} F_n^\alpha = 0 \end{cases} \tag{2.8}$$

have only the zero solution over the linear closure $\tilde{\mathcal{K}}$ of $\mathcal{K}$.

2.2.4 Association Relation

Given $H(\partial) \in \mathrm{Mat}_{\ell\times\ell}\,\mathcal{V}(\partial)$, we say that $\xi \in \mathcal{V}^{\oplus l}$ and $P \in \mathcal{V}^\ell$ are *H-associated,* and denote it by

$$\xi \stackrel{H}{\longleftrightarrow} P\,, \tag{2.9}$$

if there exist a fractional decomposition $H = AB^{-1}$ with $A, B \in \mathrm{Mat}_{\ell\times\ell}\,\mathcal{V}[\partial]$ and B non-degenerate, and an element $F \in \mathcal{K}^\ell$, such that $\xi = BF$, $P = AF$ [2].

Theorem 2.2 ([8, Thm4.12]). *Let* (2.6) *be a minimal rational expression for $H(\partial) \in \mathrm{Mat}_{\ell\times\ell}\,\mathcal{V}(\partial)$. Then, $\xi \stackrel{H}{\longleftrightarrow} P$ if and only the system of differential equations*

$$\begin{cases} B_n^\alpha F_n^\alpha = \xi\,, \ \alpha \in \mathcal{A} \\ A_i^\alpha F_i^\alpha = B_{i-1}^\alpha F_{i-1}^\alpha\,, \ 2 \le i \le n, \alpha \in \mathcal{A} \\ \sum_{\alpha \in \mathcal{A}} A_1^\alpha F_1^\alpha = P \end{cases} \tag{2.10}$$

has a solution $\{F_i^\alpha\}_{\alpha\in\mathcal{A}, i\in\{1,\dots,n\}}$ over $\mathcal{K}$.

2.2.5 Non-local Poisson Structures

To a matrix pseudodifferential operator $H = \big(H_{ij}(\partial)\big)_{i,j\in I} \in \mathrm{Mat}_{\ell\times\ell}\,\mathcal{V}((\partial^{-1}))$ we associate a map, called λ-bracket, $\{\cdot\,{}_\lambda\,\cdot\}_H : \mathcal{V}\times\mathcal{V} \to \mathcal{V}((\lambda^{-1}))$, given by the following *Master Formula* (see [2]):

$$\{f_\lambda g\}_H = \sum_{\substack{i,j\in I \\ m,n\in\mathbb{Z}_+}} \frac{\partial g}{\partial u_j^{(n)}}(\lambda+\partial)^n H_{ji}(\lambda+\partial)(-\lambda-\partial)^m \frac{\partial f}{\partial u_i^{(m)}} \in \mathcal{V}((\lambda^{-1}))\,. \tag{2.11}$$

In particular,

$$H_{ji}(\partial) = \{u_i\,{}_\partial u_j\}_{H_\rightarrow}. \tag{2.12}$$

(The arrow means that we move ∂ to the right.)

The following facts are proved in [1] and [2]. For arbitrary H, the λ-bracket (2.11) satisfies the following sesquilinearity conditions:

(i) $\{\partial f_\lambda g\} = -\lambda\{f_\lambda g\}$,
(ii) $\{f_\lambda \partial g\} = (\lambda + \partial)\{f_\lambda g\}$,

and left and right Leibniz rules ($f, g, h \in \mathcal{V}$):

(iii) $\{f_\lambda gh\} = \{f_\lambda g\}h + \{f_\lambda h\}g$,
(iv) $\{fg_\lambda h\} = \{f_{\lambda+\partial} h\}g + \{g_{\lambda+\partial} h\}f$.

Here and further an expression $\{f_{\lambda+\partial} h\}_\rightarrow g$ is interpreted as follows: if $\{f_\lambda h\} = \sum_{n=-\infty}^{N} c_n\lambda^n$, then $\{f_{\lambda+\partial} h\}_\rightarrow g = \sum_{n=-\infty}^{N} c_n(\lambda+\partial)^n g$, where we expand $(\lambda+\partial)^n$ in non-negative powers of ∂.

Skewadjointness of H is equivalent to the following skewsymmetry condition

(v) $\{f_\lambda g\} = -\{g_{-\lambda-\partial} f\}$.

The RHS of the skewsymmetry condition should be interpreted as follows: we move $-\lambda - \partial$ to the left and we expand its powers in non-negative powers of ∂, acting on the coefficients on the λ-bracket.

Let $\mathcal{V}_{\lambda,\mu} := \mathcal{V}[[\lambda^{-1},\mu^{-1},(\lambda+\mu)^{-1}]][\lambda,\mu]$, i.e. the quotient of the $\mathbb{F}[\lambda,\mu,\nu]$-module $\mathcal{V}[[\lambda^{-1},\mu^{-1},\nu^{-1}]][\lambda,\mu,\nu]$ by the submodule $(\nu-\lambda-\mu)\mathcal{V}[[\lambda^{-1},\mu^{-1},\nu^{-1}]][\lambda,\mu,\nu]$. We have the natural embedding $\iota_{\mu,\lambda} : \mathcal{V}_{\lambda,\mu} \hookrightarrow V((\lambda^{-1}))((\mu^{-1}))$ defined by expanding the negative powers of $\nu = \lambda+\mu$ by geometric series in the domain $|\mu| > |\lambda|$. In general, if H is an arbitrary matrix pseudodifferential operator, we have $\{f_\lambda\{g_\mu h\}\} \in \mathcal{V}((\lambda^{-1}))((\mu^{-1}))$ for all $f, g, h \in \mathcal{V}$. If H is a rational matrix pseudodifferential operator, we have the following admissibility condition ($f, g, h \in \mathcal{V}$):

(vi) $\{f_\lambda\{g_\mu h\}\} \in \mathcal{V}_{\lambda,\mu}$,

where we identify the space $\mathcal{V}_{\lambda,\mu}$ with its image in $\mathcal{V}((\lambda^{-1}))((\mu^{-1}))$ via the embedding $\iota_{\mu,\lambda}$.

Definition 2.1. A *non-local Poisson structure* on $\mathcal{V}$ is a skewadjoint rational matrix pseudodifferential operator H with coefficients in $\mathcal{V}$, satisfying the following Jacobi identity ($f, g, h \in \mathcal{V}$):

(vii) $\{f_\lambda\{g_\mu h\}\} - \{g_\mu\{f_\lambda h\}\} = \{\{f_\lambda g\}_{\lambda+\mu} h\}$,

where the equality is understood in the space $\mathcal{V}_{\lambda,\mu}$.

(Note that, if skewsymmetry (v) and admissibility (vi) hold, then all three terms of Jacobi identity lie in the image of $\mathcal{V}_{\lambda,\mu}$ via the appropriate embedding $\iota_{\mu,\lambda}$, $\iota_{\lambda,\mu}$ or

$\iota_{\lambda+\mu,\lambda}\cdot$.) Note that Jacobi identity (vii) holds for all $f, g, h \in \mathcal{V}$ if and only if it holds for any triple of generators u_i, u_j, u_k [1,2].

Two non-local Poisson structures $H_0, H_1 \in \mathrm{Mat}_{\ell\times\ell}\,\mathcal{V}(\partial)$ on $\mathcal{V}$ are said to be *compatible* if any their linear combination (or, equivalently, their sum) is a non-local Poisson structure. In this case we say that (H_0, H_1) form a *bi-Poisson structure* on $\mathcal{V}$.

Definition 2.2. A *non-local Poisson vertex algebra* is, by definition, a differential algebra $\mathcal{V}$ endowed with a λ-bracket $\{\cdot\,_\lambda\,\cdot\} : \mathcal{V}\times\mathcal{V} \to \mathcal{V}((\lambda^{-1}))$ satisfying conditions (i)–(vii).

We shall often drop the term "non-local", so when we will refer to Poisson structures and λ-brackets we will always mean *non-local PVA*'s and *non-local λ-brackets*. (This, of course, includes the local case as well.)

2.2.6 Hamiltonian Equations and Integrability

Recall that we have a non-degenerate pairing $(\cdot\,|\,\cdot) : \mathcal{V}^\ell \times \mathcal{V}^\ell \to \mathcal{V}/\partial\mathcal{V}$ given by $(P\,|\,\xi) = \int P\cdot\xi$ (see e.g. [1]). Let $H \in \mathrm{Mat}_{\ell\times\ell}\,\mathcal{V}(\partial)$ be a non-local Poisson structure. An evolution equation on the variables $u = \big(u_i\big)_{i\in I}$,

$$\frac{du}{dt} = P\,, \tag{2.13}$$

is called *Hamiltonian* with respect to the Poisson structure H and the Hamiltonian functional $\int h \in \mathcal{V}/\partial\mathcal{V}$ if (cf. Sect. 2.2.4)

$$\frac{\delta h}{\delta u} \overset{H}{\longleftrightarrow} P\,.$$

Equation (2.13) is called *bi-Hamiltonian* if there are two compatible non-local Poisson structures H_0 and H_1, and two local functionals $\int h_0, \int h_1 \in \mathcal{V}/\partial\mathcal{V}$, such that

$$\frac{\delta h_0}{\delta u} \overset{H_1}{\longleftrightarrow} P \;\text{ and }\; \frac{\delta h_1}{\delta u} \overset{H_0}{\longleftrightarrow} P\,. \tag{2.14}$$

An *integral of motion* for the Hamiltonian equation (2.13) is a local functional $\int f \in \mathcal{V}/\partial\mathcal{V}$ which is constant in time, i.e. such that $(P\,|\,\frac{\delta f}{\delta u}) = 0$. The usual requirement for *integrability* is to have sequences $\{\int h_n\}_{n\in\mathbb{Z}_+} \subset \mathcal{V}/\partial\mathcal{V}$ and $\{P_n\}_{n\in\mathbb{Z}_+} \subset \mathcal{V}^\ell$, starting with $\int h_0 = \int h$ and $P_0 = P$, such that

(C1) $\frac{\delta h_n}{\delta u} \overset{H}{\longleftrightarrow} P_n$ for every $n \in \mathbb{Z}_+$.
(C2) $[P_m, P_n] = 0$ for all $m, n \in \mathbb{Z}_+$.

(C3) $(P_m \mid \frac{\delta h_n}{\delta u}) = 0$ for all $m, n \in \mathbb{Z}_+$.
(C4) The elements P_n span an infinite dimensional subspace of $\mathcal{V}^\ell$.

In this case, we have an *integrable hierarchy* of Hamiltonian equations

$$\frac{du}{dt_n} = P_n \,,\; n \in \mathbb{Z}_+ \,.$$

Elements $\int h_n$'s are called *higher Hamiltonians*, the P_n's are called *higher symmetries*, and the condition $(P_m \mid \frac{\delta h_n}{\delta u}) = 0$ says that $\int h_m$ and $\int h_n$ are *in involution.* Note that (C4) implies that the elements $\frac{\delta h_n}{\delta u}$ span an infinite dimensional subspace of $\mathcal{V}^\ell$. The converse holds provided that either H_0 or H_1 is non-degenerate.

Suppose we have a bi-Hamiltonian equation (2.13), associated to the compatible Poisson structures H_0, H_1 and the Hamiltonian functionals $\int h_0, \int h_1$, in the sense of equation (2.14). The *Lenard-Magri scheme of integrability* consists in finding sequences $\{\int h_n\}_{n\in\mathbb{Z}_+} \subset \mathcal{V}/\partial\mathcal{V}$ and $\{P_n\}_{n\in\mathbb{Z}_+} \subset \mathcal{V}^\ell$, starting with $P_0 = P$ and the given Hamiltonian functionals $\int h_0, \int h_1$, satisfying the following recursive relations:

$$\frac{\delta h_{n-1}}{\delta u} \overset{H_1}{\longleftrightarrow} P_n \,,\quad \frac{\delta h_n}{\delta u} \overset{H_0}{\longleftrightarrow} P_n \quad \text{for all } n \in \mathbb{Z}_+ \,. \tag{2.15}$$

In this case, we have the corresponding bi-Hamiltonian hierarchy

$$\frac{du}{dt_n} = P_n \in \mathcal{V}^\ell \,,\quad n \in \mathbb{Z}_+ \,, \tag{2.16}$$

all Hamiltonian functionals $\int h_n$, $n \geq -1$, are integrals of motion for all equations of the hierarchy, and they are in involution with respect to both Poisson structures H_0 and H_1, and all commutators $[P_m, P_n]$ are zero, provided that one of the Poisson structures H_0 or H_1 is local (see [2, Sec.7.4]). Hence, in this situation (2.16) is an integrable hierarchy of compatible evolution equations, provided that condition (C4) holds.

2.3 Dirac Reduction for (Non-local) Poisson Structures and Hamiltonian Equations

2.3.1 Dirac Reduction of a Poisson Structure

Let $H(\partial) \in \mathrm{Mat}_{\ell\times\ell}\,\mathcal{V}(\partial)$ be a Poisson structure on $\mathcal{V}$. Let $\{\cdot\,{}_\lambda\,\cdot\}_H$ be the corresponding PVA λ-bracket on $\mathcal{V}$ given by the Master Formula (2.11). Let $\theta_1, \dots, \theta_m$ be some elements of $\mathcal{V}$, and let $\mathcal{I} = \langle\theta_1, \dots, \theta_m\rangle_\mathcal{V} \subset \mathcal{V}$ be the differential ideal generated by them. Consider the following rational matrix pseudodifferential operator

$$C(\partial) = D_\theta(\partial) \circ H(\partial) \circ D_\theta^*(\partial) \in \mathrm{Mat}_{m\times m}\,\mathcal{V}(\partial)\,, \tag{2.17}$$

where $D_\theta(\partial)$ is the $m \times \ell$ matrix differential operator of Frechet derivatives of the elements θ_i's:

$$D_\theta(\partial)_{\alpha,i} = \sum_{n\in\mathbb{Z}_+} \frac{\partial \theta_\alpha}{\partial u_i^{(n)}}\partial^n\,,\quad \alpha = 1,\dots,m,\ i = 1,\dots,\ell\,, \tag{2.18}$$

and $D_\theta^*(\partial) \in \mathrm{Mat}_{\ell\times m}\,\mathcal{V}[\partial]$ is its adjoint. Recalling the Master Formula (2.11), we get that $C(\partial)$ has matrix elements with symbol

$$C_{\alpha\beta}(\lambda) = \{\theta_{\beta\,\lambda}\,\theta_\alpha\}_H\,. \tag{2.19}$$

Note also that, by the skewadjointness of H, the corresponding λ-bracket $\{\cdot\,_\lambda\,\cdot\}_H$ is skewsymmetric, hence $C(\partial)$ is a skewadjoint pseudodifferential operator.

We shall assume that the matrix $C(\partial)$ in (2.17) is invertible in $\mathrm{Mat}_{m\times m}\,\mathcal{V}((\partial^{-1}))$, and we denote its inverse by $C^{-1}(\partial) = \big((C^{-1})_{\alpha\beta}(\partial)\big)_{\alpha,\beta=1}^m \in \mathrm{Mat}_{m\times m}\,\mathcal{V}((\partial^{-1}))$.

Definition 2.3. The *Dirac modification* of the Poisson structure $H \in \mathrm{Mat}_{\ell\times\ell}\,\mathcal{V}(\partial)$ by the *constraints* $\theta_1,\dots,\theta_m$ is the following skewadjoint $\ell\times\ell$ matrix pseudodifferential operator:

$$H^D(\partial) = H(\partial) + B(\partial)\circ C^{-1}(\partial)\circ B^*(\partial)\,, \tag{2.20}$$

where $B(\partial) = H(\partial)\circ D_\theta^*(\partial) \in \mathrm{Mat}_{\ell\times m}\,\mathcal{V}(\partial)$.

The matrix pseudodifferential operator $H^D(\partial)$ is skewadjoint and rational. The corresponding λ-bracket, given by the Master Formula (2.11), is (cf. [5])

$$\{f_\lambda g\}_H^D = \{f_\lambda g\}_H - \sum_{\alpha,\beta=1}^m \{\theta_{\beta\,\lambda+\partial}\,g\}_{H\,\to}(C^{-1})_{\beta\alpha}(\lambda+\partial)\{f_\lambda\theta_\alpha\}_H\,. \tag{2.21}$$

The following result is a special case of [5, Thm.2.2]:

Theorem 2.3. *(a) The Dirac modified λ-bracket* (2.21) *satisfies the Jacobi identity (vii). Consequently, the Dirac modification $H^D(\partial)$ is a non-local Poisson structure on $\mathcal{V}$.*

(b) All the elements θ_i, $i = 1,\dots,m$, are central with respect to the Dirac modified λ-bracket, i.e.:

$$\{f_\lambda\theta_i\}_H^D = \{\theta_{i\,\lambda}\,f\}_H^D = 0$$

for all $i = 1,\dots,m$ and $f\in\mathcal{V}$.

(c) The differential ideal $\mathcal{I} = \langle\theta_1,\dots,\theta_m\rangle_{\mathcal{V}} \subset \mathcal{V}$, generated by $\theta_1,\dots,\theta_m$, is an ideal with respect to the Dirac modified λ-bracket $\{\cdot\,_\lambda\,\cdot\}^D$, namely:

$$\{\mathcal{I}\,{}_\lambda\,\mathcal{V}\}_H^D\,,\ \{\mathcal{V}\,{}_\lambda\,\mathcal{I}\}_H^D \subset \mathcal{I}((\lambda^{-1}))\,.$$

Hence, the quotient space $\mathcal{V}/\mathcal{I}$ is a PVA, with λ-bracket induced by $\{\cdot\,{}_\lambda\,\cdot\}^D$, which we call the Dirac reduction *of $\mathcal{V}$ by the constraints $\theta_1, \dots, \theta_m$.*

Remark 2.1. If the constraints θ_i's are some generators of the algebra of differential polynomials $\mathcal{V}$, then the quotient $\mathcal{V}/\mathcal{I}$ is still an algebra of differential polynomials (in the remaining generators), and we have the induced Poisson structure $\overline{H}^D$ on this quotient (corresponding to the PVA λ-bracket of Theorem 2.3(c)).

2.3.2 Dirac Reduction of a Bi-Poisson Structure

Let (H_0, H_1) be a bi-Poisson structure on $\mathcal{V}$. Let $\theta_1, \dots, \theta_m \in \mathcal{V}$ be central elements for H_0. Suppose that the matrix pseudodifferential operator (cf. (2.17)) $C(\partial) = D_\theta(\partial) \circ H_1(\partial) \circ D_\theta^*(\partial)$ is invertible. Then we can consider the Dirac modified Poisson structure H_1^D (cf. (2.20)), and the corresponding λ-bracket $\{\cdot\,{}_\lambda\,\cdot\}_1^D$ (cf. (2.21)), and we have the following result:

Theorem 2.4 ([5, Thm.2.3]).

(a) The matrices H_0 and H_1^D form a compatible pair of Poisson structures on $\mathcal{V}$.
(b) The differential algebra ideal $\mathcal{I} = \langle\theta_1, \dots, \theta_m\rangle_\mathcal{V}$ is a PVA ideal for both the λ-brackets $\{\cdot\,{}_\lambda\,\cdot\}_0$ and $\{\cdot\,{}_\lambda\,\cdot\}_1^D$, and we have the induced compatible PVA λ-brackets on $\mathcal{V}/\mathcal{I}$.

2.3.3 Reduction of a Bi-Hamiltonian Hierarchy

Let (H_0, H_1) be a *local* bi-Poisson structure (i.e. consisting of matrix differential operators). Suppose that we have a bi-Hamiltonian hierarchy $\frac{du}{dt_n} = P_n \in \mathcal{V}^\ell$, $n \in \mathbb{Z}_+$, with respect to (H_0, H_1), and let $\int h_n \in \mathcal{V}/\partial\mathcal{V}$ be a sequence of integrals of motion satisfying the Lenard-Magri recursive condition (2.15). Let $\theta_1, \dots, \theta_m \in \mathcal{V}$ be central elements for H_0. Assume that the matrix $C(\partial) = D_\theta(\partial) \circ H_1(\partial) \circ D_\theta^*(\partial) \in \mathrm{Mat}_{m\times m}\,\mathcal{V}[\partial]$ is invertible in $\mathrm{Mat}_{m\times m}\,\mathcal{V}((\partial^{-1}))$. Then, by Theorem 2.4, $H_1^D = H_1 + B(\partial)C^{-1}(\partial)B^*(\partial)$, where $B(\partial) = H_1(\partial) \circ D_\theta^*(\partial)$, is a (non-local) Poisson structure on $\mathcal{V}$ compatible to H_0. Moroever, we have the following result:

Proposition 2.1. *Suppose that* Ker $B(\partial)$ *and* Ker $C(\partial)$ *have zero intersection over the linear closure $\tilde{\mathcal{K}}$ of $\mathcal{K}$. Then we have the Lenard-Magri recursive relations*

$$\frac{\delta h_{n-1}}{\delta u} \overset{H_1^D}{\longleftrightarrow} P_n\,,\quad \frac{\delta h_n}{\delta u} \overset{H_0}{\longleftrightarrow} P_n \quad \textit{for all } n \in \mathbb{Z}_+\,. \tag{2.22}$$

Proof. According to Theorem 2.1, the condition Ker $B(\partial)$ and Ker $C(\partial)$ have zero intersection is equivalent to saying that $H_1^D = H_1 + B(\partial)C^{-1}(\partial)B^*(\partial)$ is a minimal rational expression for H_1^D. By assumption, we have $P_n = H_0 \frac{\delta h_n}{\delta u} = H_1 \frac{\delta h_{n-1}}{\delta u}$. By Theorem 2.2 the association relation $\frac{\delta h_{n-1}}{\delta u} \overset{H_1^D}{\longleftrightarrow} P_n$ holds if there exists $F_n \in \mathcal{K}^m$ such that

$$B^*(\partial)\frac{\delta h_{n-1}}{\delta u} = C(\partial)F_n \qquad \text{and} \qquad B(\partial)F_n = 0\,.$$

Note that $B^*(\partial)\frac{\delta h_{n-1}}{\delta u} = -D_\theta(\partial)H_1(\partial)\frac{\delta h_{n-1}}{\delta u} = -D_\theta(\partial)P_n$. Since the elements θ_α's are central for the Poisson structure H_0, they are constant densities for the Hamiltonian equations (2.16) (see [5, Lem.5.2(b)]). Thus we have $D_\theta(\partial)P_n = 0$, for every $n \in \mathbb{Z}_+$. Therefore we can choose F_n to be the zero vector in $\mathcal{K}^\ell$, for every $n \in \mathbb{Z}_+$.

2.4 Dirac Reduced Homogeneous DS Hierarchy

First we review the construction of the homogeneous Drinfeld-Sokolov hierarchy, following [3].

Let $\mathfrak{g}$ be a simple finite-dimensional Lie algebra. Fix a non-degenerate symmetric invariant bilinear form $(\cdot\,|\,\cdot)$ on $\mathfrak{g}$, and a regular semisimple element $s \in \mathfrak{g}$. We have the direct sum decomposition $\mathfrak{g} = \mathfrak{h} \oplus \mathfrak{h}^\perp$, where $\mathfrak{h} = \mathrm{Ker}(\mathrm{ad}\, s)$ is a Cartan subalgebra, and $\mathfrak{h}^\perp = \mathrm{Im}(\mathrm{ad}\, s)$ is its orthogonal complement with respect to the bilinear form $(\cdot\,|\,\cdot)$, and it is the direct sum of root spaces.

Let $\mathcal{V} = S(\mathbb{F}[\partial]\mathfrak{g})$, the algebra of differential polynomials in a basis of $\mathfrak{g}$. We define a λ-bracket on $\mathcal{V}$ by

$$\{a_\lambda b\}_z = [a,b] + (a|b)\lambda + z(s|[a,b])\,, \tag{2.23}$$

for $a, b \in \mathfrak{g}$, and we extend it to a λ-bracket on $\mathcal{V}$ by (2.11), thus obtaining a PVA structure. In Eq. (2.23) $z \in \mathbb{F}$ is a parameter.

Let $\ell = \mathrm{rank}(\mathfrak{g})$ be the rank of $\mathfrak{g}$, and let Δ be the set of roots of $\mathfrak{g}$. Choose a basis $\mathfrak{g}$ as follows: $\mathcal{B} = \{x_i\}_{i=1}^\ell \cup \{e_\alpha\}_{\alpha\in\Delta}$, union of an orthonormal basis of $\mathfrak{h}$ and a collection of root vectors such that $(e_\alpha|e_{-\alpha}) = 1$. Hence $\mathcal{V} = \mathbb{C}[x_i^{(n)}, e_\alpha^{(n)} \mid i \in \{1,\dots,\ell\}, \alpha \in \Delta, n \in \mathbb{Z}_+]$ is the algebra of differential polynomials generated by the elements of the basis $\mathcal{B}$. Equation (2.23) defines a (local) bi-Poisson structure (H_0, H_1) on $\mathcal{V}$ given by $(i, j \in \{1,\dots,\ell\}, \alpha \in \Delta)$

$$\begin{cases} (H_0)_{ij}(\partial) = 0 \\ (H_0)_{\alpha i}(\partial) = 0 \\ (H_0)_{\alpha\beta}(\partial) = \delta_{\alpha,-\beta}\beta(s) \end{cases} \quad \text{and} \quad \begin{cases} (H_1)_{ij}(\partial) = \delta_{ij}\partial \\ (H_1)_{\alpha i}(\partial) = \alpha(x_i)e_\alpha \\ (H_1)_{\alpha\beta}(\partial) = [e_\beta, e_\alpha] + \delta_{\alpha,-\beta}\partial\,. \end{cases} \tag{2.24}$$

It is proved in [3] that, for any $a \in \mathfrak{h}$, we have an infinite sequence of local functionals $\{\int h_n\}_{n\geq -1} \subset \mathcal{V}/\partial\mathcal{V}$, satisfying the Lenard-Magri recursive relations (2.15), and $h_{-1} = 0$, $h_0 = a$. The next two functionals of the sequence have densities

$$h_1 = \frac{1}{2}\sum_{\alpha\in\Delta}\frac{\alpha(a)}{\alpha(s)}e_\alpha e_{-\alpha}\,,$$

and

$$\begin{aligned}h_2 = &\frac{1}{2}\sum_{\alpha\in\Delta}\frac{\alpha(a)}{\alpha(s)^2}e_{-\alpha}e'_\alpha + \frac{1}{2}\sum_{\alpha\in\Delta}\frac{\alpha(a)}{\alpha(s)^2}e_\alpha e_{-\alpha}[e_{-\alpha},e_\alpha]\\ &+\frac{1}{3}\sum_{\substack{\alpha,\beta\in\Delta\\ \alpha\neq\beta}}\frac{\alpha(a)}{\alpha(s)\beta(s)}e_\alpha e_{-\beta}[e_{-\alpha},e_\beta]\,.\end{aligned}$$

The corresponding Hamiltonian equations (2.16) are: $\frac{dx_i}{dt_0} = \frac{dx_i}{dt_1} = \frac{dx_i}{dt_2} = 0$, for $i = 1,\ldots,\ell$, and, for $\alpha \in \Delta$,

$$\begin{aligned}&\frac{de_\alpha}{dt_0} = \alpha(a)e_\alpha\,, \qquad \frac{de_\alpha}{dt_1} = \frac{\alpha(a)}{\alpha(s)}e'_\alpha + \sum_{\beta\in\Delta}\frac{\beta(a)}{\beta(s)}e_{-\beta}[e_\beta,e_\alpha]\,,\\ &\frac{de_\alpha}{dt_2} = \frac{\alpha(a)}{\alpha(s)^2}e''_\alpha + \frac{\alpha(a)}{\alpha(s)^2}(e_\alpha[e_{-\alpha},e_\alpha])' + \sum_{\beta\in\Delta}\frac{\beta(a)}{\beta(s)^2}e'_\beta[e_{-\beta},e_\alpha]\\ &+\frac{1}{3}\sum_{\beta\in\Delta\setminus\{\alpha\}}\left(\frac{\alpha(a)+\beta(a)}{\alpha(s)\beta(s)} + \frac{\beta(a)}{\beta(s)(\alpha(s)-\beta(s))}\right)(e_\beta[e_{-\beta},e_\alpha])'\\ &+\sum_{\beta\in\Delta}\frac{\beta(a)}{\beta(s)^2}\left(e_\beta[e_{-\beta},e_\beta][e_{-\beta},e_\alpha] - \frac{1}{2}(\alpha|\beta)e_\alpha e_\beta e_{-\beta}\right)\\ &+\frac{1}{3}\sum_{\substack{\beta,\gamma\in\Delta\\ \beta\neq\gamma}}\frac{2\beta(a)\gamma(s)-\gamma(a)\beta(s)}{\beta(s)\gamma(s)(\beta(s)-\gamma(s))}e_\beta e_{-\gamma}[[e_{-\beta},e_\gamma],e_\alpha]\,.\end{aligned} \tag{2.25}$$

It follows from [3, Rem.2.7] that, for every $\alpha \in \Delta$,

$$\frac{\delta h_n}{\delta e_\alpha} = (-1)^{n+1}\frac{\alpha(a)}{\alpha(s)^n}e_{-\alpha}^{(n-1)} + \text{higher polynomial order terms}\,. \tag{2.26}$$

In particular, the elements $\frac{\delta h_n}{\delta u}$ are linearly independent.

Let $\mathcal{I}$ be the differential ideal generated by the variables x_i, $i = 1,\ldots,\ell$. Clearly, as differential algebras,

$$\mathcal{V}/\mathcal{I} \simeq \overline{\mathcal{V}} = \mathbb{F}[e_\alpha^{(n)} \mid \alpha \in \Delta, n \in \mathbb{Z}_+] .$$

Note that, by Eq. (2.24), the elements x_i, $i = 1, \dots, \ell$, are central for the Poisson structure H_0. Consider the Dirac modification H_1^D of H_1 by the constraints $\{x_i\}_{i=1}^\ell$, defined by the Eq. (2.20):

$$H_1^D(\partial) = H_1(\partial) + B(\partial) \circ C^{-1}(\partial) \circ B^*(\partial) ,$$

where the matrices $B(\partial) \in \mathrm{Mat}_{(\ell+|\Delta|)\times\ell}\, \mathcal{V}[\partial]$ and $C(\partial) \in \mathrm{Mat}_{\ell\times\ell}\, \mathcal{V}[\partial]$ have entries

$$\begin{aligned} &B_{ij}(\partial) = C_{ij}(\partial) = (H_1)_{ij}(\partial) = \delta_{ij}\partial \,, \; i, j = 1, \dots, \ell \\ &B_{\alpha i}(\partial) = (H_1)_{\alpha i}(\partial) = \alpha(x_i)e_\alpha \,, \quad i = 1, \dots, \ell, \alpha \in \Delta \,. \end{aligned} \tag{2.27}$$

By Theorem 2.4, we have a bi-Poisson structure (H_0, H_1^D) on $\mathcal{V}$, and the induced bi-Poisson structure $(\overline{H}_0, \overline{H}_1^D)$ on $\overline{\mathcal{V}}$. It is given by $(\alpha, \beta \in \Delta)$

$$\begin{aligned} &(\overline{H}_0)_{\alpha\beta}(\lambda) = \delta_{\alpha,-\beta}\beta(s) \\ &(\overline{H}_1^D)_{\alpha,-\alpha}(\lambda) = \partial + (\alpha|\alpha)e_\alpha\partial^{-1} \circ e_{-\alpha} \,, \\ &(\overline{H}_1^D)_{\alpha\beta}(\lambda) = [e_\beta, e_\alpha] - (\alpha|\beta)e_\alpha\partial^{-1} \circ e_\beta \,, \text{ for } \alpha \neq -\beta \,. \end{aligned}$$

Proposition 2.2. *The Lenard-Magri recursive relations* (2.22) *hold for the bi-Poisson structure* (H_1^D, H_0). *Hence, we get an induced bi-Hamiltonian hierarchy in* $\overline{\mathcal{V}}$.

Proof. By Proposition 2.1 it suffices to show that $\mathrm{Ker}\, B(\partial) = 0$ over the linear closure $\tilde{\mathcal{K}}$ of $\mathcal{K}$. Let $F = (F_i)_{i=1}^\ell \in \tilde{\mathcal{K}}$ be an element of the kernel of $B(\partial)$. We have, for $\alpha \in \Delta$,

$$(B(\partial)F)_\alpha = (\alpha(x_1)F_1 + \dots + \alpha(x_\ell)F_\ell)\, e_\alpha \,.$$

Since Δ spans $\mathfrak{h}^*$, it follows that $F = 0$. To conclude, we just observe that, by Eq. (2.26), the images of elements $\frac{\delta h_n}{\delta u}$ in $\overline{\mathcal{V}}^{|\Delta|}$ are linearly independent. Since $\overline{H}_0$ is an invertible constant matrix, the images of the elements P_n in $\overline{\mathcal{V}}^{|\Delta|}$ are linearly independent as well.

Remark 2.2. It follows from [1, Prop.2.10] and the definition of the Lie bracket between Hamiltonian functionals given in [2, Eq.(7.8)] (using the fact that H_0 is local) that all the $\int h_n$'s, obtained by taking all possible $a \in \mathfrak{h}$, are in involution.

The first equations of the reduced bi-Hamiltonian hierarchy are ($\alpha \in \Delta$)

$$\frac{de_\alpha}{dt_0} = \alpha(a)e_\alpha\,, \qquad \frac{de_\alpha}{dt_1} = \frac{\alpha(a)}{\alpha(s)}e'_\alpha + \sum_{\beta\in\Delta\setminus\{-\alpha\}} \frac{\beta(a)}{\beta(s)}e_{-\beta}[e_\beta, e_\alpha]\,,$$

$$\frac{de_\alpha}{dt_2} = \frac{\alpha(a)}{\alpha(s)^2}e''_\alpha + \sum_{\beta\in\Delta\setminus\{\alpha\}} \frac{\beta(a)}{\beta(s)^2}e'_\beta[e_{-\beta}, e_\alpha]$$

$$+\frac{1}{3}\sum_{\beta\in\Delta\setminus\{\alpha\}} \left(\frac{\alpha(a)+\beta(a)}{\alpha(s)\beta(s)} + \frac{\beta(a)}{\beta(s)(\alpha(s)-\beta(s))}\right)(e_\beta[e_{-\beta}, e_\alpha])'$$

$$-\frac{1}{2}\sum_{\beta\in\Delta} \frac{\beta(a)}{\beta(s)^2}(\alpha|\beta)e_\alpha e_\beta e_{-\beta} + \frac{1}{3}\sum_{\substack{\beta,\gamma\in\Delta\\ \beta\neq\gamma,\gamma+\alpha}} \frac{2\beta(a)\gamma(s)-\gamma(a)\beta(s)}{\beta(s)\gamma(s)(\beta(s)-\gamma(s))}e_\beta e_{-\gamma}[[e_{-\beta}, e_\gamma], e_\alpha]\,.$$

Remark 2.3. For $\mathfrak{g} = \mathfrak{sl}_2$ we have $\Delta = \{\alpha, -\alpha\}$. Letting $s = a = [e_\alpha, e_{-\alpha}]$, the first non trivial equation of the reduced bi-Hamiltonian hierarchy is

$$\begin{cases} \dfrac{de_\alpha}{dt_2} = \dfrac{1}{2}e''_\alpha - e_\alpha^2 e_{-\alpha} \\ \dfrac{de_{-\alpha}}{dt_2} = -\dfrac{1}{2}e''_{-\alpha} + e_\alpha e_{-\alpha}^2\,. \end{cases}$$

Hence, the reduced DS homogeneous hierarchy for the Lie algebra $\mathfrak{g} = \mathfrak{sl}_2$ coincides with the *NLS hierarchy* (AKNS), and $(\overline{H}_0, \overline{H}_1^D)$ coincides with its well-known bi-Poisson structure.

2.5 Dirac Reduced Minimal DS Hierarchy

We recall here the construction of the classical $\mathcal{W}$-algebra associated to a minimal nilpotent element following [4].

Let $\mathfrak{g}$ be a simple Lie algebra with a non-degenerate symmetric invariant bilinear form $(\cdot\,|\,\cdot)$, and let $f \in \mathfrak{g}$ be a minimal nilpotent element, that is a lowest root vector of $\mathfrak{g}$. Let $\{f, h = 2x, e\} \subset \mathfrak{g}$ be an $\mathfrak{sl}_2$-triple. The ad x-eigenspace decomposition is

$$\mathfrak{g} = \mathbb{F}f \oplus \mathfrak{g}_{-\frac{1}{2}} \oplus \mathfrak{g}_0 \oplus \mathfrak{g}_{\frac{1}{2}} \oplus \mathbb{F}e\,.$$

Note that $(x|a) = 0$ for all $a \in \mathfrak{g}_0^f$. Hence, the subalgebra $\mathfrak{g}_0 \subset \mathfrak{g}$ admits the orthogonal decomposition $\mathfrak{g}_0 = \mathfrak{g}_0^f \oplus \mathbb{F}x$. For $a \in \mathfrak{g}_0$ we denote by $a^\sharp$ its projection to $\mathfrak{g}_0^f$.

We fix a basis of $\mathfrak{g}^f = \mathfrak{g}_0^f \oplus \mathfrak{g}_{-\frac{1}{2}} \oplus \mathbb{F}f$ as follows. Let $\{a_i\}_{i\in J_0^f} \subset \mathfrak{g}_0^f$ be an orthonormal basis of $\mathfrak{g}_0^f$ with respect to $(\cdot\,|\,\cdot)$. Let also $\{v_k\}_{k\in J_{-\frac{1}{2}}} \subset \mathfrak{g}_{\frac{1}{2}}$ be a basis of $\mathfrak{g}_{\frac{1}{2}}$ and let $\{v^k\}_{k\in J_{-\frac{1}{2}}} \subset \mathfrak{g}_{\frac{1}{2}}$ be the dual basis with respect to the nondegenerate

Table 2.1 λ-brackets among generators of $\mathcal{W}$ for minimal nilpotent f

$\{\cdot\,{}_\lambda\,\cdot\}_z$	L	b	u_1
L	$(\partial+2\lambda)L$ $-(x\|x)\lambda^3+4(x\|x)z\lambda$	$(\partial+\lambda)b$	$(\partial+\frac{3}{2}\lambda)u_1$
a	λa	$[a,b]+(a\|b)\lambda$	$[a,u_1]$
u	$(\frac{1}{2}\partial+\frac{3}{2}\lambda)u$	$[u,b]$	Eq. (2.28)

skewsymmetric pairing $(f|[\cdot\,,\cdot])$ on $\mathfrak{g}_{\frac{1}{2}}$. Equivalently, letting $u_k=[f,v_k]$, we have that $\{u_k\}_{k\in J_{-\frac{1}{2}}}\subset\mathfrak{g}_{-\frac{1}{2}}$ and $\{v^k\}_{k\in J_{-\frac{1}{2}}}\subset\mathfrak{g}_{\frac{1}{2}}$ are dual bases with respect to $(\cdot\,|\,\cdot)$.

An explicit description of the classical $\mathcal{W}$-algebra $\mathcal{W}=\mathcal{W}_z(\mathfrak{g},f)$, associated to the Lie algebra $\mathfrak{g}$ and the minimal nilpotent element f, is as follows. As a differential algebra, it is $\mathcal{W}=S(\mathbb{F}[\partial]\mathfrak{g}^f)$, namely the algebra of differential polynomials in the differential variables $\{a_i\}_{i\in J_0^f}\subset\mathfrak{g}_0^f$, $\{u_k\}_{k\in J_{\frac{1}{2}}}\subset\mathfrak{g}_{-\frac{1}{2}}$, and f. We also let $L=f+\frac{1}{2}\sum_{i\in J_0^f}a_i^2\in\mathcal{W}$ (which we can take as a differential generator in place of f). The λ-brackets on generators are given by Table 2.1 ($a,b\in\mathfrak{g}_0^f$, $u,u_1\in\mathfrak{g}_{-\frac{1}{2}}$).

The λ-bracket of two elements u and u_1 of $\mathfrak{g}_{-\frac{1}{2}}$ is

$$\begin{aligned}\{u_\lambda u_1\}_z=&\sum_{k\in J_{-\frac{1}{2}}}[u,v^k]^\sharp[u_1,v_k]^\sharp+(\partial+2\lambda)[u,[e,u_1]]^\sharp\\&+\frac{(e|[u,u_1])}{2(x|x)}f-\lambda^2(e|[u,u_1])+z(e|[u,u_1])\,.\end{aligned}\tag{2.28}$$

Associated to this 1-parameter family of PVA λ-brackets we have compatible Poisson structures H_0 and H_1, defined by $\{\cdot\,{}_\lambda\,\cdot\}_z=\{\cdot\,{}_\lambda\,\cdot\}_{H_1}-z\{\cdot\,{}_\lambda\,\cdot\}_{H_0}$. It follows by [3, Theorem 4.18] that we can find an infinite sequence of linearly independent local functionals $\int h_n\in\mathcal{W}/\partial\mathcal{W}$, $n\in\mathbb{Z}_+$, starting with $h_0=f$, satisfying the Lenard-Magri recursive relations (2.15). The first few integrals of motion and the corresponding equations of the integrable hierarchy are computed in [4, Section 6].

Note that, by Table 2.1, all elements of $\mathfrak{g}_0^f$ are central for the Poisson structure H_0. Therefore, by Theorem 2.4 we have a Dirac modified bi-Poisson structure (H_0,H_1^D) on $\mathcal{W}$ with respect to the constraints $\{a_i\}_{i\in J_0^f}$, and the induced Dirac reduced bi-Poisson structure $(\overline{H}_0,\overline{H}_1^D)$ on $\overline{\mathcal{W}}=\mathcal{W}/\langle a_i\rangle_{i\in J_0^f}$. Let $\ell=|J_0^f|$ and $N=|J_0^f|+|J_{-\frac{1}{2}}|+1$. By Definition 2.3 we have $H_1^D=H_1+BC^{-1}B^*$, where $C(\partial)\in\mathrm{Mat}_{\ell\times\ell}\,\mathcal{W}[\partial]$ and $B(\partial)\in\mathrm{Mat}_{N\times\ell}\,\mathcal{W}[\partial]$ are matrix differential operators with entries as follows ($i,j\in J_0^f$, $k\in J_{-\frac{1}{2}}$):

$$\begin{cases} B_{ij}(\partial) = C_{ij}(\partial) = [a_j, a_i] + \delta_{ij}\partial \\ B_{kj}(\partial) = [a_j, u_k]\,, \\ B_{Nj}(\partial) = a_j\partial\,. \end{cases} \tag{2.29}$$

Proposition 2.3. *The Lenard-Magri recursive relations* (2.22) *hold for the bi-Poisson structure* (H_1^D, H_0). *Consequently, we get an induced integrable bi-Hamiltonian hierarchy in* $\overline{\mathcal{W}}$.

Proof. For the first assertion, by Proposition 2.1 it suffices to show that $\operatorname{Ker} B(\partial) = 0$ in $\tilde{\mathcal{K}}^\ell$ (recall that $\tilde{\mathcal{K}}$ denotes the linear closure of $\mathcal{K}$, and $\mathcal{K}$ is the field of fractions of $\mathcal{W}$). Let $F = (F_i)_{i=1}^\ell \in \tilde{\mathcal{K}}$ be an element of the kernel of $B(\partial)$. Looking at the first ℓ rows of the matrix $B(\partial)$, we get the equations

$$F_i' = \sum_{j \in J_0^f} [a_i, a_j] F_j\,, \quad i \in J_0^f\,. \tag{2.30}$$

Let $\mathcal{W}_0 = \mathbb{F}[a_i^{(n)} \mid i \in J_0^f, n \in \mathbb{Z}_+] \subset \mathcal{W}$, be the algebra of differential polynomials in the differential variables $\{a_i\}_{i \in J_0^f}$, let $\mathcal{K}_0$ be its differential field of fractions, and let $\tilde{\mathcal{K}}_0$ be its linear closure. It is a differential subfield of $\tilde{\mathcal{K}}$ with the same subfield of constants $\overline{\mathbb{F}}$. The space of solutions of equation (2.30) in $\tilde{\mathcal{K}}^\ell$ is an $\overline{\mathbb{F}}$-linear subspace of $\tilde{\mathcal{K}}_0^\ell$ of dimension ℓ. Let $E \in \operatorname{Mat}_{\ell\times\ell}\tilde{\mathcal{K}}_0$ be a non-degenerate matrix, whose columns form a basis of the space of solutions of equation (2.30). Then, all solutions of equation (2.30) have the form

$$F = EC\,, \tag{2.31}$$

for some constant vector $C \in \overline{\mathbb{F}}^\ell$, see e.g. [9].

Next, consider the following $N - \ell - 1$ rows of the matrix $B(\partial)$. We get the equations

$$\sum_{i \in J_0^f, k \in J_{-\frac{1}{2}}} [u_k, a_i] E_{ij} C_j = 0\,, \quad k \in J_{-\frac{1}{2}}\,. \tag{2.32}$$

The left hand side of equation (2.32) lies in $\mathfrak{g}_{-\frac{1}{2}} \otimes \tilde{\mathcal{K}}_0$. Hence, we can apply $\xi = (v|\cdot) \in \mathfrak{g}_{\frac{1}{2}}^*$ to it (considered as a linear map $\xi : \mathfrak{g}_{-\frac{1}{2}} \otimes \tilde{\mathcal{K}}_0 \to \tilde{\mathcal{K}}_0$):

$$\sum_{i \in J_0^f, k \in J_{-\frac{1}{2}}} ([v, u_k]|a_i) E_{ij} C_j = 0\,, \tag{2.33}$$

for all $v \in \mathfrak{g}_{\frac{1}{2}}$ and $k \in J_{-\frac{1}{2}}$. Note that $[\mathfrak{g}_{\frac{1}{2}}, \mathfrak{g}_{-\frac{1}{2}}] = \mathfrak{g}_0$ (it follows by the fact that $\mathfrak{g}$ is simple and $\mathbb{F}f \oplus \mathfrak{g}_{-\frac{1}{2}} \oplus [\mathfrak{g}_{\frac{1}{2}}, \mathfrak{g}_{-\frac{1}{2}}] \oplus \mathfrak{g}_{\frac{1}{2}} \oplus \mathbb{F}e$ is clearly an ideal of $\mathfrak{g}$). Since the inner product $(\cdot\,|\,\cdot)$ is non-degenerate on $\mathfrak{g}_0^f$, Eq. (2.33) implies $EC = 0$, from which we get that $C = 0$, by the non-degeneracy of the matrix E. Hence $F = 0$, as required.

For the last assertion, we consider the images of the conserved densities h_n in $\overline{\mathcal{W}}$, and the corresponding variational derivatives $\frac{\delta \bar{h}_n}{\delta u} \in \overline{\mathcal{W}}^{N-\ell}$. It follows by [3, Lem.4.15] that they span an infinite dimensional space. Since $\overline{H}_0$ is a non-degenerate matrix differential operator, it follows that the images of the elements P_n in $\overline{\mathcal{W}}^{N-\ell}$ span an infinite dimensional space as well.

The Dirac reduced Poisson structures are explicitly as follows ($h, k \in J_{-\frac{1}{2}}$):

$$\begin{cases} (\overline{H}_0)_{hk}(\partial) = (e|[u_h, u_k]) \\ (\overline{H}_0)_{NN}(\partial) = -4(x|x)\partial \\ (\overline{H}_0)_{Nk}(\partial) = (\overline{H}_0)_{kN}(\partial) = 0 \end{cases},$$

and

$$\begin{cases} (\overline{H}_1^D)_{hk}(\partial) = \sum\limits_{i\in J_0^f} [a_i, u_h]\partial^{-1} \circ [a_i, u_k] - \dfrac{(e|[u_h, u_k])}{2(x|x)} f + (e|[u_h, u_k])\partial^2 \\ (\overline{H}_1^D)_{NN}(\partial) = f' + 2f\partial - (x|x)\partial^3 \\ (\overline{H}_1^D)_{kN}(\partial) = u_k' + \dfrac{3}{2}u_k\partial \\ (\overline{H}_1^D)_{Nk}(\partial) = \dfrac{1}{2}u_k' + \dfrac{3}{2}u_k\partial \end{cases}.$$

The first two conserved densities are $h_0 = f$ and

$$h_1 = -\frac{1}{8(x|x)} f^2 - \frac{1}{2} \sum_{k\in J_{-\frac{1}{2}}} [f, v^k] u_k',$$

and the first two equations of the reduced bi-Hamiltonian hierarchy are (for $u \in \mathfrak{g}_{-\frac{1}{2}}$) $\frac{du}{dt_0} = u'$, $\frac{df}{dt_0} = f'$, and (cf. [4, eq. (6.17)])

$$\frac{du}{dt_1} = u''' - \frac{3}{4(x|x)} fu' - \frac{3}{8(x|x)} uf' - \frac{1}{2} \sum_{i\in J_0^f, k\in J_{-\frac{1}{2}}} [a_i, u][a^i, [f, v^k]][f, v_k],$$

$$\frac{df}{dt_1} = \frac{1}{4} f''' - \frac{3}{4(x|x)} ff' + \frac{3}{2} \sum_{k\in J_{-\frac{1}{2}}} u_k [f, v^k]''.$$

Table 2.2 λ-brackets among generators of $\mathcal{W}$ for short nilpotent f

$\{\cdot_\lambda\cdot\}_z$	b	u_1
a	$[a,b]+(a\|b)\lambda$	$[a,u_1]$
u	$[u,b]$	Eq. (2.34)

2.6 Dirac Reduced Short DS Hierarchy

Let $\mathfrak{g}$ be a simple Lie algebra with a non-degenerate symmetric invariant bilinear form $(\cdot\,|\,\cdot)$. Recall that, by definition, for a short nilpotent element $f\in\mathfrak{g}$, and an $\mathfrak{sl}_2$-triple $\{f,h=2x,e\}$, we have the ad x-eigenspace decomposition $\mathfrak{g}=\mathfrak{g}_{-1}\oplus\mathfrak{g}_0\oplus\mathfrak{g}_1$. Moreover, we have the orthogonal decomposition $\mathfrak{g}_0=\mathfrak{g}_0^f\oplus[f,\mathfrak{g}_1]$, and $\mathfrak{g}_0^f=\mathfrak{g}_0^e$, $[f,\mathfrak{g}_1]=[e,\mathfrak{g}_{-1}]$. Let $^\sharp:\mathfrak{g}_0\to\mathfrak{g}_0^f$ be the corresponding orthogonal projection. Let $\{a_i\}_{i\in J_0^f}\subset\mathfrak{g}_0^f$ be an orthonormal basis of $\mathfrak{g}_0^f$. Let also $\{u_k\}_{k\in J_{-1}}\subset\mathfrak{g}_{-1}$ be a basis of $\mathfrak{g}_{-1}$ and let $\{u^k\}_{k\in J_{-1}}\subset\mathfrak{g}_1$ be the dual basis with respect to $(\cdot\,|\cdot)$.

The classical $\mathcal{W}$-algebra $\mathcal{W}=\mathcal{W}(\mathfrak{g},f)$ is, as differential algebra, the algebra of differential polynomials in the differential variables $\{a_i\}_{i\in J_0^f}\subset\mathfrak{g}_0^f$ and $\{u_k\}_{k\in J_{-1}}\subset\mathfrak{g}_{-1}$. The λ-bracket on generators are given by Table 2.2 ($a,b\in\mathfrak{g}_0^f$, $u,u_1\in\mathfrak{g}_{-1}$).

The λ-bracket of $u,u_1\in\mathfrak{g}_{-1}$ is

$$
\begin{aligned}
\{u_\lambda u_1\}_z=&\frac{1}{2}\sum_{k\in J_1}(u\circ u_k)[u_1,u^k]^\sharp-\frac{1}{2}\sum_{k\in J_1}(u_1\circ u_k)[u,u^k]^\sharp\\
&+\frac{1}{4}\sum_{h,k\in J_1}[[e,u_h],[e,u_k]][u,u^h]^\sharp[u_1,u^k]^\sharp-\frac{1}{2}(\partial+2\lambda)(u\circ u_1)\\
&+\frac{1}{4}(\partial+2\lambda)\sum_{k\in J_1}[[e,u],[e,u_k]][u_1,u^k]^\sharp+\frac{1}{4}\sum_{k\in J_1}[[e,u_1],[e,u_k]](\partial+\lambda)[u,u^k]^\sharp\\
&-\frac{1}{4}\left(3\lambda^2+3\lambda\partial+\partial^2\right)[[e,u],[e,u_1]]+\frac{1}{4}(e|u\circ u_1)\lambda^3\\
&+z[[e,u],[e,u_1]]^\sharp-(e|u\circ u_1)z\lambda\,,
\end{aligned}
\tag{2.34}
$$

where $u\circ u_1=[[e,u],u_1]$, for all $u,u_1\in\mathfrak{g}_{-1}$.

Associated to this 1-parameter family of PVA λ-brackets we have compatible Poisson structures H_0 and H_1, defined by $\{\cdot_\lambda\cdot\}_z=\{\cdot_\lambda\cdot\}_{H_1}-z\{\cdot_\lambda\cdot\}_{H_0}$. It follows by [3, Theorem 4.18] that we can find an infinite sequence of linearly independent local functionals $\int h_n\in\mathcal{W}/\partial\mathcal{W}$, $n\in\mathbb{Z}_+$, starting with $h_0=f$, satisfying the Lenard-Magri recursive relations (2.15). The first few integrals of motion and the corresponding equations of the integrable hierarchy are computed in [4, Section 7].

Note that, by Table 2.1, all elements of $\mathfrak{g}_0^f$ are central for the Poisson structure H_0. Therefore, by Theorem 2.4 we have a Dirac modified bi-Poisson structure (H_0,H_1^D) on $\mathcal{W}$ with respect to the constraints $\{a_i\}_{i\in J_0^f}$, and the induced Dirac reduced bi-Poisson structure $(\overline{H}_0, \overline{H}_1^D)$ on $\overline{\mathcal{W}} = \mathcal{W}/\langle a_i \rangle_{i \in J_0^f}$. Let $\ell = |J_0^f|$ and $N = |J_0^f| + |J_{-1}|$. By Definition 2.3 we have $H_1^D = H_1 + BC^{-1}B^*$, where $C(\partial) \in \mathrm{Mat}_{\ell\times\ell}\mathcal{W}[\partial]$ and $B(\partial) \in \mathrm{Mat}_{N\times\ell}\mathcal{W}[\partial]$ are matrix differential operators with entries as follows ($i, j \in J_0^f$, $k \in J_{-1}$):

$$\begin{cases} B_{ij}(\partial) = C_{ij}(\partial) = [a_j, a_i] + \delta_{ij}\partial \\ B_{kj}(\partial) = [a_j, u_k]\,. \end{cases} \tag{2.35}$$

Proposition 2.4. *The Lenard-Magri recursive relations* (2.22) *hold for the bi-Poisson structure* (H_1^D, H_0). *Consequently, we get an induced integrable bi-Hamiltonian hierarchy in* $\overline{\mathcal{W}}$.

Proof. It is along the same lines as the proof of Proposition 2.3.

The Dirac reduced Poisson structures are explicitly as follows ($h, k \in J_{-1}$):

$$(\overline{H}_0)_{hk}(\partial) = (e|u_h \circ u_k)\partial\,,$$

and

$$(\overline{H}_1^D)_{hk}(\partial) = \sum_{i \in J_0^f} [a_i, u_h]\partial^{-1} \circ [a_i, u_k] - \frac{1}{2}(u_h \circ u_k)' - (u_h \circ u_k)\partial + \frac{1}{4}(e|u_h \circ u_k)\partial^3\,.$$

The first two equations of the reduced bi-Hamiltonian hierarchy are (for $u \in \mathfrak{g}_{-1}$) $\frac{du}{dt_0} = u'$, and

$$\frac{du}{dt_1} = \frac{1}{4}u''' + \frac{3}{4}\sum_{h,k \in J_{-1}} (u^k * u^h|u)u_h u_k'\,,$$

where $*$ is the Jordan product on $\mathfrak{g}_1$ defined by $a * b = [[f, a], b]$, for every $a, b \in \mathfrak{g}_1$. The last equation is, after a rescaling of the variables, the Svinolupov equation associated to this Jordan product, [7]. We thus provided a bi-Hamiltonian structure for such equation and we proved its integrability.

References

1. A. Barakat, A. De Sole, V.G. Kac, Poisson vertex algebras in the theory of Hamiltonian equations. Jpn. J. Math. **4**(2), 141–252 (2009)
2. A. De Sole, V.G. Kac, Non-local Poisson structures and applications to the theory of integrable systems. Jpn. J. Math. **8**(2), 233–347 (2013)
3. A. De Sole, V.G. Kac, D. Valeri, Classical $\mathcal{W}$-algebras and generalized Drinfeld-Sokolov bi-Hamiltonian systems within the theory of Poisson vertex algebras. Commun. Math. Phys. **323**(2), 663–711 (2013)

4. A. De Sole, V.G. Kac, D. Valeri, Classical $\mathcal{W}$-algebras and generalized Drinfeld-Sokolov hierarchies for minimal and short nilpotents. Comm. Math. Phys. (2014). doi:10.1007/s00220-014-2049-2
5. A. De Sole, V.G. Kac, D. Valeri, Dirac reduction for Poisson vertex algebras. arXiv:1306.6589 [math-ph]
6. V.G. Drinfeld, V.V. Sokolov, Lie algebras and equations of KdV type. Soviet J. Math. **30**, 1975–2036 (1985)
7. S.I. Svinolupov, Jordan algebras and generalized Korteweg-de Vries equations. Teor. Mat. Fiz. **87**(3), 391–403 (1991)
8. S. Carpentier, A. De Sole, V.G. Kac, Singular degree of a rational matrix pseudodifferential operator. IMRN (2014). doi:10.1093/imrn/rnu093
9. S. Carpentier, A. De Sole, V.G. Kac, Rational matrix pseudodifferential operators. Selecta Math. (N.S.) **20**, n.2, 403–419 (2014)
10. S. Carpentier, A. De Sole, V.G. Kac, Some algebraic properties of differential operators. J. Math. Phys. **53**(6), 063501, 12pp. (2012)

Chapter 3
Some Open Problems About Aspherical Closed Manifolds

Wolfgang Lück

Abstract We discuss some open and interesting problems about aspherical closed manifolds including topologically rigidity, Poincaré duality groups and L^2-invariants.

3.1 Introduction

This article is devoted to aspherical closed manifolds and open conjectures, problems and questions about them. All the problems stated here are very interesting and hard. Any progress towards an answer is welcome and valuable. We hope that a reader may be motivated by this note to study them.

We will address the questions whether an aspherical closed manifold is topologically rigid, whether a finitely presented Poincaré duality group is the fundamental group of an aspherical closed manifold, whether an aspherical closed manifold carries an S^1-action or a Riemannian metric with positive scalar curvature, and finally state some conjectures about the possible values of L^2-Betti numbers and L^2-torsion of the universal covering of an aspherical closed manifold and the homological growth in a tower of finite coverings.

3.2 Basics About Aspherical CW-Complexes

A *CW*-complex X is called *aspherical* if it is connected and the nth homotopy group $\pi_n(X)$ vanish for $n \geq 2$, or, equivalently, it is connected and its universal covering is contractible. Two aspherical *CW*-complexes are homotopy equivalent if

W. Lück (✉)
Mathematical Institut of the University at Bonn, Endenicher Allee 60, 53115 Bonn, Germany
e-mail: wolfgang.lueck@him.uni-bonn.de

V. Ancona and E. Strickland (eds.), *Trends in Contemporary Mathematics*,
Springer INdAM Series 8, DOI 10.1007/978-3-319-05254-0_3,

and only if their fundamental groups are isomorphic. This follows from the fact that for any connected CW-complex X and any aspherical CW-complex Y two maps $f_0, f_1 \colon \pi_1(X) \to \pi_1(Y)$ are homotopic if and only if for one (and hence all points) $x \in X$ there exists path w from $f_0(x)$ to $f_1(x)$ such that the composite of the obvious map $c_w \colon \pi_1(X, f_0(x)) \to \pi_1(Y, f_1(x))$ given by conjugation with w and $\pi_1(f_0, x) \colon \pi_1(X, x) \to \pi_1(Y, f_0(x))$ is $\pi_1(f_1, x) \colon \pi_1(X, x) \to \pi_1(Y, f_1(x))$. So the homotopy theory of aspherical CW-complexes is completely determined by their fundamental groups.

Given any group G, there exists a connected aspherical CW-complex X with $\pi_1(X) \cong G$. Since X is unique up to homotopy, one often denotes such a space by BG or $K(G,1)$. One defines the homology $H_*(G)$ of a group G by $H_*(BG)$ and this definition is independent of the choice of model BG by homotopy invariance.

3.3 Basics About Aspherical Closed Manifolds

We are interested in aspherical closed (topological or smooth) manifolds. These exists in abundance.

3.3.1 Non-positive Curvature

Let M be a closed smooth manifold. Suppose that it possesses a Riemannian metric whose sectional curvature is non-positive. Then the universal covering $\tilde{M}$ inherits a complete Riemannian metric whose sectional curvature is non-positive. The Hadamard-Cartan Theorem (see [31, 3.87 on page 134]) implies that $\tilde{M}$ is diffeomorphic to $\mathbb{R}^n$. Hence M is aspherical.

3.3.2 Low-Dimensions

A connected closed 1-dimensional manifold is homeomorphic to S^1 and hence aspherical.

Let M be a connected closed 2-dimensional manifold. Then M is either aspherical or homeomorphic to S^2 or $\mathbb{RP}^2$. The following statements are equivalent: (i) M is aspherical. (ii) M admits a Riemannian metric which is *flat*, i.e., with sectional curvature constant 0, or which is *hyperbolic*, i.e., with sectional curvature constant -1. (iii) The universal covering of M is homeomorphic to $\mathbb{R}^2$.

A connected closed 3-manifold M is called *prime* if for any decomposition as a connected sum $M \cong M_0 \sharp M_1$ one of the summands M_0 or M_1 is homeomorphic to S^3. It is called *irreducible* if any embedded sphere S^2 bounds an embedded disk D^3. Every irreducible closed 3-manifold is prime. A prime closed 3-manifold is either irreducible or an S^2-bundle over S^1 (see [37, Lemma 3.13 on page 28]). A closed orientable 3-manifold is aspherical if and only if it is irreducible and has infinite fundamental group. This follows from the Sphere Theorem [37, Theorem 4.3 on page 40].

3.3.3 *Torsionfree Discrete Subgroups of Almost Connected Lie Groups*

Let L be a Lie group with finitely many path components. Let $K \subseteq L$ be a maximal compact subgroup. Let $G \subseteq L$ be a discrete torsionfree subgroup. Then $M = G\backslash L/K$ is an aspherical closed manifold with fundamental group G since its universal covering L/K is diffeomorphic to $\mathbb{R}^n$ for appropriate n (see [36, Theorem 1. in Chapter VI]). Examples for M are hyperbolic manifolds.

3.3.4 *Hyperbolization*

A very important construction of aspherical manifolds comes from the *hyperbolization technique* due to Gromov [33]. It turns a cell complex into a non-positively curved (and hence aspherical) polyhedron. The rough idea is to define this procedure for simplices such that it is natural under inclusions of simplices and then define the hyperbolization of a simplicial complex by gluing the results for the simplices together as described by the combinatorics of the simplicial complex. The goal is to achieve that the result shares some of the properties of the simplicial complexes one has started with, but additionally to produce a non-positively curved and hence aspherical polyhedron. Since this construction preserves local structures, it turns manifolds into manifolds.

We briefly explain what the *orientable hyperbolization procedure* gives. Further expositions of this construction can be found in [16, 20–22]. We start with a finite-dimensional simplicial complex Σ and assign to it a cubical cell complex $h(\Sigma)$ and a natural map $c: h(\Sigma) \to \Sigma$ with the following properties:

1. $h(\Sigma)$ is non-positively curved and in particular aspherical.
2. The natural map $c: h(\Sigma) \to \Sigma$ induces a surjection on the integral homology.
3. $\pi_1(f): \pi_1(h(\Sigma)) \to \pi_1(\Sigma)$ is surjective.
4. If Σ is an oriented closed manifold, then
   (a) $h(\Sigma)$ is an oriented closed manifold.
   (b) The natural map $c: h(\Sigma) \to \Sigma$ has degree one.
   (c) There is a stable isomorphism between the tangent bundle $Th(\Sigma)$ and the pullback $c^*T\Sigma$.

One can deduce from this construction that the condition aspherical does not impose any restrictions on the characteristic numbers of a manifold or on its bordism class, see [20, Remarks 15.1] and [22, Theorem B]. Moreover, it can be used to construct aspherical closed manifolds with rather exotics properties, for instance examples which do not possess a triangulation, whose universal covering is not homeomorphic to $\mathbb{R}^n$, whose fundamental group contains an infinite divisible abelian group or has an unsolvable word problem. For such exotic examples and more information about aspherical closed manifolds we refer for instance to [9, 20, 22, 47].

3.4 The Borel Conjecture

In this section we deal with

Conjecture 3.1 (Borel Conjecture for a group G). If M and N are aspherical closed manifolds with $\pi_1(M) \cong \pi_1(N) \cong G$, then M and N are homeomorphic and any homotopy equivalence $M \to N$ is homotopic to a homeomorphism.

The main tool to attack the Borel Conjecture is surgery theory and the Farrell-Jones Conjecture. We consider the following special version of the Farrell-Jones Conjecture.

Conjecture 3.2 (Farrell-Jones Conjecture for torsionfree groups and integer coefficients). Let G be a torsionfree group Then:

1. $K_n(\mathbb{Z}G) = 0$ for $n \leq -1$.
2. The reduced projective class group $\tilde{K}_0(\mathbb{Z}G)$ vanishes.
3. The Whitehead group $\mathrm{Wh}(G)$ vanishes.
4. For any homomorphism $w\colon G \to \{\pm 1\}$ the w-twisted L-theoretic assembly map $H_n(BG; {}^w\mathbf{L}^{\langle -\infty \rangle}) \xrightarrow{\cong} L_n^{\langle -\infty \rangle}(RG, w)$ is bijective.

The relevance of the Conjecture 3.2 for the Borel Conjecture comes from the next theorem whose proof is based on surgery theory.

Theorem 3.1 (The Farrell-Jones Conjecture and the Borel Conjecture). *Suppose that G satisfies the version of the Farrell-Jones Conjecture stated in Conjecture 3.2.*

Then the Borel Conjecture is true for aspherical closed manifolds of dimension ≥ 5 with G as fundamental group. It is true for aspherical closed manifolds of dimension 4 with G as fundamental group if G is good in the sense of Freedman (see [29, 30]).

Remark 3.1 (The Borel Conjecture in low dimensions). The Borel Conjecture is true in dimension ≤ 2 by the classification of closed manifolds of dimension ≤ 2. It is true in dimension 3 if Thurston's Geometrization Conjecture is true. This follows from results of Waldhausen (see Hempel [37, Lemma 10.1 and Corollary 13.7]) and Turaev (see [61]) as explained for instance in [42, Section 5]. A proof of Thurston's Geometrization Conjecture is given in [50] following ideas of Perelman.

Remark 3.2 (The Borel Conjecture does not hold in the smooth category). The Borel Conjecture 3.1 is false in the smooth category, i.e., if one replaces topological manifold by smooth manifold and homeomorphism by diffeomorphism. The torus T^n for $n \geq 5$ is an example (see [62, 15A]). Other counterexample involving negatively curved manifolds are constructed by Farrell-Jones [24, Theorem 0.1].

Remark 3.3 (The Borel Conjecture versus Mostow rigidity). A version of *Mostow rigidity* says for two closed hyperbolic manifolds N_0 and N_1 that they are isometrically diffeomorphic if and only if $\pi_1(N_0) \cong \pi_1(N_1)$ and any homotopy equivalence $N_0 \to N_1$ is homotopic to an isometric diffeomorphism.

One may view the Borel Conjecture as the topological version of Mostow rigidity. The conclusion in the Borel Conjecture is weaker, one gets only homeomorphisms and not isometric diffeomorphisms, but the assumption is also weaker, since there are many more aspherical closed topological manifolds than hyperbolic closed manifolds.

The following is known about the Farrell-Jones Conjecture, see for instance [3–5, 7, 8, 39, 59, 63].

Theorem 3.2. *Let $\mathcal{C}$ be the smallest class of groups satisfying:*

- *Every hyperbolic group belongs to $\mathcal{C}$.*
- *Every group that acts properly, isometrically and cocompactly on a complete proper $\mathrm{CAT}(0)$-space belongs to $\mathcal{C}$.*
- *Every lattice in an almost connected Lie group belongs to $\mathcal{C}$.*
- *Every virtually solvable group belongs to $\mathcal{C}$.*
- *Every arithmetic groups belongs to $\mathcal{C}$.*
- *The fundamental group of any 3-manifold (possibly with boundary and possibly non-compact) belongs to $\mathcal{C}$.*
- *If G_1 and G_2 belong to $\mathcal{C}$, then both $G_1 * G_2$ and $G_1 \times G_2$ belong to $\mathcal{C}$.*
- *If H is a subgroup of G and $G \in \mathcal{C}$, then $H \in \mathcal{C}$.*
- *Let $\{G_i \mid i \in I\}$ be a directed system of groups (with not necessarily injective structure maps) such that $G_i \in \mathcal{C}$ for every $i \in I$. Then the directed colimit $\mathrm{colim}_{i \in I} G_i$ belongs to $\mathcal{C}$.*

Then every group G in $\mathcal{C}$ satisfies the K- and L-theoretic Farrell-Jones Conjecture with coefficients in additive categories and with finite wreath products, and in particular the version of the Farrell-Jones Conjecture stated in Conjecture 3.2.

For more information about the Borel and the Farrell-Jones Conjecture and literature about them we refer for instance to [25, 46, 49].

3.5 Poincaré Duality Groups

The following definition is due to Johnson-Wall [38].

Definition 3.1 (Poincaré duality group). A group G is called a *Poincaré duality group of dimension n* if the following conditions holds:

1. The group G is of type FP, i.e., the trivial $\mathbb{Z}G$-module $\mathbb{Z}$ possesses a finite-dimensional projective $\mathbb{Z}G$-resolution by finitely generated projective $\mathbb{Z}G$-modules.
2. We get an isomorphism of abelian groups

$$H^i(G;\mathbb{Z}G) \cong \begin{cases} \{0\} & \text{for } i \neq n, \\ \mathbb{Z} & \text{for } i = n. \end{cases}$$

If G is the fundamental group of an aspherical closed manifold of dimension n, then it is finitely presented and a Poincaré duality group of dimension n. This leads to

Conjecture 3.3 (Poincaré duality groups). A finitely presented group is a n-dimensional Poincaré duality group if and only if it is the fundamental group of an aspherical closed n-dimensional topological manifold.

Conjecture 3.3 is known to be true if $n = 1, 2$. This is obvious for $n = 1$ and for $n = 2$ proved in [23, Theorem 2].

A topological space X is called an *absolute neighborhood retract* or briefly ANR if for every normal space Z, every closed subset $Y \subseteq Z$ and every (continuous) map $f\colon Y \to X$ there exists an open neighborhood U of Y in Z together with an extension $F\colon U \to Z$ of f to U. A *compact n-dimensional homology* ANR-*manifold* X is a compact absolute neighborhood retract such that it has a countable basis for its topology, has finite topological dimension and for every $x \in X$ the abelian group $H_i(X, X - \{x\})$ is trivial for $i \neq n$ and infinite cyclic for $i = n$. A closed n-dimensional topological manifold is an example of a compact n-dimensional homology ANR-manifold (see [19, Corollary 1A in V.26 page 191]).

The *disjoint disk property* says that for any $\epsilon > 0$ and maps $f, g\colon D^2 \to M$ there are maps $f', g'\colon D^2 \to M$ so that the distance between f and f' and the distance between g and g' are bounded by ϵ and $f'(D^2) \cap g'(D^2) = \emptyset$.

Theorem 3.3. *Let G be a finitely presented group and $n \geq 6$ be a natural number. Suppose that G satisfies the version of the Farrell-Jones Conjecture 3.2.*

Then G is the fundamental group of a compact homology ANR-*manifold of dimension n satisfying the disjoint disk property if and only if G is an n-dimensional Poincaré duality group.*

Proof. See [13, Main Theorem on page 439 and Section 8], [14, Theorem A and Theorem B], and [57, Remark 25.13 on page 297].

One would prefer if in the conclusion of Theorem 3.3 one could replace "compact homology ANR-manifold" by "closed topological manifold". The remaining obstruction is the *resolution obstruction* of Quinn which takes values in $1 + 8 \cdot \mathbb{Z}$. Any element in $1 + 8 \cdot \mathbb{Z}$ can be realized by an appropriate compact homology ANR-manifold as its *resolution obstruction*. There are compact homology ANR-manifolds that are not homotopy equivalent to closed manifolds. But no example of an aspherical compact homology ANR-manifold that is not homotopy equivalent to a closed topological manifold is known. So we could replace in the conclusion of Theorem 3.3 "compact homology ANR-manifold" by "closed topological manifold" if the following question has a positive answer.

Question 3.1 (Vanishing of the resolution obstruction in the aspherical case). Is every aspherical compact homology ANR-manifold having the DDP homotopy equivalent to a closed manifold?

We refer for instance to [13, 26, 55–57] for more information about this topic.

The question which hyperbolic groups arise as fundamental groups of aspherical closed manifolds of dimension n and which torsionfree hyperbolic groups have a sphere S^{n-1} as boundary is answered by Bartels-Lück-Weinberger [6] in dimension $n \geq 6$.

3.6 S^1-Actions

Let M be a closed aspherical manifold with a non-trivial S^1-action. Then the S^1-action has only finite isotropy groups, the inclusion of any orbit induces an injection on the fundamental group and the center of $\pi_1(X)$ contains an infinite normal cyclic subgroup. A proof can be found for instance in [17] or [44, Corollary 1.43 on page 48]. Conner-Raymond [17] conjectured that an aspherical closed manifold whose fundamental group has a non-trivial center admits a non-trivial S^1-action. This conjecture has been disproved Cappell-Weinberger-Yan [15]. One may still ask the following question

Question 3.2 (S^1-actions). If M is an aspherical closed manifold whose fundamental group has a non-trivial center, is there a finite covering which admits a non-trivial S^1-action?

3.7 Fiber Bundles

Question 3.3 (Fiber bundles). Let $f: M \to N$ be a map of aspherical closed manifolds which induces a surjection on fundamental groups.

Under which conditions is it homotopy equivalent to the projection of a locally trivial topological fiber bundle (or to a Manifold Approximate Fibration)?

A necessary condition for a positive answer is that the homotopy fiber has the homotopy type of a finite CW-complex. If the homotopy fiber is a point, or equivalently, if f is a homotopy equivalence, a positive answer (for a locally trivial fiber bundle) is equivalent to the statement that f is homotopic to a homeomorphism, in other words Question 3.3 becomes the Borel Conjecture 3.1.

3.8 L^2-Invariants

Next we mention some prominent conjectures about aspherical closed manifolds and L^2-invariants. For more information about these conjectures and their status we refer to [10,44,45]. We denote by $b_p^{(2)}(\tilde{M})$ the p-th L^2-Betti number and by $\rho^{(2)}(\tilde{M})$ the L^2-torsion of the universal covering $\tilde{M}$ of a closed manifold M.

3.8.1 *The Hopf and the Singer Conjecture*

Conjecture 3.4 (Hopf Conjecture). If M is an aspherical closed manifold of even dimension, then

$$(-1)^{\dim(M)/2} \cdot \chi(M) \geq 0.$$

If M is a closed Riemannian manifold of even dimension with sectional curvature $\sec(M)$, then

$$\begin{aligned} (-1)^{\dim(M)/2} \cdot \chi(M) &> 0 \quad \text{if } \sec(M) < 0, \\ (-1)^{\dim(M)/2} \cdot \chi(M) &\geq 0 \quad \text{if } \sec(M) \leq 0, \\ \chi(M) &\geq 0 \quad \text{if } \sec(M) \geq 0, \\ \chi(M) &> 0 \quad \text{if } \sec(M) > 0. \end{aligned}$$

Conjecture 3.5 (Singer Conjecture). If M is an aspherical closed manifold, then

$$b_p^{(2)}(\tilde{M}) = 0 \qquad \text{if } 2p \neq \dim(M).$$

If M is a closed connected Riemannian manifold with negative sectional curvature, then

$$b_p^{(2)}(\tilde{M}) \begin{cases} = 0 & \text{if } 2p \neq \dim(M), \\ > 0 & \text{if } 2p = \dim(M). \end{cases}$$

3.8.2 L^2-Torsion and Aspherical Closed Manifolds

Conjecture 3.6 (L^2-torsion for aspherical closed manifolds). If M is an aspherical closed manifold of odd dimension, then $\tilde{M}$ is det-L^2-acyclic and

$$(-1)^{\frac{\dim(M)-1}{2}} \cdot \rho^{(2)}(\tilde{M}) \geq 0.$$

If M is a closed connected Riemannian manifold of odd dimension with negative sectional curvature, then $\tilde{M}$ is det-L^2-acyclic and

$$(-1)^{\frac{\dim(M)-1}{2}} \cdot \rho^{(2)}(\tilde{M}) > 0.$$

If M is an aspherical closed manifold whose fundamental group contains an amenable infinite normal subgroup, then $\tilde{M}$ is det-L^2-acyclic and

$$\rho^{(2)}(\tilde{M}) = 0.$$

3.8.3 Simplicial Volume and L^2-Invariants

Conjecture 3.7 (Simplicial volume and L^2-invariants). Let M be an aspherical closed orientable manifold. Suppose that its simplicial volume $||M||$ vanishes. Then $\tilde{M}$ is of determinant class and

$$\begin{aligned} b_p^{(2)}(\tilde{M}) &= 0 \qquad \text{for } p \geq 0, \\ \rho^{(2)}(\tilde{M}) &= 0. \end{aligned}$$

3.8.4 Zero-in-the-Spectrum Conjecture

Conjecture 3.8 (Zero-in-the-spectrum Conjecture). Let $\tilde{M}$ be a complete Riemannian manifold. Suppose that $\tilde{M}$ is the universal covering of an aspherical closed Riemannian manifold M (with the Riemannian metric coming from M). Then for some $p \geq 0$ zero is in the Spectrum of the minimal closure

$$(\Delta_p)_{\min} : \operatorname{dom}\big((\Delta_p)_{\min}\big) \subset L^2\Omega^p(\tilde{M}) \to L^2\Omega^p(\tilde{M})$$

of the Laplacian acting on smooth p-forms on $\tilde{M}$.

3.8.5 Homological Growth

Here is a generalization of a conjecture due to Bergeron-Venkatesh [10, Conjecture 1.3].

Conjecture 3.9 (Homological growth and L^2-torsion for aspherical closed manifolds).

Let M be an aspherical closed manifold of dimension d and fundamental group $G = \pi_1(M)$. Let $G = G_0 \supseteq G_1 \supseteq \cdots$ be a descending sequence of in G normal subgroups $[G : G_i]$ with trivial intersection $\bigcap_{i \geq 0} G_i = \{1\}$. Put $M[i] = G_i \backslash \tilde{M}$, where $\tilde{M}$ is the universal covering. Let F be an field. Then

1. We get for any $p \geq 0$

$$b_p^{(2)}(\tilde{M}) = \lim_{i \to \infty} \frac{b_n(M[i]; F)}{[G : G_i]}.$$

2. We get for any natural number p with $2p + 1 \neq d$

$$\lim_{i \to \infty} \frac{\ln\left(\left|\operatorname{tors}\left(H_p(M[i])\right)\right|\right)}{[G : G_i]} = 0,$$

and we get in the case $d = 2p + 1$

$$\lim_{i \to \infty} \frac{\ln\left(\left|\operatorname{tors}\left(H_n(M[i])\right)\right|\right)}{[G : G_i]} = (-1)^p \cdot \rho^{(2)}\big(\tilde{M}\big).$$

Some evidence for Conjecture 3.9 comes from [10] and [48].

3.9 Positive Scalar Curvature

Conjecture 3.10. An aspherical closed smooth manifold does not admit a Riemannian metric of positive scalar curvature.

Some evidence comes from the following fact. Let M be an aspherical closed smooth manifold whose fundamental group $\pi = \pi_1(M)$ satisfies the *strong Novikov Conjecture*, i.e., the assembly map $K_n(B\pi) \to K_n(C_r^*(\pi))$ from the K-homology of BG to the topological K-theory of the reduced group C^*-algebra is rationally injective for all $n \in \mathbb{Z}$. Then M carries no Riemannian metric of positive scalar curvature, see [58, Theorem 3.5]. Moreover, M satisfies the Zero-in-the-Spectrum Conjecture 3.8, see [43, Corollary 4]. We refer to [49, Section 5.1.3] for a discussion about the large class of groups for which the assembly map $K_n(BG) \to K_n(C_r^*(G))$ is known to be injective or rationally injective. More information about the Novikov Conjecture can be found in for instance in [27,28,41].

3.10 Random Closed Manifolds

The idea of a random group has successfully been used to construct groups with certain properties, see for instance [2, 32, 34, 35, 51–53, 60, 64]. For example, in a precise statistical sense almost all finitely presented groups are torsionfree hyperbolic and in particular have a finite model for their classifying space.

It is not clear what it means in a precise sense to talk about a random closed manifold. Nevertheless, the author's intuition is that almost all closed manifolds are aspherical. (A related question would be whether a random closed smooth manifold admits a Riemannian metric with non-positive sectional curvature.) It is certainly true in dimension 2 since only finitely many closed surfaces are not aspherical. The characterization of closed 3-dimensional manifolds in Sect. 3.3.2 seems to fit as well.

A closed manifold M is called *asymmetric* if every finite group which acts effectively on M is trivial. This is equivalent to the statement that for any choice of Riemannian metric on M the group of isometries is trivial (see [40, Introduction]). A survey on asymmetric closed manifolds can be found in [54]. The first constructions of asymmetric aspherical closed manifolds are due to Connor-Raymond-Weinberger [18]. The first simply-connected asymmetric manifold has been constructed by Kreck [40] answering a question of Raymond and Schultz [12, page 260] which was repeated by Adem and Davis [1] in their problem list. Raymond and Schultz expressed also their feeling that a random manifold should be asymmetric. Borel has shown that an aspherical closed manifold is asymmetric if its fundamental group is centerless and its outer automorphism group is torsionfree (see the manuscript "On periodic maps of certain $K(\pi, 1)$" in [11, pages 57–60]).

This leads to the intuitive assertion:

Almost all closed manifolds are aspherical, topologically rigid in the sense of the Borel Conjecture 3.1 and asymmetric.

References

1. A. Adem, J.F. Davis, Topics in transformation groups, in *Handbook of Geometric Topology* (North-Holland, Amsterdam, 2002), pp. 1–54
2. G. Arzhantseva, T. Delzant, Examples of random groups (2008, Preprint)
3. A. Bartels, W. Lück, The Borel conjecture for hyperbolic and CAT(0)-groups. Ann. Math. (2), **175**, 631–689 (2012)
4. A. Bartels, S. Echterhoff, W. Lück, Inheritance of isomorphism conjectures under colimits, in *K-Theory and Noncommutative Geometry*, ed. by Cortinaz, Cuntz, Karoubi, Nest, and Weibel, EMS-Series of Congress Reports (European Mathematical Society, 2008), pp. 41–70
5. A. Bartels, W. Lück, H. Reich, The K-theoretic Farrell-Jones conjecture for hyperbolic groups. Invent. Math. **172**(1), 29–70 (2008)
6. A. Bartels, W. Lück, S. Weinberger, On hyperbolic groups with spheres as boundary. J. Differ. Geom. **86**(1), 1–16 (2010)
7. A. Bartels, T. Farrell, W. Lück, The Farrell-Jones Conjecture for cocompact lattices in virtually connected Lie groups. JAMS **27**(2), 339–388 (2014)

8. A. Bartels, W. Lück, H. Reich, H. Rüping, K- and L-theory of group rings over $GL_n(\mathbb{Z})$. Publ. Math. IHES **119**, 97–125 (2014)
9. I. Belegradek, Aspherical manifolds, relative hyperbolicity, simplicial volume and assembly maps. Algebr. Geom. Topol. **6**, 1341–1354 (2006) (electronic)
10. N. Bergeron, A. Venkatesh, The asymptotic growth of torsion homology for arithmetic groups. J. Inst. Math. Jussieu **12**(2), 391–447 (2013)
11. A. Borel, *Œuvres: Collected Papers. Vol. III, 1969–1982* (Springer, Berlin, 1983)
12. W. Browder, W.C. Hsiang, Some problems on homotopy theory manifolds and transformation groups, in *Algebraic and Geometric Topology (Proc. Sympos. Pure Math., Stanford Univ., Stanford, Calif., 1976), Part 2*, Proc. Sympos. Pure Math., XXXII (American Mathematical Society, Providence, 1978), pp. 251–267
13. J. Bryant, S. Ferry, W. Mio, S. Weinberger, Topology of homology manifolds. Ann. Math. (2) **143**(3), 435–467 (1996)
14. J. Bryant, S. Ferry, W. Mio, S. Weinberger, Desingularizing homology manifolds. Geom. Topol. **11**, 1289–1314 (2007)
15. S.E. Cappell, S. Weinberger, M. Yan, Closed aspherical manifolds with center. J. Topol. **6**, 1009–1018 (2014)
16. R.M. Charney, M.W. Davis, Strict hyperbolization. Topology, **34**(2), 329–350 (1995)
17. P.E. Conner, F. Raymond, Actions of compact Lie groups on aspherical manifolds, in *Topology of Manifolds (Proc. Inst., Univ. of Georgia, Athens, Ga., 1969)* (Markham, Chicago, 1970) pp. 227–264
18. P.E. Conner, F. Raymond, P.J. Weinberger, Manifolds with no periodic maps, in *Proceedings of the Second Conference on Compact Transformation Groups (Univ. Massachusetts, Amherst, Mass., 1971), Part II*, pp. 81–108. Lecture Notes in Math., Vol. 299 (Springer, Berlin, 1972)
19. R.J. Daverman, *Decompositions of Manifolds*. Volume 124 of Pure and Applied Mathematics (Academic Press Inc., Orlando, 1986)
20. M. Davis, Exotic aspherical manifolds, in ed. by T. Farrell, L. Göttsche, W. Lück, *High Dimensional Manifold Theory*, number 9 in ICTP Lecture Notes, pp. 371–404. Abdus Salam International Centre for Theoretical Physics, Trieste, 2002. Proceedings of the summer school "High dimensional manifold theory" in Trieste May/June 2001, Number 2. http://www.ictp.trieste.it/~pub_off/lectures/vol9.html
21. M.W. Davis, *The Geometry and Topology of Coxeter Groups*. Volume 32 of London Mathematical Society Monographs Series (Princeton University Press, Princeton, 2008)
22. M.W. Davis, T. Januszkiewicz, Hyperbolization of polyhedra. J. Differ. Geom. **34**(2), 347–388 (1991)
23. B. Eckmann, P.A. Linnell, Poincaré duality groups of dimension two. II. Comment. Math. Helv. **58**(1), 111–114 (1983)
24. F.T. Farrell, L.E. Jones, Negatively curved manifolds with exotic smooth structures. J. Am. Math. Soc. **2**(4), 899–908 (1989)
25. F.T. Farrell, L.E. Jones, Isomorphism conjectures in algebraic K-theory. J. Am. Math. Soc. **6**(2), 249–297 (1993)
26. S.C. Ferry, E.K. Pedersen, Epsilon surgery theory, in *Novikov Conjectures, Index Theorems and Rigidity, Vol. 2 (Oberwolfach, 1993)* (Cambridge University Press, Cambridge, 1995), pp. 167–226
27. S.C. Ferry, A.A. Ranicki, J. Rosenberg (eds.), *Novikov Conjectures, Index Theorems and Rigidity*, vol. 1. (Cambridge University Press, Cambridge, 1995). Including papers from the conference held at the Mathematisches Forschungsinstitut Oberwolfach, Oberwolfach, 6–10 Sept 1993
28. S.C. Ferry, A.A. Ranicki, J. Rosenberg (eds.), *Novikov Conjectures, Index Theorems and Rigidity*, vol. 2. (Cambridge University Press, Cambridge, 1995). Including papers from the conference held at the Mathematisches Forschungsinstitut Oberwolfach, Oberwolfach, 6–10 Sept 1993
29. M.H. Freedman, The topology of four-dimensional manifolds. J. Differ. Geom. **17**(3), 357–453 (1982)

30. M.H. Freedman, The disk theorem for four-dimensional manifolds, in *Proceedings of the International Congress of Mathematicians, Vol. 1, 2 (Warsaw, 1983)* (PWN, Warsaw, 1984), pp. 647–663
31. S. Gallot, D. Hulin, J. Lafontaine, *Riemannian Geometry* (Springer, Berlin, 1987)
32. É. Ghys, Groupes aléatoires (d'après Misha Gromov,. . .). Astérisque **294**, viii, 173–204 (2004)
33. M. Gromov, Hyperbolic groups, in *Essays in Group Theory* (Springer, New York, 1987), pp. 75–263
34. M. Gromov, Asymptotic invariants of infinite groups, in *Geometric Group Theory, Vol. 2 (Sussex, 1991)* (Cambridge University Press, Cambridge, 1993), pp. 1–295
35. M. Gromov, Random walk in random groups. Geom. Funct. Anal. **13**(1), 73–146 (2003)
36. S. Helgason, *Differential Geometry, Lie Groups, and Symmetric Spaces* (American Mathematical Society, Providence, 2001). Corrected reprint of the 1978 original
37. J. Hempel, 3-*Manifolds*. Annals of Mathematics Studies, vol. 86 (Princeton University Press, Princeton, 1976)
38. F.E.A. Johnson, C.T.C. Wall, On groups satisfying Poincaré duality. Ann. Math. (2), **96**, 592–598 (1972)
39. H. Kammeyer, W. Lück, H. Rüping, The Farrell-Jones Conjecture for arbitrary lattices in virtually connected Lie groups. (2014, in preparation)
40. M. Kreck, Simply connected asymmetric manifolds. J. Topol. **2**(2), 249–261 (2009)
41. M. Kreck, W. Lück, *The Novikov Conjecture: Geometry and Algebra*. Volume 33 of Oberwolfach Seminars (Birkhäuser Verlag, Basel, 2005)
42. M. Kreck, W. Lück, Topological rigidity for non-aspherical manifolds. Pure Appl. Math. Quart. **5**(3), 873–914 (2009). special issue in honor of Friedrich Hirzebruch
43. J. Lott, The zero-in-the-spectrum question. Enseign. Math. (2) **42**(3–4), 341–376 (1996)
44. W. Lück, *L^2-Invariants: Theory and Applications to Geometry and K-Theory*. Volume 44 of Ergebnisse der Mathematik und ihrer Grenzgebiete. 3. Folge. A Series of Modern Surveys in Mathematics [Results in Mathematics and Related Areas. 3rd Series. A Series of Modern Surveys in Mathematics] (Springer, Berlin, 2002)
45. W. Lück, L^2-invariants from the algebraic point of view, in *Geometric and Cohomological Methods in Group Theory*. Volume 358 of London Mathematical Society Lecture Note Series (Cambridge University Press, Cambridge, 2009), pp. 63–161
46. W. Lück, K- and L-theory of group rings, in *Proceedings of the 26-th ICM in Hyderabad, August 19–27, 2010*, vol. II, ed. by R. Bhatia (World Scientific, 2011), pp. 1071–1098
47. W. Lück, Aspherical manifolds. *Bulletin of the Manifold Atlas 2012*, pp. 1–17 (2012)
48. W. Lück, Approximating L^2-invariants and homology growth. Geom. Funct. Anal. **23**(2), 622–663 (2013)
49. W. Lück, H. Reich The Baum-Connes and the Farrell-Jones conjectures in K- and L-theory, in *Handbook of K-Theory. Vol. 1, 2* (Springer, Berlin, 2005), pp. 703–842
50. J. Morgan, G. Tian, Completion of the proof of the Geometrization Conjecture (2008, preprint). arXiv:0809.4040v1 [math.DG]
51. Y. Ollivier, *A January 2005 Invitation to Random Groups*. Volume 10 of Ensaios Matemáticos [Mathematical Surveys] (Sociedade Brasileira de Matemática, Rio de Janeiro, 2005)
52. A.Y. Olshanskii, Almost every group is hyperbolic. Int. J. Algebra Comput. **2**(1), 1–17 (1992)
53. P. Pansu, Groupes aléatoires, in *Groupes et géométrie*. Volume 2003 of SMF Journ. Annu. (Société Mathématique de France, Paris, 2003), pp. 37–49
54. V. Puppe, Do manifolds have little symmetry? J. Fixed Point Theory Appl. **2**(1), 85–96 (2007)
55. F. Quinn, Resolutions of homology manifolds and the topological characterization of manifolds. Invent. Math. **72**, 267–284 (1983)
56. F. Quinn, An obstruction to the resolution of homology manifolds. Michigan Math. J. **34**(2), 285–291 (1987)
57. A.A. Ranicki, *Algebraic L-Theory and Topological Manifolds* (Cambridge University Press, Cambridge, 1992)
58. J. Rosenberg, C^*-algebras, positive scalar curvature, and the Novikov conjecture. Inst. Hautes Études Sci. Publ. Math. **58**, 197–212 (1984) 1983

59. H. Rüping, The Farrell-Jones conjecture for S-arithmetic groups (2013, preprint). arXiv:1309.7236 [math.KT]
60. L. Silberman, Addendum to: "Random walk in random groups" [Geom. Funct. Anal. **13**(1), 73–146 (2003); mr1978492] by M. Gromov. Geom. Funct. Anal. **13**(1), 147–177 (2003)
61. V.G. Turaev, Homeomorphisms of geometric three-dimensional manifolds. Mat. Zametki **43**(4), 533–542, 575 (1988). Translation in Math. Notes **43**(3–4), 307–312 (1988)
62. C.T.C. Wall, *Surgery on Compact Manifolds*. Volume 69 of Mathematical Surveys and Monographs, 2nd edn., ed. with a foreword by A. A. Ranicki (American Mathematical Society, Providence, 1999)
63. C. Wegner, The Farrell-Jones conjecture for virtually solvable groups (2013, preprint). arXiv:1308.2432 [math.GT]
64. A. Żuk, Property (T) and Kazhdan constants for discrete groups. Geom. Funct. Anal. **13**(3), 643–670 (2003)

Chapter 4
Quantum Statistical Mechanics, L-Series and Anabelian Geometry I: Partition Functions

Gunther Cornelissen and Matilde Marcolli

Abstract The zeta function of a number field can be interpreted as the partition function of an associated quantum statistical mechanical (QSM) system, built from abelian class field theory.

We introduce a general notion of isomorphism of QSM-systems and prove that it preserves (extremal) KMS equilibrium states.

We prove that two number fields with isomorphic quantum statistical mechanical systems are arithmetically equivalent, i.e., have the same zeta function. If one of the fields is normal over $\mathbb{Q}$, this implies that the fields are isomorphic. Thus, in this case, isomorphism of QSM-systems is the same as isomorphism of number fields, and the *noncommutative* space built from the *abelianized* Galois group can replace the *anabelian* absolute Galois group from the theorem of Neukirch and Uchida.

4.1 Introduction

The starting point for this study is the observation that the zeta function of a number field $\mathbb{K}$ can be realized as the partition function of a quantum statistical mechanical (QSM) system in the style of Bost and Connes (cf. [3] for $\mathbb{K} = \mathbb{Q}$). The QSM-systems for general number fields that we consider are those that were

---

This paper is an updated version of part of [9]. We have split the original preprint into various parts, depending on the methods that are used in them. In the current part, these belong mainly to mathematical physics.

G. Cornelissen (✉)
Mathematisch Instituut, Universiteit Utrecht, Postbus 80.010, 3508 TA Utrecht, Nederland
e-mail: g.cornelissen@uu.nl

M. Marcolli
Mathematics Department, Mail Code 253-37, Caltech, 1200 E. California Blvd. Pasadena, CA 91125, USA
e-mail: matilde@caltech.edu

V. Ancona and E. Strickland (eds.), *Trends in Contemporary Mathematics*, Springer INdAM Series 8, DOI 10.1007/978-3-319-05254-0_4,

constructed by Ha and Paugam (see section 8 of [12], which is a specialization of their more general class of QSM-systems associated to Shimura varieties), and further studied by Laca, Larsen and Neshveyev in [16]. This quantum statistical mechanical system $(A_{\mathbb{K}}, \sigma_{\mathbb{K}})$ consists of a C^*-algebra $A_{\mathbb{K}}$ (the noncommutative analogue of a topological space) with a time evolution $\sigma_{\mathbb{K}}$ (i.e., a continuous group homomorphism $\mathbb{R} \to \mathrm{Aut}\, A_{\mathbb{K}}$). The structure of the algebra is that of a semigroup crossed product

$$A_{\mathbb{K}} := C(X_{\mathbb{K}}) \rtimes J_{\mathbb{K}}^{+}, \text{ with } X_{\mathbb{K}} := G_{\mathbb{K}}^{\mathrm{ab}} \times_{\hat{\mathcal{O}}_{\mathbb{K}}^{*}} \hat{\mathcal{O}}_{\mathbb{K}},$$

where $\hat{\mathcal{O}}_{\mathbb{K}}$ is the ring of finite integral adeles and $J_{\mathbb{K}}^{+}$ is the semigroup of ideals, which acts on the space $X_{\mathbb{K}}$ by Artin reciprocity. The time evolution is only non-trivial on elements $\mu_{\mathfrak{n}} \in A_{\mathbb{K}}$ corresponding to ideals $\mathfrak{n} \in J_{\mathbb{K}}^{+}$, where it acts by multiplication with the norm $N(\mathfrak{n})^{it}$. For exact definitions, see Sect. 4.3.

We call two general QSM-systems *isomorphic* if there is a C^*-algebra isomorphism between the algebras that intertwines the time evolutions. In Sect. 4.2, we prove that such an isomorphism induces a homeomorphism between (extremal) KMS-equilibrium states of the systems at a given temperature.

Our main result for the QSM-systems of number fields is:

Theorem (= Theorem 4.1 below). *If the QSM-systems $(A_{\mathbb{K}}, \sigma_{\mathbb{K}})$ and $(A_{\mathbb{L}}, \sigma_{\mathbb{L}})$ of two number fields $\mathbb{K}$ and $\mathbb{L}$ are isomorphic, then $\mathbb{K}$ and $\mathbb{L}$ are arithmetically equivalent, i.e., they have the same Dedekind zeta function.*

Using some other known consequences of arithmetical equivalence, we get the following ([19], Theorem 1): if number fields $\mathbb{K}$ and $\mathbb{L}$ have isomorphic QSM-systems, then, for any rational prime p, there is a bijection between the prime ideals $\mathfrak{p}$ of $\mathbb{K}$ above p and the prime ideals $\mathfrak{q}$ of $\mathbb{L}$ above p that preserves the inertia degrees: $f(\mathfrak{p}\,|\mathbb{K}) = f(\mathfrak{q}\,|\mathbb{L})$. Furthermore, the number fields have the same degree over $\mathbb{Q}$, the same discriminant, normal closure, isomorphic unit groups, and the same number of real and complex embeddings. However, it does not follow from arithmetical equivalence that $\mathbb{K}$ and $\mathbb{L}$ have the same class group (or even class number), cf. [10]. In general, arithmetic equivalence does not imply that $\mathbb{K}$ and $\mathbb{L}$ are isomorphic, as was shown by Gaßmann ([11], cf. also Perlis [19], or [14]). An example is provided by $\mathbb{K} = \mathbb{Q}(\sqrt[8]{3})$ and $\mathbb{L} = \mathbb{Q}(\sqrt[8]{3 \cdot 2^4})$ [15, 19]. However, the implication is true *if* $\mathbb{K}$ and $\mathbb{L}$ are Galois over $\mathbb{Q}$ (Theorem of Bauer [1, 2], nowadays a corollary of Chebotarev's density theorem, see, e.g., Neukirch [18] 13.9), so we find:

Corollary. *If the QSM-systems $(A_{\mathbb{K}}, \sigma_{\mathbb{K}})$ and $(A_{\mathbb{L}}, \sigma_{\mathbb{L}})$ of two number fields $\mathbb{K}$ and $\mathbb{L}$ are isomorphic and the extension $\mathbb{K}\,/\,\mathbb{Q}$ is normal, then $\mathbb{K}$ and $\mathbb{L}$ are isomorphic as fields.* □

This corollary is somewhat reminiscent of the anabelian theorem of Neukirch and Uchida [17,20], which says that number fields with isomorphic absolute Galois groups are isomorphic (Neukirch [17] proved this if one of the fields is normal over $\mathbb{Q}$, just as in our corollary). It is interesting to notice that the QSM-system involves the abelianized Galois group and the adeles, but not the absolute Galois group. In this sense, it is "not anabelian"; but of course, it is "noncommutative" (in noncommutative topology, the crossed product construction is an analog of taking quotients). The emerging philosophy seems to be that one can substitute the consideration of the "anabelian" absolute Galois group (with its difficult representation theory studied in the Langlands programme) by the dynamics of the action of Frobeniusses in the abelianized Galois group, with its "easy" representation theory given by class field theory.

One may wonder whether QSM-system isomorphism in general implies field isomorphism. In [8], this is proven for global function fields. We will discuss the number field case in the remaining instalments of this work.

The structure of this paper is as follows: first, we introduce isomorphism of QSM-systems. We deduce some basic properties, such as preservation of (extremal) KMS-states. Then we recall the QSM-system of a number field, and we prove our main theorem. In the final section, we make explicit the matching of KMS states for number fields.

4.2 Isomorphism of QSM Systems

We recall some definitions and refer to [4,5], and Chapter 3 of [6] for more information and for some physics background. After that, we introduce isomorphism of QSM-systems, and prove it preserves KMS-states.

Definition 4.1. A *quantum statistical mechanical system* (QSM-system) (A, σ) is a (unital) C^*-algebra A together with a so-called *time evolution* σ, which is a continuous group homomorphism $\sigma : \mathbb{R} \to \text{Aut } A : t \mapsto \sigma_t$. A *state* on A is a continuous positive unital linear functional $\omega : A \to \mathbb{C}$. We say ω is a *KMS*$_\beta$ *state* for some $\beta \in \mathbb{R}_{>0}$ if for all $a, b \in A$, there exists a function $F_{a,b}$, holomorphic in the strip $0 < \Im z < \beta$ and bounded continuous on its boundary, such that

$$F_{a,b}(t) = \omega(a\sigma_t(b)) \text{ and } F_{a,b}(t + i\beta) = \omega(\sigma_t(b)a) \quad (\forall t \in \mathbb{R}).$$

Equivalently, ω is a σ-invariant state with $\omega(ab) = \omega(b\sigma_{i\beta}(a))$ for a, b in a dense set of σ-analytic elements. The set $\text{KMS}_\beta(A, \sigma)$ of KMS_β states is topologized as a subspace of the convex set of states, a weak* closed subset of the unit ball in the operator norm of bounded linear functionals on the algebra. A KMS_β state is called *extremal* if it is an extremal point in the (compact convex) set of KMS_β states for the weak (i.e., pointwise convergence) topology.

Remark 4.1 (Physical origins). This notion of QSM-system is one of the possible physical theories of quantum statistical mechanics; one should think of A as the algebra of observables, represented on some Hilbert space $\mathcal{H}$ with orthonormal basis $\{\Psi_i\}$; the time evolution, in the given representation, is generated by a Hamiltonian H by

$$\sigma_t(a) = e^{itH} a e^{-itH}, \tag{4.1}$$

and (mixed) states of the system are combinations

$$a \mapsto \sum \lambda_i \langle \Psi_i | a \Psi_i \rangle$$

which will mostly be of the form $a \mapsto \mathrm{trace}(\rho a)$ for some density matrix ρ. A typical equilibrium state (here, this means stable by time evolution) is a Gibbs state

$$a \mapsto \mathrm{trace}(a e^{-\beta H}) / \mathrm{trace}(e^{-\beta H})$$

at temperature $1/\beta$, where we have normalized by the *partition function* $\mathrm{trace}(e^{-\beta H})$. The KMS-condition (named after Kubo, Martin and Schwinger) is a correct generalization of the notion of equilibrium state to more general situations, for example when the trace class condition $\mathrm{trace}(e^{-\beta H}) < \infty$, needed to define Gibbs states, no longer holds (cf. [13]).

We now introduce the following equivalence relation for QSM-systems:

Definition 4.2. An *isomorphism* of two QSM-systems (A, σ) and (B, τ) is a C^*-algebra isomorphism $\varphi : A \xrightarrow{\sim} B$ that intertwines time evolutions, i.e., such that the following diagram commutes:

$$\begin{array}{ccc} A & \xrightarrow[\sim]{\varphi} & B \\ {\scriptstyle\sigma}\downarrow & & \downarrow{\scriptstyle\tau} \\ A & \xrightarrow[\sim]{\varphi} & B \end{array}$$

Proposition 4.1. *Let $\varphi : (A, \sigma) \xrightarrow{\sim} (B, \tau)$ denote an isomorphism of QSM systems. Then for any $\beta > 0$,*

(i) *Pullback*

$$\varphi^* \; : \; \mathrm{KMS}_\beta(B, \tau) \to \mathrm{KMS}_\beta(A, \sigma) \; : \; \omega \mapsto \omega \circ \varphi$$

is a homeomorphism between the spaces of KMS_β *states on B and A.*

(ii) *φ^* induces a homeomorphism between extremal* KMS_β *states on B and A.*

Proof. The map φ obviously induces a bijection between states on B and states on A.

For (i), let $F_{a,b}$ be the holomorphic function that implements the KMS_β-condition for the state ω on (B, τ) at $a, b \in B$, so

$$F_{a,b}(t) = \omega(a \tau_t(b)) \text{ and } F_{a,b}(t + i\beta) = \omega(\tau_t(b) a).$$

The following direct computation then shows that the function $F_{\varphi(c),\varphi(d)}$ implements the KMS$_\beta$-condition for the state $\varphi^*\omega$ on (A,σ) at $c,d \in A$:

$$(\omega \circ \varphi)(c\sigma_t(d)) = \omega(\varphi(c)\tau_t(\varphi(d))) = F_{\varphi(c),\varphi(d)}(t),$$

and similarly at $t + i\beta$. Also, note that pullback is continuous, since C^*-algebra isomorphism is compatible with the topology on the set of KMS-states.

For (ii), if a KMS$_\beta$ state ω on B is not extremal, then the GNS-representation π_ω of ω is not factorial. As in Prop 3.8 of [5], there exists a positive linear functional, which is dominated by ω, namely $\omega_1 \leq \omega$, and which extends from B to the von Neumann algebra given by the weak closure $\mathscr{M}_\omega$ of B in the GNS representation. The functional ω_1 is of the form $\omega_1(b) = \omega(hb)$ for some positive element h in the center of the von Neumann algebra $\mathscr{M}_\omega$. Consider then the pullbacks

$$\varphi^*(\omega)(a) = \omega(\varphi(a))$$

and

$$\varphi^*(\omega_1)(a) = \omega_1(\varphi(a)) = \omega(h\varphi(a))$$

for $a \in A$. The continuous linear functional $\varphi^*(\omega_1)$ has norm $\|\varphi^*(\omega_1)\| \leq 1$. In fact, since we are dealing with unital algebras, $\|\varphi^*(\omega_1)\| = \varphi^*(\omega_1)(1) = \omega(h)$. The linear functional $\omega_2(b) = \omega((1-h)b)$ also satisfies the positivity property $\omega_2(b^*b) \geq 0$, since $\omega_1 \leq \omega$. The decomposition

$$\varphi^*(\omega) = \lambda\eta_1 + (1-\lambda)\eta_2,$$

with $\lambda = \omega(h)$,

$$\eta_1 = \varphi^*(\omega_1)/\omega(h) \text{ and } \eta_2 = \varphi^*(\omega_2)/\omega(1-h)$$

shows that the state $\varphi^*(\omega)$ is not extremal. Notice that η_1 and η_2 are both KMS states. To see this, it suffices to check that the state $\omega_1(b)/\omega(h)$ is KMS. In fact, one has for all analytic elements $a, b \in B$:

$$\omega_1(ab) = \omega(hab) = \omega(ahb) = \omega(hb\tau_{i\beta}(a)).$$

This proves the proposition.

4.3 A QSM-System for Number Fields

Bost and Connes [3] introduced a QSM-system for the field of rational numbers. More general QSM-systems associated to arbitrary number fields were constructed by Ha and Paugam in [12] as a special case of their more general class of systems for Shimura varieties, which in turn generalize the GL(2)-system of [5]. We recall here briefly the construction of the systems for number fields in an equivalent formulation (cf. also [16]).

We denote by $J_{\mathbb{K}}^{+}$ the semigroup of integral ideals, with the norm function

$$N : J_{\mathbb{K}}^{+} \to \mathbb{Z} : \mathfrak{n} \mapsto N(\mathfrak{n}) = N_{\mathbb{Q}}^{\mathbb{K}}(\mathfrak{n}) = N_{\mathbb{K}}(\mathfrak{n}).$$

Denote by $G_{\mathbb{K}}^{\mathrm{ab}}$ the Galois group of the maximal abelian extension of $\mathbb{K}$. The Artin reciprocity map is denoted by

$$\vartheta_{\mathbb{K}} : \mathbf{A}_{\mathbb{K}}^{*} \to G_{\mathbb{K}}^{\mathrm{ab}}.$$

By abuse of notation, we will also write $\vartheta_{\mathbb{K}}(\mathfrak{n})$ for the image under this map of an ideal $\mathfrak{n}$, which is seen as an idele by choosing a non-canonical section s of

$$\mathbf{A}_{\mathbb{K},f}^{*} \twoheadrightarrow J_{\mathbb{K}} \quad (\text{section } s) \qquad : \qquad (x_{\mathfrak{p}})_{\mathfrak{p}} \mapsto \prod_{\mathfrak{p} \text{ finite}} \mathfrak{p}^{v_{\mathfrak{p}}(x_{\mathfrak{p}})} .$$

The abuse lies in the fact that the image depends on this choice of section (thus, up to a unit in the finite ideles), but it is canonically defined in (every quotient of) the Galois group $G_{\mathbb{K},\mathfrak{n}}^{\mathrm{ab}}$ of the maximal abelian extension unramified at prime divisors of $\mathfrak{n}$: on every finite quotient of this, it is the "Frobenius element" of $\mathfrak{n}$. The notation $\vartheta_{\mathbb{K}}(\mathfrak{n})$ will only occur in situations where this ambiguity plays no role.

We consider the fibered product

$$X_{\mathbb{K}} := G_{\mathbb{K}}^{\mathrm{ab}} \times_{\hat{\mathcal{O}}_{\mathbb{K}}^{*}} \hat{\mathcal{O}}_{\mathbb{K}},$$

(where $\hat{\mathcal{O}}_{\mathbb{K}}$ is the ring of finite integral adeles), where the balancing is defined for $\gamma \in G_{\mathbb{K}}^{\mathrm{ab}}$ and $\rho \in \hat{\mathcal{O}}_{\mathbb{K}}$ by the equivalence

$$(\gamma, \rho) \sim (\vartheta_{\mathbb{K}}(u^{-1}) \cdot \gamma, u\rho) \text{ for all } u \in \hat{\mathcal{O}}_{\mathbb{K}}^{*}.$$

Definition 4.3. The *QSM-system* $(A_{\mathbb{K}}, \sigma_{\mathbb{K}})$ *associated to a number field* $\mathbb{K}$ is defined as the semigroup crossed product algebra

$$A_{\mathbb{K}} := C(X_{\mathbb{K}}) \rtimes J_{\mathbb{K}}^{+} = C(G_{\mathbb{K}}^{\mathrm{ab}} \times_{\hat{\mathcal{O}}_{\mathbb{K}}^{*}} \hat{\mathcal{O}}_{\mathbb{K}}) \rtimes J_{\mathbb{K}}^{+}, \tag{4.2}$$

where the crossed product structure is given by $\mathfrak{n} \in J_{\mathbb{K}}^{+}$ acting on $f \in C(X_{\mathbb{K}})$ as

$$(\mathfrak{n}, f) \mapsto \rho_{\mathfrak{n}}(f)(\gamma, \rho) = f(\vartheta_{\mathbb{K}}(\mathfrak{n})\gamma, s(\mathfrak{n})^{-1}\rho)e_{\mathfrak{n}},$$

with $e_{\mathfrak{n}} = \mu_{\mathfrak{n}}\mu_{\mathfrak{n}}^{*}$ the projector onto the space of $[(\gamma, \rho)]$ where $s(\mathfrak{n})^{-1}\rho \in \hat{\mathcal{O}}_{\mathbb{K}}$. Here $\mu_{\mathfrak{n}}$ is the isometry that implements the action of $J_{\mathbb{K}}^{+}$. Note that, because of the balancing over the finite idelic units $\hat{\mathcal{O}}^*_{\mathbb{K}}$, the dependence of $\vartheta_{\mathbb{K}}(\mathfrak{n})$ on s is again of no influence. The action has a partial inverse defined by

$$\sigma_{\mathfrak{n}}(f)(x) = f(\mathfrak{n} * x)$$

where we have defined the action $\mathfrak{n} * x$ of an ideal $\mathfrak{n} \in J^+_{\mathbb{K}}$ on an element $x \in X_{\mathbb{K}}$ as

$$\mathfrak{n} * [(\gamma, \rho)] = [(\vartheta_{\mathbb{K}}(\mathfrak{n})^{-1}\gamma, s(\mathfrak{n})\rho)].$$

Then the following defining relations hold in the semigroup crossed product algebra:

$$\mu_{\mathfrak{n}}\mu^*_{\mathfrak{n}} = e_{\mathfrak{n}};\ \mu^*_{\mathfrak{n}}\mu_{\mathfrak{n}} = 1;\ \rho_{\mathfrak{n}}(f) = \mu_{\mathfrak{n}} f \mu^*_{\mathfrak{n}};$$

$$\sigma_{\mathfrak{n}}(f) = \mu^*_{\mathfrak{n}} f \mu_{\mathfrak{n}};\ \sigma_{\mathfrak{n}}(\rho_{\mathfrak{n}}(f)) = f;\ \rho_{\mathfrak{n}}(\sigma_{\mathfrak{n}}(f)) = f e_{\mathfrak{n}}.$$

Finally, the time evolution is given by

$$\begin{cases} \sigma_{\mathbb{K},t}(f) = f, & \forall f \in C(G^{ab}_{\mathbb{K}} \times_{\hat{\mathcal{O}}^*_{\mathbb{K}}} \hat{\mathcal{O}}_{\mathbb{K}}); \\ \sigma_{\mathbb{K},t}(\mu_{\mathfrak{n}}) = N(\mathfrak{n})^{it}\, \mu_{\mathfrak{n}}, & \forall\, \mathfrak{n} \in J^+_{\mathbb{K}}, \end{cases} \tag{4.3}$$

where $\mu_{\mathfrak{n}}$ are the isometries that implement the semigroup action of $J^+_{\mathbb{K}}$.

4.4 Hilbert Space Representation, Partition Function, KMS-States

Let us abbreviate $\mathrm{KMS}_\beta(\mathbb{K}) := \mathrm{KMS}_\beta(A_{\mathbb{K}}, \sigma_{\mathbb{K}})$. A complete classification of the KMS states for the systems $(A_{\mathbb{K}}, \sigma_{\mathbb{K}})$ was obtained in [16], Thm. 2.1. In particular, in the low temperature range $\beta > 1$, the extremal KMS_β states are parameterized by elements $\gamma \in G^{ab}_{\mathbb{K}}$, and are in Gibbs form, given by

$$\omega_{\beta,\gamma}(f) = \frac{1}{\zeta_{\mathbb{K}}(\beta)} L_{\mathbb{K}}(\gamma, f, \beta), \text{ where } L_{\mathbb{K}}(\gamma, f, \beta) := \sum_{\mathfrak{n} \in J^+_{\mathbb{K}}} \frac{f(\mathfrak{n} * \gamma)}{N(\mathfrak{n})^\beta} \tag{4.4}$$

is a *generalized L-series* associated to $\gamma \in G^{ab}_{\mathbb{K}}$ and $f \in A_{\mathbb{K}}$.

Associated to any element $\gamma \in G^{ab}_{\mathbb{K}}$ is a natural representation π_γ of the algebra $A_{\mathbb{K}}$ on the Hilbert space $\ell^2(J^+_{\mathbb{K}})$. Namely, let $\varepsilon_{\mathfrak{m}}$ denote the canonical basis of $\ell^2(J^+_{\mathbb{K}})$. Then the action on $\ell^2(J^+_{\mathbb{K}})$ of an element $f_{\mathfrak{n}}\mu_{\mathfrak{n}} \in A_{\mathbb{K}}$ with $\mathfrak{n} \in J^+_{\mathbb{K}}$ and $f_{\mathfrak{n}} \in C(X_{\mathbb{K}})$ is given by

$$\pi_\gamma(f_{\mathfrak{n}}\mu_{\mathfrak{n}})\, \varepsilon_{\mathfrak{m}} = f_{\mathfrak{n}}(\mathfrak{n}\,\mathfrak{m} * \gamma)\, \varepsilon_{\mathfrak{n}\,\mathfrak{m}}.$$

In this picture, the time evolution is implemented (in the sense of formula (4.1)) by a Hamiltonian

$$H_{\sigma_{\mathbb{K}}} \varepsilon_{\mathfrak{n}} = \log N(\mathfrak{n})\ \varepsilon_{\mathfrak{n}}. \tag{4.5}$$

In this representation,

$$\operatorname{trace}(\pi_\gamma(f) e^{-\beta H_{\sigma_{\mathbb{K}}}}) = \sum_{\mathfrak{n} \in J_{\mathbb{K}}^+} \frac{f(\mathfrak{n} * \gamma)}{N(\mathfrak{n})^\beta}.$$

Setting $f = 1$, the Dedekind zeta function

$$\zeta_{\mathbb{K}}(\beta) = \sum_{\mathfrak{n} \in J_{\mathbb{K}}^+} N(\mathfrak{n})^{-\beta}$$

appears as the partition function

$$\zeta_{\mathbb{K}}(\beta) = \operatorname{trace}(e^{-\beta H_{\sigma_{\mathbb{K}}}})$$

of the system (convergent for $\beta > 1$).

4.5 Hamiltonians and Arithmetic Equivalence

We show that the existence of an isomorphism of the quantum statistical mechanical systems implies arithmetic equivalence; this is basically because the zeta functions of $\mathbb{K}$ and $\mathbb{L}$ are the partition functions of the respective systems. Some care has to be taken since the systems are not represented on the same Hilbert space.

Theorem 4.1. *Let $\varphi : (A_{\mathbb{K}}, \sigma_{\mathbb{K}}) \xrightarrow{\sim} (A_{\mathbb{L}}, \sigma_{\mathbb{L}})$ be an isomorphism of QSM-systems of number fields $\mathbb{K}$ and $\mathbb{L}$. Then $\mathbb{K}$ and $\mathbb{L}$ are arithmetically equivalent, i.e., they have the same Dedekind zeta function.*

Proof. The isomorphism $\varphi : (A_{\mathbb{K}}, \sigma_{\mathbb{K}}) \xrightarrow{\sim} (A_{\mathbb{L}}, \sigma_{\mathbb{L}})$ induces an identification of the sets of extremal KMS-states of the two systems, via pullback $\varphi^* : \mathrm{KMS}_\beta(\mathbb{L}) \to \mathrm{KMS}_\beta(\mathbb{K})$.

Consider the GNS representations associated to regular low temperature KMS states $\omega = \omega_\beta$ and $\varphi^*(\omega)$. We denote the respective Hilbert spaces by $\mathscr{H}_\omega$ and $\mathscr{H}_{\varphi^*\omega}$. As in Lemma 4.3 of [7], we observe that the factor $\mathscr{M}_\omega$ obtained as the weak closure of $A_{\mathbb{L}}$ in the GNS representation is of type I_∞, since we are only considering the low temperature KMS states that are of Gibbs form. Thus, the space $\mathscr{H}_\omega$ decomposes as

$$\mathscr{H}_\omega = \mathscr{H}(\omega) \otimes \mathscr{H}',$$

with an irreducible representation π_ω of $A_{\mathbb{L}}$ on $\mathscr{H}(\omega)$ and

$$\mathscr{M}_\omega = \{T \otimes 1 \mid T \in \mathscr{B}(\mathscr{H}(\omega))\}$$

($\mathscr{B}$ indicates the set of bounded operators). Moreover, we have

$$\langle (T \otimes 1)1_\omega, 1_\omega \rangle = \mathrm{Tr}(T\rho)$$

for a density matrix ρ (positive, of trace class, of unit trace).

We know that the low temperature extremal KMS states for the system $(A_{\mathbb{L}}, \sigma_{\mathbb{L}})$ are of Gibbs form and given by the explicit expression in Eq. (4.4) for some $\gamma \in G_{\mathbb{L}}^{\mathrm{ab}}$; and similarly for the system $(A_{\mathbb{K}}, \sigma_{\mathbb{K}})$. Thus, we can identify $\mathscr{H}(\omega)$ with $\ell^2(J_{\mathbb{L}}^+)$ and the density ρ correspondingly with

$$\rho = e^{-\beta H_{\sigma_{\mathbb{L}}}} / \mathrm{Tr}(e^{-\beta H_{\sigma_{\mathbb{L}}}});$$

this is the representation considered in Sect. 4.4. As in Lemma 4.3 of [7], the evolution group e^{itH_ω} generated by the Hamiltonian H_ω that implements the time evolution $\sigma_{\mathbb{L}}$ in the GNS representation on $\mathscr{H}_\omega$ agrees with $e^{itH_{\sigma_{\mathbb{L}}}}$ on the factor $\mathscr{M}_\omega$. We find

$$e^{itH_\omega} \pi_\omega(f) e^{-itH_\omega} = \pi_\omega(\sigma_{\mathbb{L}}(f)) = e^{itH_{\sigma_{\mathbb{L}}}} \pi_\omega(f) e^{-itH_{\sigma_{\mathbb{L}}}}.$$

As observed in §4.2 of [7], this gives us that the Hamiltonians differ by a constant:

$$H_\omega = H_{\sigma_{\mathbb{L}}} + \log \lambda_1 \text{ for some } \lambda_1 \in \mathbb{R}_+^* . \tag{4.6}$$

The argument for the GNS representation for $\pi_{\varphi^*(\omega)}$ is similar and it gives an identification of the Hamiltonians

$$H_{\varphi^*(\omega)} = H_{\sigma_{\mathbb{K}}} + \log \lambda_2 \text{ for some } \lambda_2 \in \mathbb{R}_+^* . \tag{4.7}$$

The algebra isomorphism φ induces a unitary equivalence Φ of the Hilbert spaces of the GNS representations of the corresponding states, and the Hamiltonians that implement the time evolution in these representations are therefore related by

$$H_{\varphi^*(\omega)} = \Phi H_\omega \Phi^*. \tag{4.8}$$

In particular the Hamiltonians $H_{\varphi^*(\omega)}$ and H_ω then have the same spectrum.

By combining (4.6)–(4.8), we find that

$$H_{\sigma_{\mathbb{K}}} = \Phi H_{\sigma_{\mathbb{L}}} \Phi^* + \log \lambda$$

for a unitary operator Φ and a $\lambda \in \mathbb{R}_+^*$. This gives at the level of zeta functions

$$\zeta_{\mathbb{L}}(\beta) = \lambda^{-\beta}\zeta_{\mathbb{K}}(\beta) \tag{4.9}$$

for sufficiently large real β, hence for all β by analytic continuation. Now consider the left hand side and right hand side as classical Dirichlet series of the form

$$\sum_{n\geq 1}\frac{a_n}{n^\beta} \text{ and } \sum_{n\geq 1}\frac{b_n}{(\lambda n)^\beta},$$

respectively. Observe that $a_1 = b_1 = 1$. Taking the limit as $\beta \to +\infty$ in (4.9), we find

$$a_1 = \lim_{\beta\to+\infty} b_1\lambda^{-\beta},$$

from which we conclude that $\lambda = 1$. Thus, we obtain $\zeta_{\mathbb{K}}(\beta) = \zeta_{\mathbb{L}}(\beta)$, which gives arithmetic equivalence of the number fields.

4.6 Matching of Generalized L-Series

Since the zeta functions are equal, the matching of extremal KMS_β states as in 4.1 implies a matching of generalized L-series, as follows:

Corollary 4.1. *Let $\varphi : (A_{\mathbb{K}}, \sigma_{\mathbb{K}}) \xrightarrow{\sim} (A_{\mathbb{L}}, \sigma_{\mathbb{L}})$ be an isomorphism of QSM-systems of number fields $\mathbb{K}$ and $\mathbb{L}$. There exists a homeomorphism $\psi\colon G_{\mathbb{L}}^{\text{ab}} \xrightarrow{\sim} G_{\mathbb{K}}^{\text{ab}}$ such that we have an identification of generalized L-series*

$$L_{\mathbb{L}}(\gamma, f, \beta) = L_{\mathbb{K}}(\psi(\gamma), \varphi^{-1}(f), \beta)$$

for all $f \in A_{\mathbb{L}}$ and all $\gamma \in G_{\mathbb{L}}^{\text{ab}}$. □

References

1. M. Bauer, Über einen Satz von Kronecker. Arch. der Math. u. Phys. (3) **6**, 218–219 (1903)
2. M. Bauer, Über zusammengesetzte Körper. Arch. der Math. u. Phys. (3) **6**, 221–222 (1903)
3. J.-B. Bost, A. Connes, Hecke algebras, type III factors and phase transitions with spontaneous symmetry breaking in number theory. Selecta Math. (N.S.) **1**, 411–457 (1995)
4. O. Bratteli, D.W. Robinson, *Operator Algebras and Quantum Statistical Mechanics 2*, 2nd edn. Texts and Monographs in Physics (Springer, Berlin, 1997)
5. A. Connes, M. Marcolli, From physics to number theory via noncommutative geometry, I: quantum statistical mechanics of $\mathbb{Q}$-lattices, in *Frontiers in Number Theory, Physics, and Geometry I*, ed. by P.E. Cartier, B. Julia, P. Moussa, P. Vanhove (Springer, Berlin, 2006), pp. 269–347

6. A. Connes, M. Marcolli, *Noncommutative Geometry, Quantum Fields and Motives*. American Mathematical Society Colloquium Publications, vol. 55 (American Mathematical Society, Providence, 2008)
7. A. Connes, C. Consani, M. Marcolli, Noncommutative geometry and motives: the thermodynamics of endomotives. Adv. Math. **214**, 761–831 (2007)
8. G. Cornelissen, Curves, dynamical systems, and weighted point counting. Proc. Natl. Acad. Sci. USA **110**(24), 9669–9673 (2013)
9. G. Cornelissen, M. Marcolli, Quantum statistical mechanics, L-series and anabelian geometry. Preprint, arXiv:1009.0736 (2010)
10. B. de Smit, R. Perlis, Zeta functions do not determine class numbers. Bull. Am. Math. Soc. (N.S.) **31**, 213–215 (1994)
11. F. Gaßmann, Bemerkungen zur Vorstehenden Arbeit von Hurwitz: Über Beziehungen zwischen den Primidealen eines algebraischen Körpers und den Substitutionen seiner Gruppen. Math. Z. **25**, 661(665)–675 (1926)
12. E. Ha, F. Paugam, Bost-Connes-Marcolli systems for Shimura varieties. I. Definitions and formal analytic properties. IMRP Int. Math. Res. Pap. **5**, 237–286 (2005)
13. R. Haag, N.M. Hugenholtz, M. Winnink, On the equilibrium states in quantum statistical mechanics. Comm. Math. Phys. **5**, 215–236 (1967)
14. N. Klingen, *Arithmetical Similarities*. Oxford Mathematical Monographs (The Clarendon Press/Oxford University Press, New York, 1998)
15. K. Komatsu, On the adele rings of algebraic number fields. Kōdai Math. Sem. Rep. **28**, 78–84 (1976)
16. M. Laca, N.S. Larsen, S. Neshveyev, On Bost-Connes types systems for number fields. J. Number Theory **129**, 325–338 (2009)
17. J. Neukirch, Kennzeichnung der p-adischen und der endlichen algebraischen Zahlkörper. Invent. Math. **6**, 296–314 (1969)
18. J. Neukirch, *Algebraic Number Theory*. Grundlehren der Mathematischen Wissenschaften, vol. 322 (Springer, Berlin, 1999)
19. R. Perlis, On the equation $\zeta_K(s) = \zeta_{K'}(s)$. J. Number Theory **9**, 342–360 (1977)
20. K. Uchida, Isomorphisms of Galois groups. J. Math. Soc. Jpn. **28**, 617–620 (1976)

Chapter 5
Exploring Noncommutative Algebras via Deformation Theory

Pavel Etingof

Abstract This is an expository paper which explains how one can use deformation theory to construct new algebras from known ones, and study their properties.

5.1 Introduction

In this paper I would like to address the following question: given an associative algebra A_0, what are the possible ways to deform it? Consideration of this question for concrete algebras often leads to interesting mathematical discoveries. I will discuss several approaches to this question, and examples of applying them.

5.2 Deformation Theory

5.2.1 Formal Deformations

The most general approach to the question "how to deform A_0?" is the theory of formal deformations.

Let k be a field and $K := k[[\hbar_1, \ldots, \hbar_\ell]]$ the ring of formal power series in variables $\hbar_i$. Let $\mathfrak{m}$ be the maximal ideal in K.

A K-module M is said to be **topologically free** if it is isomorphic to $M_0[[\hbar_1, \ldots, \hbar_\ell]]$ for some vector space M_0.

P. Etingof (✉)
Department of Mathematics, Massachusetts Institute of Technology, Cambridge, MA 02139, USA
e-mail: etingof@math.mit.edu

V. Ancona and E. Strickland (eds.), *Trends in Contemporary Mathematics*, Springer INdAM Series 8, DOI 10.1007/978-3-319-05254-0_5,

Let A_0 be an algebra over k.[1]

Definition 5.1. An **ℓ-parameter flat formal deformation** of A_0 is an algebra A over K which is topologically free as a K-module, together with an isomorphism of algebras $\phi : A/\mathfrak{m} \to A_0$.[2]

For simplicity we will mostly consider 1-parameter deformations. If A is a 1-parameter flat formal deformation of A_0 then we can choose an identification $A \to A_0[[\hbar]]$ as K-modules, which reduces to ϕ modulo $\hbar$. Then the algebra structure on A transforms into a new K-linear multiplication law μ on $A_0[[\hbar]]$. Such a multiplication law is determined by the product $\mu(a,b)$, $a,b \in A_0 \subset A_0[[\hbar]]$, which is given by the formula

$$\mu(a,b) = \mu_0(a,b) + \hbar\mu_1(a,b) + \hbar^2\mu_2(a,b) + \ldots, a, b \in A_0,$$

where $\mu_i : A_0 \otimes A_0 \to A_0$ are linear maps, and $\mu_0(a,b)$ is the undeformed product ab in A_0. Thus, to find formal deformations of A_0 means to find all such series μ which satisfy the associativity equation, modulo the automorphisms of the K-module $A_0[[\hbar]]$ which are the identity modulo $\hbar$. [3]

The associativity equation $\mu \circ (\mu \otimes Id) = \mu \circ (Id \otimes \mu)$ reduces to a hierarchy of linear equations:

$$\sum_{s=0}^{N} \mu_s(\mu_{N-s}(a,b),c) = \sum_{s=0}^{N} \mu_s(a, \mu_{N-s}(b,c)). \tag{5.1}$$

(These equations are linear in μ_N if μ_i, $i < N$, are known).

5.2.2 *Hochschild Cohomology*

Equations (5.1) can be analyzed using Hochschild cohomology. Let us recall its definition. Let M be a bimodule over A_0. A Hochschild n-cochain of A_0 with coefficients in M is a linear map $A_0^{\otimes n} \to M$. The space of such cochains is denoted by $C^n(A_0, M)$. The differential $d : C^n(A_0, M) \to C^{n+1}(A_0, M)$ is defined by the formula

---

[1]By "an algebra" we always mean an associative algebra with unit.

[2]The word "flat" refers to the fact that A is a (topologically) flat module over K, i.e. the functor of completed tensor product with this module is exact.

[3]Note that we don't have to worry about the existence of a unit in A since a flat formal deformation of an algebra with unit always has a unit.

$$df(a_1,\ldots,a_{n+1}) = f(a_1,\ldots,a_n)a_{n+1} - f(a_1,\ldots,a_na_{n+1})$$
$$+f(a_1,a_2a_3,\ldots,a_{n+1}) - \ldots + (-1)^n f(a_1a_2,\ldots,a_{n+1})$$
$$+(-1)^{n+1}a_1 f(a_2,\ldots,a_{n+1}).$$

It is easy to show that $d^2 = 0$, and one defines the Hochschild cohomology $H^\bullet(A_0, M)$ to be the cohomology of the complex $(C^\bullet(A_0, M), d)$. If $M = A_0$, the algebra itself, then we will denote $H^\bullet(A_0, M)$ by $H^\bullet(A_0)$ (it is an algebra). For example, $H^0(A_0)$ is the center of A_0, and $H^1(A_0)$ is the quotient of the Lie algebra of derivations of A_0 by inner derivations.

The following are standard facts from deformation theory (due to Gerstenhaber [13]), which can be checked directly.

1. The linear equation for μ_1 says that μ_1 is a Hochschild 2-cocycle. Thus algebra structures on $A_0[\hbar]/\hbar^2$ deforming μ_0 are parametrized by the space $Z^2(A_0)$ of Hochschild 2-cocycles of A_0 with values in $M = A_0$.
2. If μ_1, μ_1' are two 2-cocycles such that $\mu_1 - \mu_1'$ is a coboundary, then the algebra structures on $A_0[\hbar]/\hbar^2$ corresponding to μ_1 and μ_1' are equivalent by a transformation of $A_0[\hbar]/\hbar^2$ that equals the identity modulo $\hbar$, and vice versa. Thus equivalence classes of multiplications on $A_0[\hbar]/\hbar^2$ deforming μ_0 are parametrized by the cohomology $H^2(A_0)$.
3. The linear equation for μ_N says that $d\mu_N$ is a certain quadratic expression b_N in $\mu_0, \mu_1, \ldots, \mu_{N-1}$. This expression is always a Hochschild 3-cocycle, and the equation is solvable iff it is a coboundary. Thus the cohomology class of b_N in $H^3(A_0)$ is the only obstruction to solving this equation.

5.2.3 *Universal Deformation*

In particular, if $H^3(A_0) = 0$ then the equation for μ_n can be solved for all n, and for each n the freedom in choosing the solution, modulo equivalences, is the space $H := H^2(A_0)$. Thus there exists an algebra structure over $k[[H]]$ on the space $A_u := A_0[[H]]$ of formal functions from H to A_0, $a, b \mapsto \mu_u(a,b) \in A_0[[H]]$, $(a, b \in A_0)$, such that $\mu_u(a,b)(0) = ab \in A_0$, and every 1-parameter flat formal deformation A of A_0 is given by the formula $\mu(a,b)(\hbar) = \mu_u(a,b)(\gamma(\hbar))$ for a unique formal series $\gamma \in \hbar H[[\hbar]]$, with the property that $\gamma'(0)$ is the cohomology class of the cocycle μ_1.

Such an algebra A_u is called a universal deformation of A_0. It is unique up to an isomorphism.

Thus in the case $H^3(A_0) = 0$, deformation theory allows us to completely classify 1-parameter flat formal deformations of A_0. In particular, we see that the "moduli space" parametrizing formal deformations of A_0 is a smooth space – it is the formal neighborhood of zero in H.

5.2.4 Quantization of Poisson Structures

If $H^3(A_0)$ is nonzero then in general the universal deformation parametrized by H does not exist, as there are obstructions to deformations. In this case, the moduli space of deformations will be a closed subscheme of H, which is often singular. On the other hand, even when $H^3(A_0) \neq 0$, the universal deformation parametrized by H may exist (although it may be more difficult to prove than in the vanishing case). In this case one says that the deformations of A_0 are **unobstructed** (since all obstructions vanish even though the space of obstructions doesn't).

To illustrate these statements, consider the quantization theory of Poisson manifolds. Let M be a smooth C^∞-manifold or a smooth affine algebraic variety over $\mathbb{C}$, and A_0 the structure algebra of M.

Remark. In the C^∞-case, we will consider only local maps $A_0^{\otimes n} \to A_0$, i.e. those given by polydifferential operators, and all deformations and the Hochschild cohomology is defined using local, rather than general, cochains.

Theorem 5.1 (Hochschild-Kostant-Rosenberg [15]). $H^i(A_0) = \Gamma(M, \wedge^i TM)$ *as a module over* $A_0 = H^0(A_0)$.

In particular, H^2 is the space of bivector fields, and H^3 the space of trivector fields. So the cohomology class of μ_1 is a bivector field; in fact, it is $\pi(a,b) := \mu_1(a,b) - \mu_1(b,a)$, since any 2-coboundary in this case is symmetric. The equation for μ_2 says that $d\mu_2$ is a certain trivector field that depends quadratically on π. It is easy to show that this is the Schouten bracket $[\pi, \pi]$. Thus, for the existence of μ_2 it is necessary that $[\pi, \pi] = 0$, i.e. that π be a **Poisson bracket**.

Suppose now that π is a Poisson bracket, i.e. $[\pi, \pi] = 0$. In this case the algebra $A = A_0[[\hbar]]$ with the product μ is said to be **a quantization** of π, and (M, π) the **quasiclassical limit** of (A, μ). So, is it possible to construct a quantization of π?

By the above arguments, μ_2 exists (and a choice of μ_2 is unique up to adding an arbitrary bivector). So there arises the question of existence of μ_3 etc., i.e. the question whether there are other obstructions.

The answer to this question is yes and no. Namely, if you don't pick μ_2 carefully, you may be unable to find μ_3, but you can always pick μ_2 so that μ_3 exists, and there is a similar situation in higher orders. This subtle fact is a consequence of the following deep theorem of Kontsevich:

Theorem 5.2 ([16]). *Any Poisson structure* π *on* A_0 *can be quantized. Moreover, there is a natural bijection between products* μ *up to an isomorphism and Poisson brackets* $\pi_0 + \hbar\pi_1 + \hbar^2\pi_2 + \ldots$, *such that the quasiclassical limit of* μ *is* π_0.

Remark. Note that, as was shown by O. Mathieu, [17], a Poisson bracket on a general commutative $\mathbb{C}$-algebra may fail to admit a quantization.

Let us consider the special case of symplectic manifolds, i.e. the case when π is a nondegenerate bivector. In this case we can consider $\pi^{-1} = \omega$, which is a closed, nondegenerate 2-form (= symplectic structure) on M. In this case, Kontsevich's theorem is easier, and was proved by De Wilde – Lecomte, and later Deligne and

Fedosov (see e.g. [12]). Moreover, in this case there is the following additional result, see [18].

Theorem 5.3. *If M is symplectic and A is a quantization of M, then the Hochschild cohomology $H^i(A[\hbar^{-1}])$ is isomorphic to $H^i(M, \mathbb{C}((\hbar)))$.*

Remark. Here the algebra $A[\hbar^{-1}]$ is regarded as a (topological) algebra over the field of Laurent series $\mathbb{C}((\hbar))$, so Hochschild cochains are, by definition, linear maps $A_0^{\otimes n} \to A_0((\hbar))$.

Example 5.1. The algebra $B = A[\hbar^{-1}]$ provides an example of an algebra with possibly nontrivial $H^3(B)$, for which the universal deformation parametrized by $H = H^2(B)$ exists. Namely, this deformation is attached through the correspondence of Theorem 5.2 (and inversion of $\hbar$) to the Poisson bracket $\pi = (\omega + t_1\omega_1 + \ldots + t_r\omega_r)^{-1}$, where $\omega_1, \ldots, \omega_r$ are closed 2-forms on M which represent a basis of $H^2(M, \mathbb{C})$, and $t_1, \ldots, t_r$ are the coordinates on H corresponding to this basis.

5.2.5 *Examples*

Example 5.2. Let V be a symplectic vector space over $\mathbb{C}$ with symplectic form ω. Let Weyl(V) denote the Weyl algebra of V, which is the quotient of the free (= tensor) algebra on V by the ideal generated by elements $xy - yx - \omega(x, y)$.

Let G be a finite group acting symplectically on V. Then G acts on Weyl(V), and one can form a semidirect product algebra $A_0 = G \ltimes \mathrm{Weyl}(V)$. Let us study deformations of A_0.

We say that an element $g \in G$ is a symplectic reflection in V if $\mathrm{rank}(g-1)|_V = 2$. Let S be the set of symplectic reflections in G.

Proposition 5.1 ([1]). *$H^i(A_0)$ is the space of functions on the set of conjugacy classes of elements $g \in G$ such that* $\mathrm{rank}(g-1)|_V = i$. *In particular, $H^i(A_0) = 0$ if i is odd, and $H^2(A_0) = \mathbb{C}[S]^G$.*

Corollary 5.1. *There exists a universal deformation $A_u = \mathbf{H}_c(V, G)$ of A_0, which is parametrized by $c \in \mathbb{C}[S]^G$.*

The algebra $\mathbf{H}_c(V, G)$ is called the symplectic reflection algebra (see [9]). Such algebras were first considered by Drinfeld in 1986, [6]. If $V = \mathfrak{h} \oplus \mathfrak{h}^*$, where $\mathfrak{h}$ is a representation of G, and the symplectic form on G is the pairing between $\mathfrak{h}$ and $\mathfrak{h}^*$, then $\mathbf{H}_c(V, G)$ is called the rational Cherednik algebra. We will later construct $\mathbf{H}_c(V, G)$ explicitly.

Example 5.3. Let X be a smooth affine algebraic variety over $\mathbb{C}$, with an action of a finite group G. Let $D(X)$ be the algebra of algebraic differential operators on X. Let $A_0 = G \ltimes D(X)$. Let us study deformations of A_0.

For every $g \in G$, the fixed set X^g of g in Y is a smooth affine variety, which consists of connected components X_j^g, possibly of different dimensions. Such a component is said to be a **reflection hypersurface** if it has codimension 1 in X. Let S be the set of pairs (g, Y), where $g \in G$, and $Y \subset X^g$ is a connected component which is a reflection hypersurface (i.e., has codimension 1).

Proposition 5.2 ([8]). *One has $H^2(A_0) = (H^2(X, \mathbb{C}) \oplus \mathbb{C}[S])^G$. Moreover, there exists a universal deformation of A_0 parametrized by $H = H^2(A_0)$.*

This deformation $\mathbf{H}_c[X, G]$ is called the rational Cherednik algebra attached to (X, G), and is described in [8]. If X is a vector space $\mathfrak{h}$ and G acts linearly, then $\mathbf{H}_c[\mathfrak{h}, G] = \mathbf{H}_c(\mathfrak{h} \oplus \mathfrak{h}^*, G)$ is the rational Cherednik algebra discussed above.

Example 5.4. The following example from the paper [5] (conjecturally) generalizes examples 5.1, 5.2, and 5.3.

Let M be a symplectic C^∞-manifold (or affine complex algebraic variety). Let G be a finite group acting on M by symplectic transformations, and B be a quantization of M which is equivariant under G (such a quantization always exists). Let $A_0 = G \ltimes B[\hbar^{-1}]$. Let us study deformations of A_0.

The Hochschild cohomology of A_0 is given by the following theorem. Let the fixed set M^g be the union of connected components M_i^g, $i = 1, \ldots, N_g$.

Theorem 5.4 (see [5]). *$H^*(A_0)$ equals, as a vector space, the orbifold cohomology of M/G with coefficients in $\mathbb{C}((\hbar))$. Namely,*

$$H^p(A_0) = (\oplus_{g \in G} \oplus_{i=1}^{N_g} H^{p-\operatorname{codim} M_i^g}(M_i^g))^G.$$

(where the coefficients on the RHS are $\mathbb{C}((\hbar))$).

Remark. Let S be the set of pairs (g, Y), where $g \in G$, and $Y \subset M^g$ is a connected component of codimension 2. Theorem 5.4 implies that $H^2(A_0) = (H^2(M) \oplus \mathbb{C}[S])^G$.

Thus, we see that $H^3(A_0)$ does not always vanish. Nevertheless, we make the following conjecture.

Conjecture 5.1 ([5]). The deformations of the algebra A_0 are unobstructed. Thus there exists a universal deformation H_c of this algebra parametrized by $c \in H^2(A_0)$.

Thus the conjecture implies that if $S \neq \emptyset$, then there exist "interesting" deformations of A_0, i.e., ones not coming from G-invariant deformations of B.

Let us give a few examples in which this conjecture is true.

1. $H^3(A_0) = 0$. This includes the following interesting case considered in [10]: Σ is a smooth affine algebraic surface such that $H^1(\Sigma, \mathbb{C}) = 0$, and $M = \Sigma^n$, $G = S_n$. In this case there is one interesting deformation parameter corresponding to reflections in S_n.
2. G is trivial (Example 5.1).
3. $M = T^*Y$, where Y is a smooth affine variety, and G acts on Y (Example 5.3).

4. If $M = V$ is a symplectic vector space and G acts linearly (Example 5.2).
5. Let $M = V/L$, where V is a symplectic vector space and L a lattice in V (i.e., L is the abelian group generated by a basis of V). Thus M is an algebraic torus with a symplectic form. We assume that the symplectic form is integral and unimodular on L. Let $G \subset Sp(L)$ be a finite subgroup; then G acts naturally on M. In this case H_c is an "orbifold Hecke algebra" defined in [8] (it will be discussed below).
6. In the case when $G = \mathbb{Z}/2$, the conjecture was proved, under some assumptions, in [14].

5.3 Algebras Given by Generators and Relations

5.3.1 *Giving Formal Deformations by Generators and Relations*

Another approach to exploring deformations of A_0 is defining deformations by generators and relations.

Let us first consider the setting of formal deformations, which we have discussed in the previous section. Namely, let A_0 be an algebra over a field k, generated by $a_1, a_2, \ldots$ with defining relations $R_j^0(a_1, a_2, \ldots) = 0$ (here R_j^0 are elements in the free k-algebra F generated by a_i). Let us now define a formal deformation of A_0 as the algebra over $K = k[[\hbar]]$ with the same generators and deformed relations $R_j = R_j^0 + \hbar R_j^1 + \hbar^2 R_j^2 + \ldots$. That is, A is the quotient of the free algebra $F[[\hbar]]$ by the $\hbar$-adically closed ideal generated by the relations R_j.

Example 5.5 (The Weyl algebra). Let $A_0 = \mathbb{C}[x, y]$ be the algebra generated by x, y with the defining relation $yx - xy = 0$. We can then define A by the same generators and the deformed relation $yx - xy = \hbar$ (the Heisenberg indeterminacy relation). Then A is indeed a 1-parameter flat formal deformation of A_0, which provides a quantization of the standard Poisson bracket $\{y, x\} = 1$.

So, is A always a 1-parameter flat formal deformation of A_0? In general the answer is **no**: the flatness property can fail. The following typical example of this is obtained by adding just one relation to the relations above.

Example 5.6. Assume the algebra A_0 is defined by generators x, y and defining relations

$$yx - xy = 0, \; x = 0,$$

and A is defined by generators x, y and relations

$$yx - xy = \hbar, \; x = 0.$$

Then A is not topologically free, as it contains $\hbar$-torsion. Indeed, $\hbar \cdot 1 = yx - xy = 0$ since $x = 0$. On the other hand, $1 \neq 0$, since the algebra $A_0 = \mathbb{C}[y]$ is nonzero.

In fact, it is easy to show that if we add any relation to $xy - yx = \hbar$, it will produce a non-flat deformation (unless the algebra to be deformed is zero to begin with). This shows that if one wants to secure flatness, one has to deform the relations in a very special way. In fact, it is usually rather difficult to do so, as well as to check that the resulting deformations are actually flat. Below I would like to show several situations when this task can be successfully completed.

5.3.2 Deformations of Quadratic Algebras

The first situation is deformation theory of quadratic algebras.

Let R be a finite dimensional semisimple algebra (say over $\mathbb{C}$). Let A be a $\mathbb{Z}_+$-graded algebra, $A = \oplus_{i \geq 0} A[i]$, such that $A[0] = R$. For simplicity assume that the spaces $A[i]$ are finite dimensional for all i.

Definition 5.2. (i) The algebra A is said to be quadratic if it is generated over R by $A[1]$, and has defining relations in degree 2.
(ii) A is Koszul if all elements of $\mathrm{Ext}^i(R, R)$ (where R is the augmentation module over A) have grade degree precisely i.

Remarks. 1. Thus, in a quadratic algebra, $A[2] = A[1] \otimes_R A[1]/E$, where E is the subspace (R-subbimodule) of relations.
2. It is easy to show that a Koszul algebra is quadratic, since the condition to be quadratic is just the Koszulity condition for $i = 1, 2$.
3. Many important algebras, e.g. the free algebra, the polynomial algebra and the exterior algebra are Koszul.

Now let A_0 be a quadratic algebra, $A_0[0] = R$. Let E_0 be the space of relations for A_0. Let $E \subset A_0[1] \otimes_R A_0[1][[\hbar]]$ be a topologically free (over $\mathbb{C}[[\hbar]]$) R-subbimodule which reduces to E_0 modulo $\hbar$ ("deformation of the relations"). Let A be the ($\hbar$-adically complete) algebra generated over $R[[\hbar]]$ by $A[1] = A_0[1][[\hbar]]$ with the space of defining relations E. Thus A is a $\mathbb{Z}_+$-graded algebra.

Then we have the following fundamental result

Theorem 5.5 (Koszul deformation principle,[2, 3, 7, 19]). *If A_0 is Koszul then A is a topologically free $\mathbb{C}[[\hbar]]$ module if and only if so is $A[3]$.*

Remark. Note that $A[i]$ for $i < 3$ are obviously topologically free.

5.3.3 Symplectic Reflection Algebras

We will now demonstate by an example how the Koszul deformation principle works.

Let V be a finite dimensional symplectic vector space over $\mathbb{C}$ with a symplectic form ω, and G be a finite group acting symplectically on V. For simplicity let us assume that $(\wedge^2 V)^G = \mathbb{C}\omega$.

If $s \in G$ is a symplectic reflection, then let $\omega_s(x, y)$ be the form ω applied to the projections of x, y to the image of $1 - s$ along the kernel of $1 - s$; thus ω_s is a skewsymmetric form of rank 2 on V.

Let $S \subset G$ be the set of symplectic reflections, and $c : S \to \mathbb{C}$ be a function which is invariant under the action of G. Let $t \in \mathbb{C}$.

Definition 5.3. The symplectic reflection algebra $H_{t,c} = H_{t,c}(V, G)$ is the quotient of the algebra $G \ltimes \mathbf{T}(V)$ by the ideal generated by the relation

$$[x, y] = t\omega(x, y) - 2\sum_{s\in S} c_s\omega_s(x, y)s. \tag{5.2}$$

The following theorem shows that the algebras $H_{t,c}(V, G)$ satisfy a flatness property, and moreover, they are the only ones satisfying this property within a certain natural class.

Theorem 5.6. *Let $\kappa : \wedge^2 V \to \mathbb{C}[G]$ be a G-equivariant function (G acts on the target by conjugation). Define the algebra H_κ to be the quotient of the algebra $G \ltimes \mathbf{T}(V)$ by the relation $[x, y] = \kappa(x, y)$, $x, y \in V$. Put an increasing filtration on H_κ by setting* $\deg(V) = 1$, $\deg(G) = 0$, *and define $\xi : G \ltimes SV \to \mathrm{gr} H_\kappa$ to be the natural surjective homomorphism. Then ξ is an isomorphism if and only if κ has the form*

$$\kappa(x, y) = t\omega(x, y) - 2\sum_{s\in S} c_s\omega_s(x, y)s,$$

for some $t \in \mathbb{C}$ and G-invariant function $c : S \to \mathbb{C}$.

Before proving this theorem, let us point out a corollary. Denote by $\mathbf{H}_c(V, G)$ the algebra defined as $H_{t,c}(V, G)$, but with $t = 1$ and c being a formal parameter.

Corollary 5.2. *The algebra $\mathbf{H}_c(V, G)$ is a flat formal deformation of $G \ltimes \mathrm{Weyl}(V)$, parametrized by $\mathbb{C}[S]^G$.*

In fact, it turns out (see [9]) that $\mathbf{H}_c(V, G)$ is the universal deformation of $G \ltimes \mathrm{Weyl}(V)$, whose existence was proved in Example 5.2.

Proof (of Theorem 5.6). Let $\kappa : \wedge^2 V \to \mathbb{C}[G]$ be an equivariant map. We write $\kappa(x, y) = \sum_{g\in G} \kappa_g(x, y)g$, where $\kappa_g(x, y) \in \wedge^2 V^*$. To apply Theorem 5.5, let us homogenize our algebras. Namely, let $A_0 = (G \ltimes SV) \otimes \mathbb{C}[u]$. Also let $\hbar$ be a formal parameter, and consider the deformation $A = H_{\hbar u^2 \kappa}$ of A_0. That is, A is the quotient of $G \ltimes \mathbf{T}(V)[u][[\hbar]]$ by the relations $[x, y] = \hbar u^2 \kappa(x, y)$. This is a deformation of the type considered in Theorem 5.5, and it is easy to see that its flatness in $\hbar$ is equivalent to Theorem 5.6. Also, the algebra A_0 is Koszul, because the polynomial algebra SV is a Koszul algebra. Thus by Theorem 5.5, it suffices to show that A is flat in degree 3.

The flatness condition in degree 3 is "the Jacobi identity"

$$[\kappa(x, y), z] + [\kappa(y, z), x] + [\kappa(z, x), y] = 0,$$

which must be satisfied in $G \ltimes V$. In components, this equation transforms into the system of equations

$$\kappa_g(x, y)(z - z^g) + \kappa_g(y, z)(x - x^g) + \kappa_g(z, x)(y - y^g) = 0$$

for every $g \in G$ (here z^g denotes the result of the action of g on z).

This equation, in particular, implies that if x, y, g are such that $\kappa_g(x, y) \neq 0$ then for any $z \in V$ $z - z^g$ is a linear combination of $x - x^g$ and $y - y^g$. Thus $\kappa_g(x, y)$ is identically zero unless the rank of $(1 - g)|_V$ is at most 2, i.e. $g = 1$ or g is a symplectic reflection.

If $g = 1$ then $\kappa_g(x, y)$ has to be G-invariant, so it must be of the form $t\omega(x, y)$, where $t \in \mathbb{C}$.

If g is a symplectic reflection, then $\kappa_g(x, y)$ must be zero for any x such that $x - x^g = 0$. Indeed, if for such an x there had existed y with $\kappa_g(x, y) \neq 0$ then $z - z^g$ for any z would be a multiple of $y - y^g$, which is impossible since $\mathrm{Im}(1 - g)|_V$ is 2-dimensional. This implies that $\kappa_g(x, y) = -2c_g\omega_g(x, y)$, and c_g must be invariant.

Thus we have shown that if A is flat (in degree 3) then κ must have the form given in Theorem 5.6. Conversely, it is easy to see that if κ does have such form, then the Jacobi identity holds. So Theorem 5.6 is proved.

5.3.4 *Deformation of Representations*

Another method of estabishing flatness of a deformation A of A_0 defined by generators and relations is showing that a given faithful representation M_0 of the algebra A_0 (for example, the regular representation) can be deformed (flatly) to a representation M of A. In this case it follows automatically that A is flat. Let us give two examples of situations where this method can be applied.

Example 5.7 (see [8]). Let X be a connected, simply connected complex manifold, and G a discrete group of automorphisms of X. In this case the quotient X/G is a complex orbifold. Let $X' \subset X$ be the set of points having trivial stabilizer (it is a nonempty open subset of X). Define the braid group $\tilde{G}$ of the orbifold X/G to be the fundamental group of the manifold X'/G with some base point x_0. We have a surjective homomorphism $\phi : \tilde{G} \to G$, which corresponds to gluing back the points which have a nontrivial stabilizer. Let K be the kernel of this homomorphism.

The kernel K can be described by simple relations, corresponding to reflection hypersurfaces in X. Namely, given a reflection hypersurface $Y \subset X$, we have a conjugacy class C_Y in $\tilde{G}$ which corresponds to the loop in X'/G which goes counterclockwise around Y. Let T_Y be a representative of C_Y. Also, let $G_Y \subset G$ be the stabilizer of a generic point on Y; this is a cyclic group of some order n_Y. Then it follows from basic topology that the elements $T_Y^{n_Y}$ belong to K, and K is the smallest normal subgroup of $\tilde{G}$ containing all of them. In other words, the group G is the quotient of the braid group $\tilde{G}$ by the relations

$$T_Y^{n_Y} = 1. \tag{5.3}$$

Now let $A_0 = \mathbb{C}[G]$, and let us define a deformation A of A_0 to be the quotient of the group algebra of the braid group $\tilde{G}$ by a deformation of relations (5.3). Namely, for every reflection hypersurface $Y \subset X$ we introduce formal parameters $\tau_{Y,j}$, $j = 1, \ldots, n_Y$ (which are conjugation invariant), and replace relations (5.3) by the relations

$$\prod_{j=1}^{n_Y} (T_Y - e^{2\pi ij/n_Y} e^{\tau_{Y,j}}) = 0. \tag{5.4}$$

The quotient A of $\mathbb{C}[\tilde{G}][[\tau]]$ by these relations is called the **orbifold Hecke algebra** of X/G, and denoted by $\mathcal{H}_\tau(X, G)$.

Theorem 5.7 ([8]). *If $H^2(X, \mathbb{C}) = 0$ then $\mathcal{H}_\tau(X, G)$ is a flat deformation of $\mathbb{C}[G]$.*

Remark. If X is $\mathbb{C}^n$ and $G = G_0 \cdot L$, where L is a lattice of rank $2n$ and G_0 is a finite group acting on L then $\mathcal{H}_\tau(X, G)$ is, essentially, the algebra which was mentioned in Example 5.4.

To illustrate the relevance of the condition $H^2(X, \mathbb{C}) = 0$, let us consider the special case when G is the triangle group $F_{p,q,r}$, generated by a, b, c with defining relations

$$a^p = 1,\ b^q = 1,\ c^r = 1,\ abc = 1,$$

where $p, q, r > 1$ are positive integers. The group G is the group generated by rotations around the vertices of a triangle with angles $\pi/p, \pi/q, \pi/r$, by twice the angle at the vertex. Let $S = \frac{1}{p} + \frac{1}{q} + \frac{1}{r}$. The triangle lies on the sphere, Euclidean plane, or hyperbolic plane X when $S > 1$, $S = 1$, and $S < 1$, respectively. The deformation $\mathcal{H}_\tau(X, G)$ is generated by a, b, c with defining relations

$$\prod_{j=1}^{p} (a - \alpha_j) = 0,\ \prod_{j=1}^{q} (b - \beta_j) = 0,\ \prod_{j=1}^{r} (c - \gamma_j) = 0,\ abc = 1,$$

where

$$\alpha_j = e^{2\pi ij/p} e^{\tau_{1j}}, \beta_j = e^{2\pi ij/q} e^{\tau_{2j}}, \gamma_j = e^{2\pi ij/r} e^{\tau_{3j}}.$$

Theorem 5.7 says that the deformation is flat for the Euclidean and hyperbolic plane, but says nothing about the sphere, i.e. the triples (p, q, r) equal to $(2, 2, n)$, $(2, 3, 3)$, $(2, 3, 4)$, $(2, 3, 5)$, in which case the group G is finite. And indeed, in this case $\mathcal{H}_\tau(X, G)$ is actually not flat! To see this, note that in the sphere case $\mathcal{H}_\tau$, if it were flat, would have dimension $|G|$ (over $\mathbb{C}[[\tau]]$). So we may take the determinant of the relation $abc = 1$ (using the fact that the eigenvalues of a, b, c are $\alpha_j, \beta_j, \gamma_j$, with equal multiplicities). This yields a nontrivial relation on τ:

$$(\prod_{j=1}^{p} \alpha_j)^{|G|/p} (\prod_{j=1}^{q} \beta_j)^{|G|/q} (\prod_{j=1}^{r} \gamma_j)^{|G|/r} = 1,$$

which rules out flatness of $\mathcal{H}_\tau$.

Example 5.8 ([11]). Let W be a Coxeter group of rank r with generators s_i and defining relations

$$s_i^2 = 1,\ (s_i s_j)^{m_{ij}} = 1 \text{ for } m_{ij} < \infty,\ i, j = 1, \dots, r,\ i \neq j,$$

where $m_{ij} = m_{ji}$ are integers ≥ 2 or ∞, defined for $i \neq j$. Let W_+ be the subgroup of even elements of W. It is easy to see that W_+ is generated by the elements $a_{ij} := s_i s_j$, with defining relations

$$a_{ij} a_{ji} = 1,\ a_{ij} a_{jk} a_{ki} = 1,\ a_{ij}^{m_{ij}} = 1.$$

Define a deformation of $A_0 = \mathbb{C}[W_+]$ as follows. Introduce invertible parameters $t_{ij,k} = t_{ji,-k}^{-1}$, $k \in \mathbb{Z}/m_{ij}\mathbb{Z}$ for $m_{ij} < \infty$. Let $R = \mathbb{C}[t_{ij,k}]$, and A be the R-algebra generated by a_{ij} with defining relations

$$a_{ij} a_{ji} = 1,\ a_{ij} a_{jk} a_{ki} = 1,\ \prod_{k=1}^{m_{ij}} (a_{ij} - t_{ij,k}) = 0.$$

For any $x \in W_+$, fix a reduced word $w(x)$ representing x. Let $T_{w(x)}$ be the element of A corresponding to this word.

Theorem 5.8 ([11]).

(i) The elements $T_{w(x)}$ for $x \in W_+$ span A over R.
(ii) These elements form a basis of A over R if and only if W has no finite parabolic subgroups of rank 3, i.e. iff for each i, j, l,

$$\frac{1}{m_{ij}} + \frac{1}{m_{jl}} + \frac{1}{m_{li}} \leq 1.$$

Corollary 5.3. *Let $\hat{A}$ be the completion of A with respect to the ideal generated by $t_{ij,k} - e^{2\pi k\sqrt{-1}/m_{ij}}$. Then $\hat{A}$ is a flat deformation of A_0 iff W has no finite parabolic subgroups of rank 3.*

Remark. Note that triangle groups $F_{p,q,r}$ are groups W_+ for Coxeter groups of rank 3 (with $m_{12} = p, m_{23} = q, m_{31} = r$), so the "only if" part of Theorem 5.8 (and the "if" part in rank 3) follow from Example 5.7.

In both of these examples, flatness is established by showing, using geometric methods (D-modules or constructible sheaves), that the regular representation of A_0 can be flatly deformed to a representation of the deformation. Let us conclude by illustrating this in Example 5.7, in the case when $X = E$ is a complex vector space, and G is a finite group acting linearly on E. In this case, Theorem 5.7 was proved by Broué, Malle, and Rouquier [4], following an idea of Cherednik. Let us sketch their proof.

The main idea of the proof is to introduce Dunkl operators D_a, $a \in E$, which act on functions on E (with poles on the reflection hyperplanes Y):

$$D_a = \partial_a + \sum_Y \frac{\alpha_Y(a)}{\alpha_Y} \Big(\sum_{g \in G_Y} c_{Y,g} g\Big),$$

where the summation is over all reflection hyperplanes Y, α_Y is the nonzero element of E^* vanishing on Y, and $c_{Y,g}$ is a conjugation invariant function of Y, g.

It can be shown that the Dunkl operators commute: $[D_a, D_b] = 0$. This implies that the system of equations $D_a\psi = 0$, $a \in E$, can be regarded as a local system with fiber $\mathbb{C}G$ on $(E \setminus \cup Y)/G$. The fundamental group of $(E \setminus \cup Y)/G$, by definition, is $\tilde{G}$, so we may consider the corresponding monodromy representation of this group. If $c = 0$, the monodromy representation is the standard homomorphism $\mathbb{C}\tilde{G} \to \mathbb{C}G$. One may show that if $c \neq 0$, then the monodromy representation is a deformation of this standard homomorphism, which factors through the Hecke algebra $\mathcal{H}_\tau(E, G)$, for an appropriate linear change of variables $c \to \tau$. This implies the flatness of $\mathcal{H}_\tau(E, G)$.

Acknowledgements This paper is based on my lecture at "Giornata IndAM", Naples, June 7, 2005. I would like to thank the organizers, in particular Corrado De Concini and Paolo Piazza for this wonderful opportunity. I am also grateful to J. Stasheff for useful comments.

References

1. J. Alev, M.A. Farinati, T. Lambre, A.L. Solotar, Homologie des invariants d'une algèbre de Weyl sous l'action d'un groupe fini. J. Algebra **232**, 564–577 (2000)

2. A. Braverman, D. Gaitsgory, Poincaré-Birkhoff-Witt theorem for quadratic algebras of Koszul type. J. Algebra **181**(2), 315–328 (1996). hep-th/9411113
3. A. Beilinson, V. Ginzburg, W. Soergel, Koszul duality patterns in representation theory. J. Am. Math. Soc. **9**, 473–527 (1996)
4. M. Broué, G. Malle, R. Rouquier, Complex reflection groups, braid groups, Hecke algebras. J. Reine und Angew. Math. **500**, 127–190 (1998)
5. V. Dolgushev, P. Etingof, Hochschild cohomology of quantized symplectic orbifolds and the Chen-Ruan cohomology, math.QA/0410562. Int. Math. Res. Not. **27**, 1657–1688 (2005)
6. V.G. Drinfeld, Degenerate affine Hecke algebras and Yangians. Func. Anal. Appl. **20**, 62–64 (1986)
7. V. Drinfeld, On quadratic commutation relations in the quasiclassical case, Mathematical physics, functional analysis (Russian), 25–34, 143, "Naukova Dumka", Kiev, 1986. Selecta Math. Sovietica. **11**, 317–326 (1992)
8. P. Etingof, Cherednik and Hecke algebras of varieties with a finite group action. math.QA/0406499
9. P. Etingof, V. Ginzburg, Symplectic reflection algebras, Calogero-Moser systems, and a deformed Harish-Chandra isomorphism. Invent. Math. **147**, 243–348 (2002)
10. P. Etingof, A. Oblomkov, Quantization, orbifold cohomology, and Cherednik algebras. math.QA/0311005, Contemp. Math. **417**, 171–182 (American Mathematical Society, Providence, 2006)
11. P. Etingof, E. Rains, New deformations of group algebras of Coxeter groups. math.QA/0409261, Int. Math. Res. Not. **2005**(10), 635–646
12. B.V. Fedosov, A simple geometrical construction of deformation quantization. J. Diff. Geom. **40**, 213–238 (1994)
13. M. Gerstenhaber, On the deformation of rings and algebras. Ann. Math. (2) **79**, 59–103 (1964)
14. G. Halbout, X. Tang, Dunkl operator and quantization of Z/2-singularity. J. für die Reine und Ang. Mat. **2012**(673), 209–235
15. G. Hochschild, B. Kostant, A. Rosenberg, Differential forms on regular affine algebras. Trans. Am. Math. Soc. **102**, 383–408 (1962)
16. M. Kontsevich, Deformation quantization of Poisson manifolds. Lett. Math. Phys. **66**, 157–216 (2003). q-alg/9709040
17. O. Mathieu, Homologies associated with Poisson structures, in *Deformation theory and symplectic geometry (Ascona, 1996)*. Mathematical Physics Studies, vol. 20 (Kluwer Academic, Dordrecht, 1997), pp. 177–199
18. R. Nest, B. Tsygan, Algebraic index theorem. Comm. Math. Phys. **172**(2), 223–262 (1995)
19. A. Polishchuk, L. Positselski, Quadratic algebras. Preprint, 1996 (to be published by the AMS, 2005)

Chapter 6
Mathematical Models and Solutions for the Analysis of Human Genotypes

Giuseppe Lancia

Abstract The past few years have seen the birth and the growth of a new reseach area in bioinformatics, called haplotyping. Haplotyping problems are combinatorial and optimization problems concerned with the analysis of human polymorphisms in populations, and with the study of common patterns for such polymorphisms. In this chapter we review the most important haplotyping problems, and describe the mathematical models and algorithmic approaches employed for their solution.

6.1 Introduction

Over the last two decades, the field of computational biology (also known as *Bioinformatics*) has experienced a tremendous growth. Computational biology problems are mostly concerned with the interpretation and analysis of large volumes of genomic data. More precisely, the simulation of biological processes in the cell, the discovery of important signals in genomic data, or even the detection and removal of experimental errors, are cast as computational problems over some suitable mathematical models. Once the genomic data have been translated into mathematical objects (such as graphs and permutations), the original biology questions become computational problems to be solved by standard techniques. Most of the times these problems turn out to be very hard optimization problems, whose hardness not only stems from their theoretical complexity but also from the large size of the instances of interest in real-life applications.

The availability of genomic data has increased at almost exponential rate over the past 30 years. For example, the EMBL repository for nucleotide sequences [22] has increased from roughly 1,000 entries of 1982 to about 100,000,000 of today.

G. Lancia (✉)
Dipartimento di Matematica e Informatica
University of Udine, via delle Scienze 206, 33100 Udine, Italy
e-mail: giuseppe.lancia@uniud.it

V. Ancona and E. Strickland (eds.), *Trends in Contemporary Mathematics*, Springer INdAM Series 8, DOI 10.1007/978-3-319-05254-0_6,

Notice that each entry itself is a sequence that can range from a few hundred to several hundred thousand of nucleotides. This increase in both volume and size of genomic data has posed new challenging problems which require sophisticated lines of attack.

Without any doubts, mathematical programming techniques rank amongst the most powerful approaches for hard optimization problems [29, 30], and hence their use in bioinformatics has also steadily increased over the past years. In the mathematical programming approach, a problem is typically modeled as the search of the optimal values for a set of integer and/or continuous variables satisfying a set of linear (or, more rarely, non-linear) constraints. The objective function to optimize can either be linear or non-linear. Mathematical approaches have been applied to a large number of computational biology problems (for a survey, we refer the reader to [23]). In this chapter we focus on a particular area which has received a great deal of attention in the recent years. The area is concerned with the analysis of human polymorphisms in populations, and of the study of common patterns for such polymorphisms (these patterns are also called *haplotypes*).

The remainder of the chapter is organized as follows. In Sect. 6.2 we describe the biological aspects of the problems, starting from the concepts of Single Nucleotide Polymorphisms, haplotypes and genotypes. In Sect. 6.3 we describe Clark's rule and the first combinatorial haplotyping problem. In Sect. 6.4 we focus on the parsimony haplotyping problem. In Sect. 6.5 we review haplotyping for perfect phylogenies. Some conclusions are drawn in Sect. 6.6.

6.2 Single Nucleotide Polymorphisms

The analysis of large quantities of genomic data has confirmed that the genetic makeup of humans is remarkably well-conserved. Generally speaking, the differences at DNA level between any two individuals amount to less than 5 % of their genomic sequences, so that the differences at the phenotype level (i.e., in the way the individuals look) are caused by small regions of differences in the genomes. The smallest possible region consists of a single nucleotide and is called *Single Nucleotide Polymorphism* or SNP (pronounced "snip"). SNPs are the predominant form of human genetic variation, and they find use, e.g., in medical, drug-design, diagnostic, and forensic applications. A SNP is almost always a polymorphism with only two alleles (i.e., only two bases, out of the four possible, are observed in the population). The two alleles can be different for different SNPs.

In diploid organisms (such as humans) each genome is organized in pairs of homologous chromosomes, and a single copy of each chromosome pair is inherited from each of the two parents. For a diploid organism, at each SNP an individual can either be *homozygous* (i.e., possess the same allele on both chromosomes) or *heterozygous* (i.e., possess two different alleles). The values of a set of SNPs on a particular chromosome copy define a *haplotype*.

Individual 1, paternal: taggtccCgatttCccaggcgcCgtatacttcgacgggTctat
Individual 1, maternal: taggtccGgatttAccaggcgcGgtatacttcgacgggTctat

Individual 2, paternal: taggtccCgatttAccaggcgcGgtatacttcgacgggTctat
Individual 2, maternal: taggtccGgatttCccaggcgcGgtatacttcgacgggCctat

Individual 3, paternal: taggtccCgatttAccaggcgcGgtatacttcgacgggTctat
Individual 3, maternal: taggtccGgatttAccaggcgcCgtatacttcgacgggCctat

Fig. 6.1 A chromosome in three individuals. There are four SNPs

In Fig. 6.1, we illustrate a simplistic example, showing a specific chromosome in three individuals. For each individual, the pair of his chromosome copies are reported. There are four SNPs. The alleles for SNP 1, in this example, are **C** and **G**, while for SNP 4 they are **T** and **C**. Individual 1 is heterozygous for SNPs 1, 2 and 3, and homozygous for SNP 4. His haplotypes are CCCT and GAGT. The haplotypes of individual 3 are CAGT and GACC.

Haplotyping an individual consists in determining his two haplotypes, for a given chromosome. The direct experiment for determining an individual's haplotypes is called *sequencing*. By sequencing, we are able to read the sequence of base pairs along the DNA molecule, but the experiment has several limitations: (i) there is a relatively high probability of reading errors; (ii) only short DNA fragments can be sequenced at a time, so that "regular" sequences have to be broken into smaller pieces which must then be somehow assembled back together; (iii) the cost for sequencing is quite high.

When the individuals for which we seek to determine the haplotypes are many, we talk of *population haplotyping*, and in this case the third limitation above becomes the most important. Instead of directly sequencing the haplotypes, biologists have then devised a cheap experiment which can determine a less informative and often ambiguous type of data, i.e., the *genotypes*, from which the haplotypes can then be retrieved computationally.

A genotype of an individual contains the information about the two (possibly identical) alleles at each SNP, but without specifying their paternal or maternal origin. Given a genotype, there may be many possible pairs of haplotypes that are consistent with that genotype. For example, assume we only know that an individual is heterozygous for the alleles $\{\mathtt{C}, \mathtt{T}\}$ at SNP 1 and for the alleles $\{\mathtt{A}, \mathtt{G}\}$ at SNP 2. Then, either one of these alternatives may be true:

(i) One parent gave the alleles C and A, the other gave the alleles T and G.
(ii) One parent gave the alleles C and G, the other gave the alleles T and A.

Both possibilities are plausible. Associating the alleles to the parents is called *phasing* the alleles. For k heterozygous SNPs there are 2^k possible phasings, which makes choosing the correct one a difficult problem. The two haplotypes that are obtained by phasing the alleles are said to *resolve*, or to *explain*, the genotype.

The most general population haplotyping problem can be stated as follows:

Given a set G of genotypes, corresponding to an existing, unknown, set H of haplotypes, retrieve H.

The goal is to compute a set H of haplotypes which contains, for each genotype $g \in G$, the two haplotypes h_1, h_2 obtained by the correct phasing of g. It is not easy to describe constraints, based only on the knowledge of G, that define precisely which of the exponentially many phasings of a genotype is the correct one. Biologists have therefore described several sensible criteria for "good" phasings. For instance, under a widely accepted parsimony principle, a good solution may be one which minimizes the number of distinct haplotypes inferred.

Once it has been mathematically modeled, haplotyping gives rise to several nice and challenging problems (for surveys on population haplotyping problems see, e.g., [4, 18]). In this chapter we address some of the most interesting combinatorial haplotyping models proposed in the literature. Each model and objective function has specific biological motivations, which are discussed in the following sections.

Given a set of n SNPs, fix arbitrarily a binary encoding of the two alleles for each SNP (i.e., call one of the two alleles `0` and the other `1`). Once the encoding has been fixed, each haplotype becomes a binary vector of length n. For a haplotype h, denote by $h[i]$ the value of its i-th component. Given two haplotypes h' and h'', we define a special sum whose result is a vector $g := h' \oplus h''$. The vector g has components in $\{\mathtt{0}, \mathtt{1}, \mathbf{X}\}$, defined by

$$g[i] := \begin{cases} \mathtt{0} & \text{if } h'[i] = h''[i] = \mathtt{0} \\ \mathtt{1} & \text{if } h'[i] = h''[i] = \mathtt{1} \\ \mathbf{X} & \text{if } h'[i] \neq h''[i] \end{cases}.$$

We call a vector g with entries in $\{\mathtt{0}, \mathtt{1}, \mathbf{X}\}$ a *genotype*. Each position i such that $g[i] = \mathbf{X}$ is called an *ambiguous position* (or *ambiguous site*). Denote by $A(g) \subseteq \{1, \ldots, n\}$ the set of ambiguous positions of g. Biologically, genotype entries of value `0` or `1` correspond to homozygous SNP sites, while entries of value **X** correspond to heterozygous sites (see Fig. 6.2).

A *resolution* of a genotype g is given by a pair of haplotypes h' and h'' such that $g = h' \oplus h''$. The haplotypes h' and h'' are said to resolve g. A genotype is *ambiguous* if it has more than one possible resolution, i.e., if it has at least two ambiguous positions. A haplotype h is said to be *compatible* with a genotype g if h can be used in a resolution of g, namely, if and only if at each position where $g[i] \neq \mathbf{X}$ it is $g[i] = h[i]$. Two genotypes g and g' are compatible if there exists at least one haplotype compatible with both of them, otherwise, they are *incompatible*. Note that g and g' are compatible if and only if at each i where they are both non-ambiguous, it is $g[i] = g'[i]$.

Haplotype 1, paternal: 0 1 0 1	X X X 1 Genotype 1
Haplotype 1, maternal: 1 0 1 1	
Haplotype 2, paternal: 0 0 1 1	X X 1 X Genotype 2
Haplotype 2, maternal: 1 1 1 0	
Haplotype 3, paternal: 0 0 1 1	X 0 X X Genotype 3
Haplotype 3, maternal: 1 0 0 0	

Fig. 6.2 Haplotypes and corresponding genotypes for three individuals

The experiment yielding each genotype is such that, at each SNP, it is known if an individual is homozygous for allele 0 (so that $g[i] = 0$) or homozygous for allele 1 (so that $g[i] = 1$) or heterozygous (so that $g[i] = \mathbf{X}$). For the haplotyping problems described in this chapter, the input data consist in a set G of m genotypes $g_1, \dots, g_m$, corresponding to m individuals in a population. The output is set H of haplotypes such that, for each $g \in G$, there is at least one pair of haplotypes $h', h'' \in H$ with $g = h' \oplus h''$. Such a set H of haplotypes is said to explain G. In addition to explaining G, the set H is also required to satisfy some particular constraints, which are different for specific types of haplotyping problems.

6.3 Clark's Rule

The geneticist Clark proposed in [9] a rule to derive new haplotypes by inference from known ones:

Clark's Inference Rule: Given a genotype g and a compatible haplotype h, obtain a new haplotype q by setting $q[j] := 1 - h[j]$ at all positions $j \in A(g)$ and $q[j] := h[j]$ at the remaining positions.

Notice that q and h resolve g. In order to resolve all genotypes of G, Clark suggested the following procedure, based on successive applications of the inference rule:

Clark's Algorithm: Let $G' \subset G$ be the set of non-ambiguous genotypes, and let H be the set of haplotypes obtained from G'. Reset $G := G - G'$. Then, repeat the following. If they exist, take a $g \in G$ and a compatible $h \in H$ and apply the inference rule, obtaining q. Set $H := H \cup \{q\}$, $G := G - \{q \oplus h', h' \in H\}$, and iterate. When no such g and h exist, the algorithm has succeeded if $G = \emptyset$ and has failed otherwise.

Notice that the procedure is nondeterministic since it does not specify how to choose the pair (g, h) whenever there are more candidates. For example, suppose $G = \{\mathbf{X}000, \mathbf{XX}00, 1\ 1\mathbf{XX}\}$. The algorithm starts by setting $H = \{0000, 1000\}$

and $G = \{$**XX**00, 11**XX**$\}$. The inference rule can be used to resolve **XX**00 from 0000, obtaining 1100, which can, in turn, be used to resolve 11**XX**, obtaining 1111. However, one could have started by using 1000 to resolve **XX**00 obtaining 0100. At that point, there would be no way to resolve 11**XX**. The non-determinism in the choice of the pair (g, h) can be settled by fixing a deterministic rule based on a sorting of the data. In [9], a large number of random sortings is used to run the algorithm, and the best solution overall is reported. Computational tests showed that, although most times the algorithm could resolve all genotypes, many times it still failed.

The problem of finding an ordering of application of Clark's rule that eventually leaves the fewest number of unresolved genotypes was defined and studied by Gusfield [14,15], who proved it is NP-hard and APX-hard (i.e, there is a value $\delta > 1$ for which it is NP-hard even to give a δ-approximation algorithm). As for practical algorithms, Gusfield proposed an Integer Linear Programming (ILP) approach for a graph-theoretic reformulation of the problem, and noticed that the solution of the LP relaxation was very often integer for the real-life instances tested. The model was applied to real data as well as random instances, with up to 80 genotypes, 60 of which were ambiguous, over 15 SNPs.

6.4 Pure Parsimony

Pure parsimony haplotyping (PPH) has the objective of minimizing the size of H. This objective has several biological motivations. For instance, today there are many human beings, but they all descend from a small number of ancestors, so that their haplotypes should be the same as their ancestors' (if it were not for some recombination events and mutations).

The PPH problem is NP-hard. In fact, Lancia et al. [26] showed that the problem is APX-hard, even when each genotype is restricted to have at most three ambiguous sites. Notice that, although the problem is APX-hard, there exist constant-ratio approximation algorithms when genotype has at most k ambiguous sites, for each constant k [26].

Pure parsimony haplotyping has been attacked by using several algorithmic approaches, many of them employing Mathematical Programming techniques. In particular, we recall the following approaches:

ILP formulations of exponential size. Let $H(G)$ be the set of all haplotypes compatible with at least one genotype of G, and let $\chi(G) := |H(G)|$. The first ILP formulation for PPH is called TIP, and was given by Gusfield in [17]. TIP has $\chi(G)$ variables and $O(m2^n)$ constraints. There ia a binary variable x_h associated to every $h \in H(G)$ (where $x_h = 1$ means that h is taken in the solution, whereas $x_h = 0$ means that h is not taken). For every $g \in G$, let $\mathcal{P}_g := \{\{h_1, h_2\} \in H(G) \times H(G) \mid h_1 \oplus h_2 = g\}$. For every $g \in G$ and $\{h_1, h_2\} \in \mathcal{P}_g$, there is a binary variable y_{h_1,h_2} used to select a pair $\{h_1, h_2\}$ in order to resolve the genotype g.

Gusfield managed to employ some preprocessing rules to get rid of variables that can be proved to be non-essential in the formulation. The resulting model is called RTIP (standing for Reduced TIP). Although practically useful, the preprocessing still leaves an exponential model, whose size grows quite quickly with respect to the instance size. The experimental results of [17] showed that RTIP can be used to tackle problems with up to 50 genotypes, over 30 SNPs, but there must be relatively small levels of heterozygosity in the input genotypes.

Lancia and Rizzi showed in [24] that when each genotype has at most two ambiguous sites, Gusfield's ILP is naturally integer and hence the problem can be solved in polynomial time.

In [25], Lancia and Serafini proposed an exponential ILP model which exploits an interpretation of PPH as a particular Set Covering (SC), based on the following observation: *if H is a set of haplotypes resolving G, then for each genotype g, ambiguous position $i \in A(g)$ and value $a \in \{\mathtt{0}, \mathtt{1}\}$, there is a haplotype $h \in H \cap H(g)$ such that $h[i] = a$.* This condition represents the covering constraints for a set cover problem in which the universe is the set of all triples (g, i, a), for $g \in G$, $i \in A(g)$ and $a \in \{0, 1\}$, and each haplotype h represents a subset of the universe, namely $h \leftrightarrow \{(g, i, a) \mid h \in H \cap H(g), h_i = a\}$. The condition is only necessary, but not sufficient, for H to be a feasible solution of PPH. Consider, for example, the following "diagonal" instance $G = \{\mathtt{1XXX}, \mathtt{X1XX}, \mathtt{XX1X}, \mathtt{XXX1}\}$. The set $\{\mathtt{0111}, \mathtt{1011}, \mathtt{1101}, \mathtt{1110}\}$ satisfies the covering condition but does not resolve G.

The SC associated to PPH seeks to minimize the number of haplotypes needed to satisfy all covering conditions for the given set of genotypes G. This SC is in fact a relaxation of PPH. The formulation of the SC model has an exponential number of variables and constraints, and can be solved by Branch-and-Cut-and-Price, i.e., via the generation of variables and constraints at run-time. If the optimal solution of SC resolves G, it is an optimal PPH solution as well. If not, one can try to obtain a good feasible PPH solution from the optimal cover by adding only a small number of haplotypes. This idea is exploited by an effective heuristic presented in [25]. The computational results show that the SC approach can be orders of magnitude faster than RTIP and can be applied to instances with $n = m = 50$ and $\chi(G)$ up to 10^9. These are among the largest-size instances for which optimal solutions have been obtained in the literature.

ILP formulations of polynomial size. Many authors have independently proposed polynomial-size ILP formulations [2, 6, 19, 26]. These formulations have $O(mn)$ variables, representing the bits of the haplotypes in the solution. The basic idea is that for each genotype $g^i \in G$ one must determine two haplotypes h_1^i and h_2^i such that $h_1^i \oplus h_2^i = g^i$. This implies that, for each position j such that $g^i[j] = \mathbf{X}$, one needs to decide a variable $x_{ij} \in \{0, 1\}$, and set $h_1^i[j] = x_{ij}$ and $h_2^i[j] = 1 - x_{ij}$. The polynomial formulations for PPH express the objective function and constraints in terms of the x variables and possibly a set of additional variables.

The LP relaxation of these formulations is generally quite weak. The addition of some extra valid inequalities [2, 6] improves the quality of the bound, but the integrality gap between the integer optimum and the LP relaxation remains large. Brown and Harrower [7] also proposed an hybrid model in which variables for a fixed subset of haplotypes are explicitly present, while the rest of the haplotypes are implicitly represented by polynomially many variables and constraints. These polynomial/hybrid formulations were successfully used for problems of similar size as those solvable by TIP. Furthermore, some tests were conducted on larger problems (30 genotypes over up to 75 SNPs), on which the exponential formulation could not be applied successfully due to the IP size.

The last polynomial model for PPH was proposed by Catanzaro et al. [8] and is based on the "class representatives with smallest index" technique for the Vertex Color problem. This ILP model turns out to be quite effective and to outperform other polynomial models for the PPH problem.

Quadratic, semi-definite programming approaches. A quadratic formulation, solved by semi-definite programming, was proposed by Kalpakis and Namjoshi in [21]. The formulation has a variable for each possible haplotype and hence it cannot be used when $\chi(G)$ is too large. According to the computational results, the size of the problems solved is comparable to that of RTIP and of the best polynomial ILP models. Based on a similar formulation, an (exponential-time) approximation algorithm was presented in [20].

Combinatorial branch-and-bound approaches. In [32], Wang and Xu proposed a simple combinatorial branch-and-bound approach, which implicitly enumerates all possible resolutions for each genotype. The lower bound is the number of haplotypes used so far. Since the search space is exponential and the bound is weak, the method is not able to solve instances of size comparable to the other approaches. Even the solution for 20 genotypes over 20 SNPs can sometimes take an extremely long time to be found.

Boolean Optimization. Among the approaches that were applied to PPH there is Pseudo Boolean Optimization (PBO), a technique by which a problem can be modeled via integer linear constraints over a set of boolean variables. The goal is to find an assignment of the variables which satisfies all constraints and minimizes a given objective function. The model is then solved by a SAT solver.

The PBO models for PPH are mostly based on the following feasibility question: "given a tentative cardinality k, does there exist as set H of k haplotypes that resolves G?" The feasibility question is expressed in terms of boolean variables similar to the variables of the polynomial ILP models, and is repeated for increasing k until it yields a positive answer.

The first PBO algorithm for PPH was presented by Lynce and Marques-Silva [27]. Later works aimed at breaking the symmetries in the original PBO model, and at computing tight lower and upper bounds to be used for pruning the search space [13,28]. The SAT approaches showed to be competitive with the best mathematical programming approaches for PPH.

Heuristics. Generally speaking, all exact models run into troubles when trying to solve "large" instances (where the most critical parameter is the number of ambiguous positions per genotype). Therefore, in order to tackle large instances of PPH (e.g., instances with hundreds of genotypes over hundreds of SNPs with high levels of heterozygosity) one needs to resort to the use of effective heuristics.

One such heuristic procedure is `CollHaps`, proposed by Tininini et al. [31]. `CollHaps` is based on the representation of the solution as a matrix M' of $2m$ rows (the solving haplotypes, also called *symbolic haplotypes* because they may contain variables), in which some entries are fixed, while the others, corresponding to ambiguous positions in the input genotypes, are variables. At each step, `CollHaps` fixes the variables in a greedy way, trying to maintain as few distinct symbolic haplotypes as possible and, eventually, to end up with as few distinct actual haplotypes as possible. Experimental results have shown that `CollHaps` is a very effective and accurate heuristic for PPH, and ranks amongst the best available procedures for this problem.

In another heuristic approach for PPH, Di Gaspero and Roli [10] proposed the use of stochastic local search, which they considered to yield a reasonable compromise between solutions quality and procedure's running time. In their work, Di Gaspero and Roli utilized a family of local search strategies, such as best improvement (BI), stochastic first improvement (SFI), simulated annealing (SA), and tabu search (TS).

6.5 Perfect Phylogeny

One limitation of the previous haplotyping models is that they do not take into account the fact that haplotypes evolve over time. Because of mutations and recombinations, for long enough haplotypes, there can be a a haplotype of an individual which was not possessed by any of his ancestors. Therefore, when the haplotypes studied are long, different models of haplotyping should be considered. One of these is haplotyping for *perfect phylogeny*.

The perfect phylogeny model is used under the hypothesis that no recombination events happened, but there were only mutations. It is assumed that at the beginning there existed only one ancestral haplotype, and new haplotypes were derived over time from existing haplotypes as follows. If at some point there existed a haplotype h in the population and then a mutation of $h[i]$ happened, a new haplotype h' started to exist, with $h'[j] = h[j]$ for $j \neq i$, and $h'[i] = 1 - h[i]$. We say that h is the "father" of h' in the tree of haplotype evolution. The evolution of the haplotypes can be described by a rooted arborescence, in which the haplotypes are the vertices, and each arc is directed from father to child. A perfect phylogeny is such an arborescence. Given a set H of haplotypes, and a haplotype h^* from which all other haplotypes have evolved, a perfect phylogeny for H is a rooted binary tree T such that:

1. The root of T corresponds to h^*.
2. The leaves of T are in 1-to-1 correspondence with H.

3. Each position $i \in 1, \ldots, n$ labels at most one edge in T.
4. For each leaf $h \in H$ and edge e along the path from h^* to h, if e is labeled with position i, then $h[i] \neq h^*[i]$.

Without loss of generality, it can be assumed that $h^* = \texttt{00} \cdots \texttt{0}$. It can be shown that a perfect phylogeny for H exists if and only if there are no four haplotypes $h^1, \ldots, h^4 \in H$ and two positions i, j such that

$$\{h^a[i]\, h^a[j], 1 \leq a \leq 4\} = \{\texttt{00}, \texttt{01}, \texttt{10}, \texttt{11}\}.$$

The *Haplotyping for Perfect Phylogeny* problem can then be stated as follows: Given a set G of genotypes, find a set H of haplotypes such that H resolves G and there is a perfect phylogeny for H.

Haplotyping for Perfect Phylogeny was introduced by Gusfield [16], who first showed that the problem is polynomial and conjectured the existence of a linear-time algorithm for its solution. To prove that it is polynomial, Gusfield reduced the problem to a well-known but complex graph theory problem, i.e., the *graph realization*, with an $O(nm\,\alpha(n, m))$ algorithm, where α is the slow-growing inverse Ackerman function. Much simpler to implement while still very effective algorithms were designed by Bafna et al. [1] and Eskin et al. [12]. The complexity for both algorithms is $O(n^2 m)$. Eventually, Ding et al. [11] were able to obtain an algorithm for perfect phylogeny haplotyping of complexity $O(nm)$, i.e., a linear-time algorithm. Almost at the same time as Ding et al. obtained their result, another linear-time algorithm for perfect phylogeny haplotyping was proposed by Bonizzoni [3], who showed that the problem solution can be obtained via the recognition of special posets of width two.

The perfect phylogeny model does not take into account events such as back mutations. In [5], Bonizzoni et al. considered a variant of the perfect phylogeny model which they called Persistent Perfect Phylogeny (referred as P-PP). In the P-PP model, each SNP site can mutate to a new value and then back to its original value only once. They developed an exact algorithm for solving the P-PP problem that is exponential in the number of SNPs and polynomial in the number of individuals.

6.6 Conclusions

In this chapter we have reviewed some combinatorial optimization problems originated in the context of genotype analysis for populations.

For one problem (perfect phylogeny haplotyping) there exist extremely fast (i.e., linear time) exact algorithms, while the remaining problems have been attacked by many algorithmic approaches, among which mathematical programming modeling has played a crucial role. From the computational results reported, it can be argued that the use of such sophisticated optimization procedures has proved quite successful for these problems, at least when the instances are not too large. On the other hand, heuristic procedures are the key to the solution of large instances for the NP-hard problems that we have described.

References

1. V. Bafna, D. Gusfield, G. Lancia, S. Yooseph, Haplotyping as perfect phylogeny: a direct approach. J. Comput. Biol. **10**, 323–340 (2003)
2. P. Bertolazzi, A. Godi, M. Labbé, L. Tininini, Solving haplotyping inference parsimony problem using a new basic polynomial formulation. Comput. Math. Appl. **55**, 900–911 (2008)
3. P. Bonizzoni, A linear-time algorithm for the perfect phylogeny haplotype problem. Algorithmica **48**, 267–285 (2007)
4. P. Bonizzoni, G. Della Vedova, R. Dondi, J. Li, The haplotyping problem: an overview of computational models and solutions. J. Comput. Sci. Technol. **19**, 1–23 (2004)
5. P. Bonizzoni, C. Braghin, R. Dondi, G. Trucco, The binary perfect phylogeny with persistent characters. Theor. Comput. Sci. **454**, 51–63 (2012)
6. D.G. Brown, I.M. Harrower, A new integer programming formulation for the pure parsimony problem in haplotype analysis, in *Annual Workshop on Algorithms in Bioinformatics – WABI*, Bergen. LNCS 3240 (Springer, Berlin/Heidelberg, 2004), pp. 254–265
7. D.G. Brown, I.M. Harrower, A new formulation for haplotype inference by pure parsimony. Technical report CS-2005-03, Department of Computer Science, University of Waterloo, Canada (2005)
8. D. Catanzaro, A. Godi, M. Labbè, A class representative model for pure parsimony haplotyping. INFORMS J. Comput. **22**, 195–209 (2010)
9. A. Clark, Inference of haplotypes from PCR amplified samples of diploid populations. Mol. Biol. Evol. **7**, 111–122 (1990)
10. L. Di Gaspero, A. Roli, Stochastic local search for large-scale instances of the haplotype inference problem by pure parsimony. J. Algebra **63**, 55–69 (2008)
11. Z. Ding, V. Filkov, D. Gusfield, A linear-time algorithm for the perfect phylogeny haplotyping problem. J. Comput. Biol. **13**, 522–553 (2006)
12. E. Eskin, E. Halperin, R. Karp, Efficient reconstruction of haplotype structure via perfect phylogeny. J. Bioinformatics Comput. Biol. **1**, 1–20 (2003)
13. A. Graca, J. Marques-Silva, I. Lynce, A.L. Oliviera, Efficient haplotype inference with pseudo-Boolean optimization, in *2nd International Conference on Algebraic Biology – AB*, Castle of Hagenberg. LNCS 4545 (Springer, Berlin/Heidelberg/New York, 2007), pp. 125–139
14. D. Gusfield, A Practical algorithm for optimal inference of haplotypes from diploid populations, in *Annual International Conference on Intelligent Systems for Molecular Biology (ISMB)*, La Jolla/San Diego (AAAI, Menlo Park, 2000), pp. 183–189
15. D. Gusfield, Inference of haplotypes from samples of diploid populations: complexity and algorithms. J. Comput. Biol. **8**, 305–324 (2001)
16. D. Gusfield, Haplotyping as perfect phylogeny: conceptual framework and efficient solutions, in *International Conference on Computational Molecular Biology – RECOMB*, Washington (ACM, New York, 2002), pp. 166–175
17. D. Gusfield, Haplotype inference by pure parsimony, in *Annual Symposium on Combinatorial Pattern Matching – CPM*, Morelia, Michocán. LNCS 2676 (Springer, Berlin/Heidelberg/New York, 2003), pp. 144–155
18. D. Gusfield, S.H. Orzack, Haplotype inference, in *Handbook of Computational Molecular Biology*, ed. by S. Aluru. (Champman and Hall/CRC, 2005), pp. 1–28
19. B. Halldorsson, V. Bafna, N. Edwards, R. Lippert, S. Yooseph, S. Istrail, A survey of computational methods for determining haplotypes, in *Computational Methods for SNP and Haplotype Inference: DIMACS/RECOMB Satellite Workshop*, Piscataway. LNCS 2983 (Springer, Berlin, 2004), pp. 26–47

20. Y.T. Huang, K.M. Chao, T. Chen, An approximation algorithm for haplotype inference by maximum parsimony, in *ACM Symposium on Applied Computing – SAC*, Santa Fe, pp. 146–150 (2005)
21. K. Kalpakis, P. Namjoshi, Haplotype phasing using semidefinite programming, in *5th IEEE Symposium on Bioinformatics and Bioengineering – BIBE*, Minneapolis, pp. 145–152 (2005)
22. C. Kanz et al., The EMBL nucleotide sequence database. Nucl. Acid Res. **33**, D29–D33 (2005)
23. G. Lancia, Applications to computational molecular biology, in *Handbook on Modeling for Discrete Optimization*, eds. by G. Appa, P. Williams, P. Leonidas, H. Paul. International Series in Operations Research and Management Science, vol. 88 (Springer, New York, 2006), pp. 270–304
24. G. Lancia, R. Rizzi, A polynomial case of the parsimony haplotyping problem. Oper. Res. Lett. **34**, 289–295 (2006)
25. G. Lancia, P. Serafini, A set covering approach with column generation for parsimony haplotyping. INFORMS J. Comput. **21**, 151–166 (2009)
26. G. Lancia, C. Pinotti, R. Rizzi, Haplotyping populations by pure parsimony: complexity, exact and approximation algorithms. INFORMS J. Comput. **16**, 17–29 (2004)
27. I. Lynce, J. Marques-Silva, SAT in bioinformatics: making the case with haplotype inference, in *Theory and Applications of Satisfiability Testing – SAT*, Seattle, LNCS 4121 (Springer, Berlin/Heidelberg, 2006), pp. 136–141
28. J. Marques-Silva, I. Lynce, A. Graca, A.L. Oliveira, Efficient and tight upper bounds for haplotype inference by pure parsimony using delayed haplotype selection, in *Progress in Artificial Intelligence*, ed. by J. Neves, M.F. Santos, J.M. Machado. LNCS 4874 (Springer, Berlin, 2007), pp. 621–632
29. G.L. Nemhauser, L.A. Wolsey, *Integer and Combinatorial Optimization* (Wiley, New York, 1988)
30. A. Schrijver, *Theory of Linear and Integer Programming* (Wiley, New York, 1986)
31. L. Tininini, P. Bertolazzi, A. Godi, G. Lancia, CollHaps: a heuristic approach to haplotype inference by parsimony. IEEE/ACM Trans. Comput. Biol. Bioinformatics **7**, 511–523 (2010)
32. L. Wang, Y. Xu, Haplotype inference by maximum parsimony. Bioinformatics **19**, 1773–1780 (2003)

Chapter 7
Kodaira-Spencer Formality of Products of Complex Manifolds

Marco Manetti

Abstract We shall say that a complex manifold X is *Kodaira-Spencer formal* if its Kodaira-Spencer differential graded Lie algebra $A_X^{0,*}(\Theta_X)$ is formal; if this happen, then the deformation theory of X is completely determined by the graded Lie algebra $H^*(X, \Theta_X)$ and the base space of the semiuniversal deformation is a quadratic singularity. Determine when a complex manifold is Kodaira-Spencer formal is generally difficult and we actually know only a limited class of cases where this happen. Among such examples we have Riemann surfaces, projective spaces, holomorphic Poisson manifolds with surjective anchor map $H^*(X, \Omega_X^1) \to H^*(X, \Theta_X)$ [4] and every compact Kähler manifold with trivial or torsion canonical bundle, see [9] and references therein. In this short note we investigate the behavior of this property under finite products. Let X, Y be compact complex manifolds; we prove that whenever X and Y are Kähler, then $X \times Y$ is Kodaira-Spencer formal if and only if the same holds for X and Y (Corollary 7.2). A revisit of a classical example by Douady shows that the above result fails if the Kähler assumption is dropped.

7.1 Review of Differential Graded (Lie) Algebras and Formality

In this section every vector space and tensor product is intended over a fixed field $\mathbb{K}$ of characteristic 0. In rational homotopy theory, an important role is played by the notion of formality of a differential graded algebra [2, p. 260]. A similar role in deformation theory is played by the notion of formality of a differential graded Lie algebra [5, p. 52].

M. Manetti (✉)
Dipartimento di Matematica "Guido Castelnuovo", Università degli Studi di Roma "La Sapienza", Piazzale Aldo Moro 5, 00185 Roma, Italy
e-mail: manetti@mat.uniroma1.it

V. Ancona and E. Strickland (eds.), *Trends in Contemporary Mathematics*, Springer INdAM Series 8, DOI 10.1007/978-3-319-05254-0_7,

Definition 7.1. A DG-algebra (short for differential graded commutative algebra) is the data of a $\mathbb{Z}$-graded vector space $A = \oplus_{n\in\mathbb{Z}} A^n$, equipped with a differential $d\colon A^n \to A^{n+1}$, $d^2 = 0$, and a product

$$A^n \times A^m \to A^{n+m}, \qquad (a,b) \mapsto ab,$$

which satisfy the following conditions:

1. (Associativity) $(ab)c = a(bc)$.
2. (Graded commutativity) $ab = (-1)^{\deg(a)\deg(b)} ba$.
3. (Graded Leibniz) $d(ab) = d(a)b + (-1)^{\deg(a)} a d(b)$.

In particular every DG-algebra is also a cochain complex and its cohomology inherits a structure of graded commutative algebra. A morphism of DG-algebras is simply a morphism of graded algebras commuting with differentials. A DG-algebra A is called unitary if there exists a unit $1 \in A^0$ such that $1a = a$ for every $a \in A$.

Typical examples of DG-algebras are the de Rham complex $A_X^{*,*}$ and the Dolbeault complex $A_X^{0,*}$ of a holomorphic manifold X, equipped with the usual wedge product of differential forms.

Definition 7.2. A morphism $f\colon A \to B$ of DG-algebras is called a quasi-isomorphism if it is a quasi-isomorphism of the underlying cochain complexes. Two DG-algebras are said to be quasi-isomorphic if they are equivalent under the equivalence relation generated by quasi-isomorphisms.

A DG-algebra A is called formal if it is quasi-isomorphic to its cohomology algebra $H^*(A)$.

Example 7.1 (The Iwasawa DG-algebra). Probably the simplest example of non formal DG-algebra is the Iwasawa algebra: consider the vector space V with basis e_1, e_2, e_3 and the unique differential on the exterior algebra $R = \oplus_i R^i$, $R^i := \bigwedge^i V$ such that

$$de_1 = de_2 = 0, \qquad de_3 = -e_1 \wedge e_2 \,.$$

According to Leibniz rule we have

$$d(e_1 \wedge e_2) = d(e_2 \wedge e_3) = d(e_1 \wedge e_3) = d(e_1 \wedge e_2 \wedge e_3) = 0$$

and there exists an obvious injective morphism $j\colon H^*(R) \hookrightarrow R$ of cochain complexes whose image is the graded vector subspace spanned by the six linearly independent vectors $1, e_1, e_2, e_1\wedge e_3, e_2\wedge e_3, e_1\wedge e_2\wedge e_3$; however j is not a morphism of algebras.

Whenever $\mathbb{K} = \mathbb{R}$ the algebra R can be identified with the algebra of right-invariant differential forms on the Lie group of real matrices of type

$$\begin{pmatrix} 1 & x_1 & x_3 \\ 0 & 1 & x_2 \\ 0 & 0 & 1 \end{pmatrix},$$

by setting $e_1 = dx_1$, $e_2 = dx_2$ and $e_3 = dx_3 - x_1 dx_2$. The non formality of R may be easily checked, as in [7], by computing the triple Massey products; here we obtain again this result as a consequence of Proposition 7.2.

Definition 7.3. A DG-Lie algebra (short for differential graded Lie algebra) is the data of a $\mathbb{Z}$-graded vector space $L = \oplus_{n\in\mathbb{Z}} L^n$, equipped with a differential $d\colon L^n \to L^{n+1}$, $d^2 = 0$, and a bracket

$$L^n \times L^m \to L^{n+m}, \qquad (a,b) \mapsto [a,b],$$

which satisfy the following conditions:

1. (Graded anti commutativity) $[a,b] = -(-1)^{\deg(a)\deg(b)}[b,a]$.
2. (Graded Leibniz) $d[a,b] = [da,b] + (-1)^{\deg(a)}[a,db]$.
3. (Graded Jacobi) $[[a,b],c] = [a,[b,c]] - (-1)^{\deg(a)\deg(b)}[b,[a,c]]$.

As above, every DG-Lie algebra is also a cochain complex and its cohomology inherits a structure of graded Lie algebra. A morphism of DG-Lie algebras is simply a morphism of graded Lie algebras commuting with differentials.

Example 7.2. The Kodaira-Spencer DG-Lie algebra KS_X of a complex manifold X is defined as the Dolbeault complex $A_X^{0,*}(\Theta_X)$ of the holomorphic tangent sheaf equipped with the natural extension of the usual bracket on smooth sections of Θ_X, see e.g. [6].

If L is a DG-Lie algebra and A is a DG-algebra, then the tensor product $L \otimes A$ has a natural structure of DG-Lie algebra, where:

$$d(x\otimes a) = dx\otimes a + (-1)^{\deg(x)} x\otimes da, \quad [x\otimes a, y\otimes b] = (-1)^{\deg(a)\deg(y)}[x,y]\otimes ab\,.$$

Let's denote by **Art** the category of Artin local $\mathbb{K}$-algebras with residue field $\mathbb{K}$ and by **Set** the category of sets. Unless otherwise specified, for every $A \in$ **Art** we shall denote by $\mathfrak{m}_A$ its maximal ideal. Every DG-Lie algebra L gives a functor

$$\mathrm{MC}_L\colon \mathbf{Art} \to \mathbf{Set}, \qquad \mathrm{MC}_L(A) = \left\{ x \in L^1 \otimes \mathfrak{m}_A \;\middle|\; dx + \frac{1}{2}[x,x] = 0 \right\}.$$

The equation $dx + [x,x]/2 = 0$ is called the Maurer-Cartan equation and MC_L is called the Maurer-Cartan functor associated to L. Two elements $x, y \in \mathrm{MC}_L(A)$ are said to be gauge equivalent if there exists $a \in L^0 \otimes \mathfrak{m}_A$ such that

$$y = e^a * x := x + \sum_{n=0}^{\infty} \frac{[a,-]^n}{(n+1)!}([a,x] - da).$$

Then we define the functor $\mathrm{Def}_L\colon \mathbf{Art} \to \mathbf{Set}$ defined as (we refer to [5, 12, 13] for details):

$$\mathrm{Def}_L(A) = \frac{\mathrm{MC}_L(A)}{\text{gauge equivalence}}.$$

The projection $\mathrm{MC}_L \to \mathrm{Def}_L$ is a formally smooth natural transformation: this means that, given a surjective morphism $A \xrightarrow{\alpha} B$ in the category **Art**, an element $x \in \mathrm{MC}_L(B)$ can be lifted to $\mathrm{MC}_L(A)$ if and only if its equivalence class $[x] \in \mathrm{Def}_L(B)$ can be lifted to $\mathrm{Def}_L(A)$.

In this paper we shall need several times the following results (for a proof see e.g. Theorem 5.71 of [13]). A morphism of DG-Lie algebras $f\colon L \to M$ is called a quasi-isomorphism if the induced map in cohomology $f\colon H^*(L) \to H^*(M)$ is an isomorphism of graded Lie algebras.

Theorem 7.1 (Schlessinger-Stasheff [18]). *Let $L \to M$ be a morphism of differential graded Lie algebras. Assume that:*

1. *$H^0(L) \to H^0(M)$ is surjective.*
2. *$H^1(L) \to H^1(M)$ is bijective.*
3. *$H^2(L) \to H^2(M)$ is injective.*

Then the induced natural transformation $\mathrm{Def}_L \to \mathrm{Def}_M$ is an isomorphism of functors.

Corollary 7.1. *Let $L \to M$ be a quasi-isomorphism of differential graded Lie algebras. Then the induced natural transformation $\mathrm{Def}_L \to \mathrm{Def}_M$ is an isomorphism of functors.*

The notion of formality extends immediately to differential graded Lie algebras. A DG-Lie algebra L is called formal if it is connected to the graded Lie algebra $H^*(L)$ by a finite chain of quasi-isomorphisms of DG-Lie algebras.

As a first application of Theorem 7.1 we have therefore that for a formal DG-Lie algebra L the functor Def_L is determined by the graded Lie algebra structure on $H^*(L)$.

Proposition 7.1. *If a differential graded Lie algebra L is formal, then the two maps*

$$\mathrm{Def}_L(\mathbb{K}[t]/(t^3)) \to \mathrm{Def}_L(\mathbb{K}[t]/(t^2)), \qquad \mathrm{Def}_L(\mathbb{K}[t]/(t^n)) \to \mathrm{Def}_L(\mathbb{K}[t]/(t^2))$$

have the same image for every $n \geq 3$.

Proof. We may assume that L is a graded Lie algebra and therefore its Maurer-Cartan equation becomes $[x,x] = 0$, $x \in L^1$. Therefore $tx_1 \in$

$\mathrm{Def}_L(\mathbb{K}[t]/(t^2))$ lifts to $\mathrm{Def}_L(\mathbb{K}[t]/(t^3))$ if and only if there exists $x_2 \in L^1$ such that

$$t^2[x_1,x_1] \equiv [tx_1+t^2x_2, tx_1+t^2x_2] \equiv 0 \pmod{t^3} \iff [x_1,x_1]=0$$

and $[x_1,x_1]=0$ implies that $tx_1 \in \mathrm{Def}_H(\mathbb{K}[t]/(t^n))$ for every $n \geq 3$. □

An example of non formal DG-Lie algebra is provided by the next proposition.

Proposition 7.2. *Let $\mathfrak{n}_3(\mathbb{K})$ be the Lie algebra of strictly upper triangular 3×3 matrices and let R the Iwasawa DG-algebra defined above. Then:*

1. *The differential graded Lie algebra $\mathfrak{n}_3(\mathbb{K}) \otimes R$ is formal and the functor $\mathrm{Def}_{\mathfrak{n}_3(\mathbb{K})\otimes R}$ is smooth.*
2. *The differential graded Lie algebra $\mathfrak{sl}_2(\mathbb{K}) \otimes R$ is not formal and the functor $\mathrm{Def}_{\mathfrak{sl}_2(\mathbb{K})\otimes R}$ is not smooth.*

Proof. Let's denote by $C \subset R$ the DG-vector subspace spanned by $e_3, e_1 \wedge e_2$ and by $I \subset \mathfrak{n}_3(\mathbb{K})$ the Lie ideal of matrices of type

$$\begin{pmatrix} 0 & 0 & t \\ 0 & 0 & 0 \\ 0 & 0 & 0 \end{pmatrix}, \qquad t \in \mathbb{K}\,.$$

Since $I = [\mathfrak{n}_3(\mathbb{K}), \mathfrak{n}_3(\mathbb{K})]$ and $[I, \mathfrak{n}_3(\mathbb{K})] = 0$, the subcomplex $I \otimes C$ is an acyclic Lie ideal of $\mathfrak{n}_3(\mathbb{K}) \otimes R$. The formality of $\mathfrak{n}_3(\mathbb{K}) \otimes R$ is now an immediate consequence of the easy facts that, the projection

$$\pi\colon \mathfrak{n}_3(\mathbb{K}) \otimes R \to \frac{\mathfrak{n}_3(\mathbb{K}) \otimes R}{I \otimes C}$$

is a quasi-isomorphism and

$$\pi \circ (\mathrm{Id} \otimes j)\colon \mathfrak{n}_3(\mathbb{K}) \otimes H^*(R) \to \frac{\mathfrak{n}_3(\mathbb{K}) \otimes R}{I \otimes C}$$

is a morphism of differential graded Lie algebras. The smoothness of $\mathrm{Def}_{\mathfrak{n}_3(\mathbb{K})\otimes R}$ follows from the fact that the Maurer-Cartan equation in $H^*(\mathfrak{n}_3(\mathbb{K}) \otimes R) = \mathfrak{n}_3(\mathbb{K}) \otimes H^*(R)$ is trivial.

Next, we shall use Proposition 7.1 in order to prove that $M = \mathfrak{sl}_2(\mathbb{K}) \otimes R$ is not formal. More precisely we shall prove that there exists an element in $\mathrm{MC}_M(\mathbb{K}[t]/(t^2))$ which lifts to $\mathrm{MC}_M(\mathbb{K}[t]/(t^3))$ but does not lift to $\mathrm{MC}_M(\mathbb{K}[t]/(t^4))$. Denote by u, v, h the standard basis of $\mathfrak{sl}_2(\mathbb{K})$:

$$[u,v]=h, \qquad [h,u]=2u, \qquad [h,v]=-2v,$$

and consider the element $\xi = ue_1t + ve_2t - he_3t^2 \in \mathrm{MC}_M(\mathbb{K}[t]/(t^3)) \subset M^1 \otimes \mathbb{K}[t]/(t^3)$. A generic element of $M^1 \otimes \mathbb{K}[t]/(t^4)$ lifting $ue_1t + ve_2t \in \mathrm{MC}_M(\mathbb{K}[t]/(t^2))$ may be written as

$$\eta = ue_1t+ve_2t+(ae_1+be_2+ce_3)t^2+(\alpha e_1+\beta e_2+\gamma e_3)t^3,\ a,b,c,\alpha,\beta,\gamma \in \mathfrak{sl}_2(\mathbb{K}).$$

Assume that η satisfies the Maurer-Cartan equation. Since

$$d\eta = ce_1 \wedge e_2t^2 + \gamma e_1 \wedge e_2t^3, \qquad \frac{1}{2}[\eta,\eta] = he_1 \wedge e_2t^2 + (\cdots)t^3$$

we must have $c = -h$; therefore the coefficient of $e_1 \wedge e_3t^3$ in $\frac{1}{2}[\eta,\eta]$ is equal to $[u,c] = [u,-h] = [h,u] = 2u \neq 0$ and this gives a contradiction. □

Lemma 7.1. *Let L, M be DG-Lie algebras and B a DG-algebra:*

1. *If L and B are formal, then $L \otimes B$ is a formal DG-Lie algebra.*
2. *If B is unitary, $H^*(B) \neq 0$ and $L \otimes B$ is a formal, then also L is formal.*
3. *The DG-Lie algebra $L \times M$ is formal if and only if L and M are formal.*

Proof. The first item is clear, while the second and the third are exactly Corollaries 3.5 and 3.6 of [14]. □

7.2 Deformations of Products of Compact Complex Manifolds

From now on we work over the field $\mathbb{C}$ of complex numbers; every complex manifold is assumed compact and connected.

By a general and extremely fruitful principle, introduced by Schlessinger-Stasheff [18], Deligne [1], Drinfeld and developed by many others, over a field of characteristic 0, every "reasonable" deformation problem is controlled by a differential graded Lie algebra, with quasi-isomorphic DG-Lie algebras giving the same deformation theory.

For instance, deformations of a compact complex manifold X are controlled by the quasi-isomorphism class of the Kodaira-Spencer differential graded Lie algebra $KS_X = A_X^{0,*}(\Theta_X)$ of differential forms valued in the holomorphic tangent sheaf [6, 17]. This means that the functor $\mathrm{Def}_X\colon \mathbf{Art} \to \mathbf{Set}$ of infinitesimal deformations of X is isomorphic to the functors Def_{KS_X}.

Here we must pay attention to the fact that the corresponding cohomology graded Lie algebra $H^*(A_X^{0,*}(\Theta_X)) = H^*(X,\Theta_X)$ is not a complete invariant under quasi-isomorphisms and, in general, its knowledge is not sufficient to determine the deformation theory of X, although $H^1(X,\Theta_X)$ is the space of first order deformations, $H^2(X,\Theta_X)$ is an obstruction space and the quadratic bracket

$$q\colon H^1(X,\Theta_X)\to H^2(X,\Theta_X),\qquad q(\xi)=\frac{1}{2}[\xi,\xi],$$

is the obstruction to lifting a first order deformation of X up to second order. In particular the vanishing of the bracket on $H^*(X,\Theta_X)$ does not imply that X is unobstructed.

Whenever the Kodaira-Spencer algebra KS_X is formal, the deformations of X are determined by the graded Lie algebra $H^*(X,\Theta_X)$ and the base space of the Kuranishi family is analytically isomorphic to the germ at 0 of the nullcone of the quadratic map q.

As noticed above, in general the Kodaira-Spencer algebra is not formal, even for projective manifolds. For example, Vakil proved [19, Thm. 1.1] that for every analytic singularity $(U,0)$ defined over $\mathbb{Z}$ there exists a complex surface S with very ample canonical bundle such that its local moduli space is analytically isomorphic to the germ at 0 of $U\times\mathbb{C}^n$ for some integer $n\geq 0$. Choosing $U=\{(x,y)\in\mathbb{C}^2\mid xy(x-y)=0\}$ and taking S as above, the Kodaira-Spencer algebra of S cannot be formal. As a warning against possible mistakes, we note that such a surface S is obstructed although the bracket on $H^*(S,\Theta_S)$ is trivial.

Consider now two compact connected complex manifolds X,Y; given two deformations $X_A\to\operatorname{Spec}(A)$, $Y_A\to\operatorname{Spec}(A)$, of X,Y over the same basis, their fibred product

$$X_A\times_{\operatorname{Spec}(A)}Y_A\to\operatorname{Spec}(A)$$

is a deformation of the product $X\times Y$. Therefore it is well defined a natural transformation of functors

$$\alpha\colon \operatorname{Def}_X\times\operatorname{Def}_Y\to\operatorname{Def}_{X\times Y}.$$

It is easy to describe α in terms of morphisms of differential graded Lie algebras: denote by $p\colon X\times Y\to X$ and $q\colon X\times Y\to Y$ the projections; since

$$p_*p*\Theta_X=\Theta_X\otimes p_*\mathcal{O}_{X\times Y}=\Theta_X,\qquad q_*q*\Theta_Y=\Theta_Y\otimes q_*\mathcal{O}_{X\times Y}=\Theta_Y$$

and $\Theta_{X\times Y}=p^*\Theta_X\oplus q^*\Theta_Y$, we may define two natural injective morphisms of differential graded Lie algebras

$$p^*\colon KS_X\to KS_{X\times Y},\qquad q^*\colon KS_Y\to KS_{X\times Y}.$$

Since $[p^*\eta,q^*\mu]=0$ for every $\eta\in KS_X$, $\mu\in KS_Y$, we get a morphism of differential graded Lie algebras

$$p^*\times q^*\colon KS_X\times KS_Y\to KS_{X\times Y}\tag{7.1}$$

inducing α at the level of associated deformation functors.

Lemma 7.2. *Assume X, Y compact and connected. Then the morphism α is an isomorphism if and only if*

$$H^0(X,\Theta_X)\otimes H^1(Y,\mathcal{O}_Y)=H^1(X,\mathcal{O}_X)\otimes H^0(Y,\Theta_Y)=0\,.$$

Proof. By Künneth formula ([8, Thm. 6.7.8], [10, Thm. 14]) we have:

$$\begin{aligned}
H^i(X\times Y,\Theta_{X\times Y})&=H^i(X\times Y,p^*\Theta_X)\oplus H^i(X\times Y,q^*\Theta_Y),\\
H^i(X\times Y,p^*\Theta_X)&=\bigoplus_j H^j(X,\Theta_X)\otimes H^{i-j}(Y,\mathcal{O}_Y),\\
H^i(X\times Y,q^*\Theta_Y)&=\bigoplus_j H^j(X,\mathcal{O}_X)\otimes H^{i-j}(Y,\Theta_Y).
\end{aligned}\tag{7.2}$$

The morphism $p^*\colon KS_X\to KS_{X\times Y}$ is injective in cohomology and the image of $H^i(X,\Theta_X)$ is the subspace $H^i(X,\Theta_X)\otimes H^0(Y,\mathcal{O}_Y)\subset H^i(X\times Y,p^*\Theta_X)$; similarly for the morphism q^*. Thus, $H^0(KS_{X\times Y})=H^0(KS_Y)\oplus H^0(KS_Y)$,

$$\begin{aligned}
H^1(KS_{X\times Y})=H^1(KS_X)\oplus H^1(KS_Y)&\oplus(H^0(X,\Theta_X)\otimes H^1(Y,\mathcal{O}_Y))\\
&\oplus(H^1(X,\mathcal{O}_X)\otimes H^0(Y,\Theta_Y))
\end{aligned}$$

and we have an injective map $H^2(KS_X)\oplus H^2(KS_Y)\to H^2(KS_{X\times Y})$.

If α is an isomorphism then, looking at first order deformations, we have

$$H^0(X,\Theta_X)\otimes H^1(Y,\mathcal{O}_Y)=H^1(X,\mathcal{O}_X)\otimes H^0(Y,\Theta_Y)=0\,.$$

Conversely, it is sufficient to apply Theorem 7.1 to the DG-Lie morphism $p^*\times q^*$. □

The assumption of Lemma 7.2 is satisfied in most cases; for instance, a theorem of Matsumura [15] implies that $H^0(X,\Theta_X)=0$ for every compact manifold of general type X. If $H^1(X,\mathcal{O}_X)\otimes H^0(Y,\Theta_Y)\neq 0$, then it is easy to describe deformations of $X\times Y$ that are not a product. Assume that X is a Kähler manifold, then $b_1(X)\neq 0$ and there exists at least one surjective homomorphism $\pi_1(X)\xrightarrow{g}\mathbb{Z}$. Since $H^0(Y,\Theta_Y)\neq 0$, there exists at least a nontrivial one parameter subgroup $\{\theta_t\}\subset \mathrm{Aut}(Y)$, $t\in\mathbb{C}$, of holomorphic automorphisms of Y. Therefore we get a family of representations

$$\rho_t\colon\pi_1(X)\to\mathrm{Aut}(Y),\qquad \rho_t(\gamma)=\theta_t^{g(\gamma)},\qquad t\in\mathbb{C}$$

inducing a family of locally trivial analytic Y-bundles over X. Moreover, Kodaira and Spencer proved that projective spaces $\mathbb{P}^n$ and complex tori $(\mathbb{C}^q/\Gamma)$ have unobstructed deformations, while the product $(\mathbb{C}^q/\Gamma)\times\mathbb{P}^n$ has obstructed deformations for every $q\geq 2$ and every $n\geq 1$ [11, page 436]. This was the first example of obstructed manifold.

Let's denote by $B_X^* = \{\phi \in A_X^{0,*} \mid \partial\phi = 0\}$ the DG-algebra of antiholomorphic differential forms on a complex manifold X. In the above setup we can define two morphisms

$$h_1\colon KS_X \otimes B_Y^* \to KS_{X\times Y}, \qquad h_1(\phi\otimes\eta) = p^*(\phi)\wedge q^*(\eta),$$

$$h_2\colon B_X^* \otimes KS_Y \to KS_{X\times Y}, \qquad h_2(\phi\otimes\eta) = p^*(\phi)\wedge q^*(\eta).$$

It is straightforward to check that h_1, h_2 are morphisms of differential graded Lie algebras and that the image of h_1 commutes with the image of h_2. This implies that the morphism (7.1) extends naturally to a morphism of differential graded Lie agebras

$$h\colon (KS_X \otimes B_Y^*) \times (B_X^* \otimes KS_Y) \to KS_{X\times Y} \tag{7.3}$$

Theorem 7.2. *For every pair of compact connected Kähler manifolds X, Y the morphism* (7.3) *is an injective quasi-isomorphism of differential graded Lie algebras. In particular, considering $H^*(X, \mathcal{O}_X)$ and $H^*(Y, \mathcal{O}_Y)$ as graded commutative algebras (with the usual cup product), there exists an isomorphism of functors*

$$\mathrm{Def}_{X\times Y} \cong \mathrm{Def}_{KS_X\otimes H^*(Y,\mathcal{O}_Y)} \times \mathrm{Def}_{KS_Y\otimes H^*(X,\mathcal{O}_X)}\,.$$

Proof. If X is compact Kähler, the $\partial\overline{\partial}$-lemma implies that $B_X^i \subset A_X^{0,i}$ is a set of representative for the Dolbeault cohomology group $H^i(X, \mathcal{O}_X)$ and therefore B_X^* is isomorphic to $H^*(X, \mathcal{O}_X)$ as a DG-algebra. Now, the formulas (7.2) imply immediately that the morphism (7.3) is a quasi-isomorphism. □

Corollary 7.2. *Let X, Y be compact Kähler manifolds. Then $KS_{X\times Y}$ is a formal DG-Lie algebra if and only if KS_X and KS_Y are formal.*

Proof. Immediate consequence of Lemma 7.1 and Theorem 7.2. □

7.3 A DG-Lie Revisitation of an Example by Douady

We want to prove, by a deeper study of a classical example by Douady [3, p. 18] that Corollary 7.2 fails without the Kähler assumption. The non Kähler manifold involved in this example is the Iwasawa manifold X, defined as the quotient of the group of complex matrices of type

$$\begin{pmatrix} 1 & z_1 & z_3 \\ 0 & 1 & z_2 \\ 0 & 0 & 1 \end{pmatrix}$$

by the right action of the cocompact subgroup of matrices with coefficients in the Gauss integers. By a (non trivial) result by Nakamura [16, p. 96] (cf. also [6, Lemma 6.5]), the morphism of DG-algebras

$$j\colon R\to A_X^{0,*},\qquad j(e_1)=d\overline{z_1},\quad j(e_2)=d\overline{z_2},\quad j(e_3)=d\overline{z_3}-\overline{z_1}d\overline{z_2},$$

is a quasi-isomorphism. Being X parallelizable the morphism of DG-Lie algebras $H^0(X,\Theta_X)\otimes R\to A_X^{0,*}(\Theta_X)$ is a quasi-isomorphism; in view of the isomorphism of Lie algebras $\mathfrak{n}_3(\mathbb{C})\simeq H^0(X,\Theta_X)$:

$$\begin{pmatrix}0&a&c\\0&0&b\\0&0&0\end{pmatrix}\mapsto a\frac{\partial}{\partial z_1}+b\left(\frac{\partial}{\partial z_2}+z_1\frac{\partial}{\partial z_3}\right)+c\frac{\partial}{\partial z_3}\,.$$

we get that the Kodaira-Spencer algebra of the Iwasawa manifold X is quasi-isomorphic to the formal DG-Lie algebra $\mathfrak{n}_3(\mathbb{C})\otimes R$.

Consider now $Y=\mathbb{P}^1$, then $H^*(Y,\Theta_Y)=H^0(Y,\Theta_Y)\simeq\mathfrak{sl}_2(\mathbb{C})$ and therefore the Kodaira-Spencer algebra KS_Y is quasi-isomorphic to the Lie algebra $\mathfrak{sl}_2(\mathbb{C})$.

Since every differential form in the image of j is antiholomorphic, as above we can define a morphism of DG-Lie algebras

$$(KS_X\otimes B_Y^*)\times(R\otimes KS_Y)\to KS_{X\times Y}\tag{7.4}$$

which, by Künneth formula is a quasi-isomorphism. Thus the Kodaira-Spencer algebra of $X\times Y$ is quasi-isomorphic to $(\mathfrak{n}_3(\mathbb{C})\otimes R)\times(\mathfrak{sl}_2(\mathbb{C})\otimes R)$.

Since $\mathfrak{sl}_2(\mathbb{C})\otimes R$ is not formal, by Lemma 7.1, also the Kodaira-Spencer algebra of $X\times Y$ is not formal. It is possible to prove, using the above results, that the base space of the Kuranishi family of $X\times Y$ is isomorphic to $(\mathbb{C}^6\times U,0)$, where $U\subset\mathbb{C}^6$ is a cone defined by six homogeneous polynomials of degree 3.

References

1. P. Deligne, Letter to J. Millson, April 24, 1986
2. P. Deligne, P. Griffiths, J. Morgan, D. Sullivan, Real homotopy theory of Kähler manifolds. Invent. Math. **29**, 245–274 (1975)
3. A. Douady, *Obstruction primaire à la déformation.* Sém. Cartan **13** (1960/1961), Exp. 4.
4. D. Fiorenza, M. Manetti, Formality of Koszul brackets and deformations of holomorphic Poisson manifolds. Homol. Homotopy Appl. **14**(2), 63–75 (2012)
5. W.M. Goldman, J.J. Millson, The deformation theory of representations of fundamental groups of compact Kähler manifolds. Inst. Hautes Études Sci. Publ. Math. **67**, 43–96 (1988)
6. W.M. Goldman, J.J. Millson, The homotopy invariance of the Kuranishi space. Ill. J. Math. **34**, 337–367 (1990)
7. P.H. Griffiths, J.W. Morgan, *Rational Homotopy Theory and Differential Forms.* Progress in Mathematics, vol. 16 (Birkhäuser, Boston, 1981)

8. A. Grothendieck, Éléments de géométrie algébrique. III. Étude cohomologique des faisceaux cohérents. II. Inst. Hautes Études Sci. Publ. Math. **17** (1963)
9. D. Iacono, M. Manetti, An algebraic proof of Bogomolov-Tian-Todorov theorem. Deform. Spaces **39**, 113–133 (2010). Vieweg Verlag, arXiv:0902.0732
10. G.R. Kempf, Some elementary proofs of basic theorems in the cohomology of quasi-coherent sheaves. Rocky Mt. J. Math. **10**, 637–646 (1980)
11. K. Kodaira, D.C. Spencer, On deformations of complex analytic structures, II. Ann. Math. (2) **67**, 403–466 (1958)
12. M. Manetti, Deformation theory via differential graded Lie algebras, in *Seminari di Geometria Algebrica 1998–1999* (Scuola Normale Superiore, Pisa, 1999). arXiv:math.AG/0507284
13. M. Manetti, Lectures on deformations on complex manifolds. Rend. Mat. Appl. (7) **24**, 1–183 (2004). arXiv:math.AG/0507286
14. M. Manetti, On some formality criteria for DG-Lie algebras. Submitted, arXiv:1310.3048
15. H. Matsumura, On algebraic groups of birational transformations. Rend. Accad. Lincei Ser. 8 **34**, 151–155 (1963)
16. I. Nakamura, Complex parallelisable manifolds and their small deformations. J. Differ. Geom. **10**, 85–112 (1975)
17. A. Nijenhuis, R.W. Richardson, Cohomology and deformations of algebraic structures. Bull. Am. Math. Soc. **70**(3), 406–411 (1964)
18. M. Schlessinger, J. Stasheff, Deformation theory and rational homotopy type (1979). Preprint, arXiv:1211.1647 [math.QA]
19. R. Vakil, Murphy's law in algebraic geometry: badly-behaved deformation spaces. Invent. Math. **164**, 569–590 (2006). arXiv:math.AG/0411469

Chapter 8
Monomial Transformations of the Projective Space

Olivier Debarre and Bodo Lass

Abstract We prove that, over any field, the dimension of the indeterminacy locus of a rational map $f : \mathbf{P}^n \dashrightarrow \mathbf{P}^n$ defined by monomials of the same degree d with no common factors is at least $(n-2)/2$, provided that the degree of f as a map is not divisible by d. This implies upper bounds on the multidegree of f and in particular, when f is birational, on the degree of f^{-1}.

2010 Mathematics Subject Classification (MSC2010): 14E07.

8.1 Introduction

We denote by $\mathbf{P}^n$ the n dimensional projective space over a fixed field. A monomial transformation of $\mathbf{P}^n$ is a rational map $f : \mathbf{P}^n \dashrightarrow \mathbf{P}^n$ whose components $f_0, \dots, f_n$ are monomials (of the same positive degree $d(f)$ and with no common factors) in the variables $x_0, \dots, x_n$.

Monomial transformations are of course very special among all rational transformations, but also much easier to study. For this reason, they have recently attracted some attention. In particular, there is a description of all *birational* monomial transformations f with $d(f) = 2$ in [2], §2, from which it follows that $d(f^{-1})$

O. Debarre (✉)
Département de Mathématiques et Applications – CNRS UMR 8553, École normale supérieure, 45 rue d'Ulm, 75230 Paris Cedex 05, France
e-mail: olivier.debarre@ens.fr

B. Lass
Institut Camille Jordan – CNRS UMR 5208, Université Claude Bernard – Lyon 1, 69622 Villeurbanne Cedex, France
e-mail: lass@math.univ-lyon1.fr

V. Ancona and E. Strickland (eds.), *Trends in Contemporary Mathematics*, Springer INdAM Series 8, DOI 10.1007/978-3-319-05254-0_8,

is then at most equal to n ([2], Theorem 2.6). Extensive computer calculations were then performed by Johnson in [6] and led him to suggest that the largest possible value for $d(f^{-1})$ should be $\frac{(d(f)-1)^n-1}{d(f)-2}$ when $d(f) \geq 3$.

These values should be compared with the (larger) optimal bound $d(g^{-1}) \leq d(g)^{n-1}$ for all birational transformations g of $\mathbf{P}^n$. This maximal value for $d(g^{-1})$ is attained if and only if the indeterminacy locus of g is finite (see Sect. 8.4), hence one is led to think that the indeterminacy locus of a monomial map should be rather large. This is what we prove in Theorem 8.1: *the dimension of the indeterminacy locus of a monomial map* $f : \mathbf{P}^n \dashrightarrow \mathbf{P}^n$ *is at least* $(n-2)/2$*, provided that the degree of* f *as a map is not divisible by* $d(f)$.

We show in Sect. 8.4 that this implies a bound on $d(f^{-1})$ for all birational monomial transformations f, which is however not as good as the one suggested by Johnson.

8.2 Monomial Transformations

We represent a monomial transformation $f : \mathbf{P}^n \dashrightarrow \mathbf{P}^n$, with components $f_0, \ldots, f_n$, by the $(n+1) \times (n+1)$ matrix $A = (a_{ij})_{0 \leq i,j \leq n}$ whose ith row lists the exponents of f_i. With this notation, one has $f_A \circ f_B = f_{AB}$.

The following proposition is elementary ([5]; [7], Lemma 1.2; [6]).

Proposition 8.1. *With the notation above, we have*

$$|\det(A)| = d(f)\deg(f).$$

In particular, f *is birational if and only if* $|\det(A)| = d(f)$.

Proof. The condition that all monomials $f_0, \ldots, f_n$ have the same degree $d := d(f)$ means that in each row of A, the sum of the entries is d. Adding all columns to the 0th column, then subtracting the 0th row from all other rows we obtain

$$\det(A) = \begin{vmatrix} d & a_{01} & \cdots & a_{0n} \\ \vdots & \vdots & & \vdots \\ d & a_{n1} & \cdots & a_{nn} \end{vmatrix} = d \begin{vmatrix} 1 & a_{01} & \cdots & a_{0n} \\ \vdots & \vdots & & \vdots \\ 1 & a_{n1} & \cdots & a_{nn} \end{vmatrix} = d \det(M),$$

where $M := (m_{ij})_{1 \leq i,j \leq n}$ is defined by $m_{ij} := a_{ij} - a_{0j}$. If $T \simeq (\mathbf{C}^*)^n \subset \mathbf{P}^n$ is the torus defined by $x_0 \cdots x_n \neq 0$, the map f induces a morphism $f_T : T \to T$ given by

$$f_T(x_1, \ldots, x_n) = (x_1^{m_{11}} \cdots x_n^{m_{1n}}, \ldots, x_1^{m_{n1}} \cdots x_n^{m_{nn}}).$$

The induced map $\hat{f_T} : \hat{T} \to \hat{T}$ between algebraic character groups (where $\hat{T}$ is the free abelian group $\mathbf{Z}^n$) is given by the transposed matrix $M^T : \mathbf{Z}^n \to \mathbf{Z}^n$. Performing elementary operations on M amounts to composing f_T with monomial automorphisms, so we can reduce to the case where M is diagonal, in which case it is obvious that the degree of the morphism f_T (which is the same as the degree of f) is $|\det(M)|$. □

Corollary 8.1. *With the notation above, f is birational if and only if $|\det(A)| = d(f)$. Its inverse is then also a monomial transformation.*

Proof. It is clear from the proof above that f is birational if and only if $|\det(M)| = 1$, i.e., if and only if $M \in \mathrm{GL}_n(\mathbf{Z})$ or, equivalently, if and only if f_T is an isomorphism, whose inverse is then given by the matrix M^{-1}. It is therefore a monomial transformation. □

8.3 Indeterminacy Locus

In this section, $f : \mathbf{P}^n \dashrightarrow \mathbf{P}^n$ is again a monomial transformation. *We assume that its components $f_0, \dots, f_n$ have no common factors.* In terms of the matrix A defined in Sect. 8.2, this means that each column of A has at least one 0 entry.

The indeterminacy locus B of f is then the subscheme of $\mathbf{P}^n$ defined by the equations $f_0 = \cdots = f_n = 0$. Its blow-up is the graph $\Gamma_f \to \mathbf{P}^n$ of f ([3], §1.4).

For each nonempty subset $J \subsetneq \{0, \dots, n\}$ such that the $(n+1) \times |J|$ matrix A_J constructed from the columns of A corresponding to the elements of J has no zero rows, we obtain a linear space contained in B by setting $x_j = 0$ for all $j \in J$. Its codimension in $\mathbf{P}^n$ is $|J|$. Moreover, B_{red} is the union of all such linear spaces.

Theorem 8.1. *Let $f : \mathbf{P}^n \dashrightarrow \mathbf{P}^n$ be a dominant transformation defined by monomials of degree d with no common factors. If the degree of f is not divisible by d, the dimension of the indeterminacy locus of f is at least* $(n-2)/2$.

The condition on the degree is necessary, as shown by the morphism $(x_0, \dots, x_n) \mapsto (x_0^d, \dots, x_n^d)$ of degree d^n (whose indeterminacy locus is empty).

Proof. Since the determinant of the matrix A is nonzero we may assume, upon permuting its rows and columns, that we have $a_{xx} \neq 0$ for all $x \in \{0, \dots, n\}$.

We then define an oriented graph on the set of vertices $\{0, \dots, n\}$ by adding an oriented edge from x to y whenever $a_{xy} \neq 0$. We then say that a vertex x is equivalent to a vertex y if and only if there exists an oriented path from x to y and an oriented path from y to x. This defines a partition of the set $\{0, \dots, n\}$ into equivalence classes (note that x is equivalent to x since $a_{xx} \neq 0$).

Say that an equivalence class X is greater than or equal to an equivalence class Y if there is an oriented path from an element of X to an element of Y (there exists then an oriented path from any element of X to any element of Y). This defines a partial order on the set of equivalence classes.

Choose a class X minimal for this order. Entries of A in a row corresponding to an element of X which are not in a column corresponding to an element of X are then 0 (otherwise, at least one oriented edge should come out of X to an element not in X, contradicting the minimality of X). It follows that the determinant of the submatrix A_X of A corresponding to rows and columns of X divides the determinant of A. The sum of all entries in a row of A_X is d hence, by the same reasoning used in the proof of Proposition 8.1, the determinant of A_X is nonzero, divisible by d.

Because of the condition $d \nmid \deg(f)$ and Proposition 8.1, the determinant of A is not divisible by d^2. In particular, our partial order has a unique minimal element X. Without loss of generality, we may assume $0 \in X$. Every other vertex then has an oriented path to 0. In particular, we may define an acyclic function

$$\varphi : \{1, \dots, n\} \to \{0, 1, \dots, n\}$$

such that $(x, \phi(x))$ is an oriented edge of our graph for all $x \in \{1, \dots, n\}$ ("acyclic" means that for all $x \in \{1, \dots, n\}$, there exists $k > 0$ such that $\phi^k(x) = 0$).

We keep only the n edges of the type $(x, \phi(x))$; since $a_{x\phi(x)} \neq 0$, they correspond to n nonzero entries, off the diagonal, in each row $1, \dots, n$. Since our new graph on $\{0, \dots, n\}$ has n edges and no cycles, we may color its vertices in black or white in such a way that x and $\phi(x)$ have different colors, for all $x \in \{0, \dots, n\}$.

We select the vertices of the color which has been used less often (if both colors have been used the same number of times, we select the vertices with the same color as 0). If 0 is not selected, we add it to the selection. We end up with at most $(n+2)/2$ selected vertices which are all on one of our n edges or the loop at 0.

Consider the submatrix of A formed by the $\leq (n+2)/2$ columns corresponding to the selected vertices. In each row, there is a nonzero entry: in the row 0, because 0 was selected and $a_{00} \neq 0$; in any other row x, because either x was selected and $a_{xx} \neq 0$, or $\phi(x)$ was selected and $a_{x\phi(x)} \neq 0$. This proves the theorem. □

Example 8.1. The bound in the theorem is sharp: for $d \geq 2$, one easily checks that the indeterminacy locus of the birational automorphism ([6], Example 2)

$$f_{n,d} : (x_0, \dots, x_n) \mapsto (x_0^d, x_0^{d-1} x_1, x_1^{d-1} x_2, \dots, x_{n-1}^{d-1} x_n) \tag{8.1}$$

of $\mathbf{P}^n$ has dimension exactly $\lceil (n-2)/2 \rceil$. But there are many other examples of birational monomial automorphisms of $\mathbf{P}^n$ with indeterminacy locus of dimension exactly $\lceil (n-2)/2 \rceil$, such as monomial maps defined by matrices

$$A = \begin{pmatrix}
d & 0 & \cdots & \cdots & \cdots & \cdots & \cdots & \cdots & 0 \\
d-1 & 1 & 0 & \cdots & \cdots & \cdots & \cdots & \cdots & 0 \\
0 & d-1 & 1 & 0 & \cdots & \cdots & \cdots & \cdots & 0 \\
a_{30} & a_{31} & a_{32} & 1 & 0 & \cdots & \cdots & \cdots & 0 \\
0 & 0 & 0 & d-1 & 1 & 0 & \cdots & \cdots & 0 \\
a_{50} & a_{51} & a_{52} & a_{53} & a_{54} & 1 & 0 & \cdots & 0 \\
\vdots & & & & & \ddots & \ddots & \ddots & \vdots \\
\vdots & & & & & & \ddots & \ddots & 0 \\
 & & & & & & & & 1
\end{pmatrix}$$

where, for each odd i, we choose $a_{i0} \neq 0$ and $\sum_{j=0}^{i-1} a_{ij} = d - 1$. The (reduced) indeterminacy locus is then defined by the equations

$$x_0 = x_1x_2 = x_3x_4 = \cdots = 0.$$

It has dimension $n - 1 - \lfloor n/2 \rfloor = \lceil (n-2)/2 \rceil$.

Another set of examples is provided by matrices of the form

$$\begin{pmatrix} 1 & 1 & 1 & d-3 & 0 & \cdots & \cdots & \cdots & \cdots & 0 \\ 0 & 1 & 1 & d-2 & 0 & \cdots & \cdots & \cdots & \cdots & 0 \\ d-1 & 0 & 1 & 0 & 0 & \cdots & \cdots & \cdots & \cdots & 0 \\ a_{30} & a_{31} & 0 & a_{33} & 0 & \cdots & \cdots & \cdots & \cdots & 0 \\ a_{40} & a_{41} & a_{42} & a_{43} & 1 & 0 & \cdots & \cdots & \cdots & 0 \\ 0 & 0 & 0 & 0 & d-1 & 1 & 0 & \cdots & \cdots & 0 \\ a_{60} & a_{61} & a_{62} & a_{63} & a_{64} & a_{65} & 1 & 0 & \cdots & 0 \\ \vdots & & & & & & \ddots & \ddots & \ddots & \vdots \\ \vdots & & & & & & & \ddots & \ddots & 0 \\ & & & & & & & & & 1 \end{pmatrix}$$

where $a_{30} + a_{31} + a_{33} = d \geq 3$, $a_{31}a_{33} \neq 0$, and, for each even i, we choose $a_{i0}a_{i2} \neq 0$ and $\sum_{j=0}^{i-1} a_{ij} = d - 1$.

8.4 Degrees of a Monomial Map

Let $g : \mathbf{P}^n \dashrightarrow \mathbf{P}^n$ be a rational map. One defines the ith degree $d_i(g)$ as the degree of the image by g of a general $\mathbf{P}^i \subset \mathbf{P}^n$ (more precisely, $d_i(g) := \mathbf{P}^{n-i} \cdot f_*\mathbf{P}^i$). One has $d_0(g) = 1$, $d_n(g) = \deg(g)$, and $d_1(g)$ is the integer $d(g)$ defined earlier (i.e., the common degree of the components $g_0, \ldots, g_n$ of g, *provided they have no common factors*). An alternative definition of the $d_i(g)$ is as follows: if $\Gamma_g \subset \mathbf{P}^n \times \mathbf{P}^n$ is the graph of g,

$$d_i(g) = \Gamma_g \cdot p_1^*\mathbf{P}^i \cdot p_2^*\mathbf{P}^{n-i}. \tag{8.2}$$

The sequence $d_0(g), \ldots, d_n(g)$ is known to be a log-concave sequence: it satisfies

$$\forall i \in \{1, \ldots, n-1\} \qquad d_i(g)^2 \geq d_{i+1}(g)d_{i-1}(g)$$

(this is a direct consequence of the Hodge Index Theorem; [3], (1.6)). This implies $d_i(g) \leq d_1(g)^i$.

Proposition 8.2. *Let $f : \mathbf{P}^n \dashrightarrow \mathbf{P}^n$ be a dominant map defined by monomials of degree d with no common factors. Set $c := \lfloor n/2 \rfloor + 1$. If the degree of f is not divisible by d, we have, for all $i \in \{c, \dots, n\}$,*

$$d_i(f) \le (1 - d^{-c})^{\frac{i-1}{c-1}} d^i .$$

Proof. The degrees of f can be expressed in terms of the Segre class of its indeterminacy locus B. In particular, if B is nonempty and $c' := \operatorname{codim}(B)$, one has ([3], Proposition 2.3.1)

$$d_i(f) = \begin{cases} d^i & \text{for } i < c', \\ d^i - \deg_s(B) & \text{for } i = c', \end{cases} \tag{8.3}$$

where $\deg_s(B)$ is the sum of the degrees of the top-dimensional components of B, counted with their *Samuel multiplicity* (this is larger than the "usual" multiplicity ([4], Examples 4.3.4 and 4.3.5.(c)); in particular $\deg_s(B)$ is a positive integer). Since in our case $c' \le c = n - \lceil (n-2)/2 \rceil$ (Theorem 8.1), it follows from log-concavity that we have

$$d_c(f) < d^c .$$

By log-concavity again, this implies that for $i \ge c$, one has

$$d_i(f) \le d_c(f)^{\frac{i-1}{c-1}} d^{1-\frac{i-1}{c-1}} \le (d^c - 1)^{\frac{i-1}{c-1}} d^{\frac{c-i}{c-1}} = (1 - d^{-c})^{\frac{i-1}{c-1}} d^i .$$

This proves the proposition. □

When g is birational, i.e., when $d_n(g) = 1$, it follows from (8.2) that $d_i(g^{-1}) = d_{n-i}(g)$ for all $i \in \{0, \dots, n\}$. In particular,

$$d(g^{-1}) = d_{n-1}(g) \le d(g)^{n-1} .$$

By (8.3), equality occurs exactly when the indeterminacy locus of g is finite.

When f is a monomial birational transformation of $\mathbf{P}^n$, Proposition 8.2 gives the stronger bound:

$$d(f^{-1}) \le (1 - d^{-c})^{\frac{n-2}{c-1}} d^{n-1} = d^{n-1} - \frac{n-2}{\lfloor n/2 \rfloor} d^{\lfloor (n-3)/2 \rfloor} + O(d^{-2}), \tag{8.4}$$

where $d := d(f)$. However, as mentioned in the introduction, this is not optimal.

When $d(f) = 2$, the set of possible values for $d(f^{-1})$ is $\{2, \dots, n\}$ and the maximal value n is obtained only (up to permutation of the factors) for the birational map $f_{n,2}$ of (8.1) ([2], Theorem 2.6). In particular, the other degrees of f are then fixed.

Johnson's calculations. When $d := d(f) > 2$, Johnson's computer calculations in [6] suggest that the maximal possible value for $d(f^{-1})$ should be

$$d(f_{n,d}^{-1}) = \frac{(d-1)^n - 1}{d-2} = d^{n-1} - (n-2)d^{n-2} + O(d^{n-3})$$

and that equality should only be attained when (up to permutation of the factors) $f = f_{n,d}$. More precisely, Johnson checks that when $n = 4$ and $3 \leq d \leq 5$, one has $d(f^{-1}) \leq d(f_{n,d}^{-1}) - d + 1$ if (up to permutation of the factors) $f \neq f_{n,d}$. There are also further gaps in the list of possible values for $d(f^{-1})$.

Mixed volumes. The degrees $d_i(f)$ of a monomial map f can be interpreted in terms of mixed volumes of polytopes in $\mathbf{R}^n$ as follows. Let $\Delta \subset \mathbf{R}^n$ be the standard n-dimensional simplex $\operatorname{conv}(0, \mathbf{e}_1, \dots, \mathbf{e}_n)$. Let $f : \mathbf{P}^n \dashrightarrow \mathbf{P}^n$ be a monomial map with associated matrix $A = (a_{ij})_{0 \leq i,j \leq n}$, and let $\Delta_f \subset \mathbf{R}^n$ be the simplex which is the convex hull of the points $\mathbf{a}_i = (a_{i1}, \dots, a_{in}) \in \mathbf{N}^n$, for $i \in \{0, \dots, n\}$. Then ([3], §3.5)

$$d_i(f) = \mathrm{MV}(\underbrace{\Delta, \dots, \Delta}_{n-i \text{ times}}, \underbrace{\Delta_f, \dots, \Delta_f}_{i \text{ times}}).$$

The right-hand side of this equality is a *mixed volume:* if the n-dimensional volume is normalized so that $\operatorname{vol}(\Delta) = 1/n!$, this is $(n-i)!i!$ times the coefficient of $u^{n-i}v^i$ in the polynomial $\operatorname{vol}(u\Delta + v\Delta_f)$, where $u\Delta + v\Delta_f$ is the Minkowski sum $\{u\mathbf{x} + v\mathbf{y} \mid \mathbf{x} \in \Delta, \mathbf{y} \in \Delta_f\}$.

Although mixed volumes are notoriously difficult to compute, there are computer programs such as `PHCpack` (available on Jan Verschelde's webpage) that can do that. We should also mention the article [1], which expresses the degrees of a monomial rational transformation in terms of integrals over an associated Newton region.

References

1. P. Aluffi, Multidegrees of monomial rational maps. arXiv:1308.4152 [math.AG]
2. B. Costa, A. Simis, Cremona maps defined by monomials. J. Pure Appl. Algebra **216**, 202–215 (2012)
3. I. Dolgachev, Lectures on Cremona transformations (2011). Available at http://www.math.lsa.umich.edu/~idolga/cremonalect.pdf
4. W. Fulton, *Intersection Theory* (Springer, Berlin, 1984)
5. G. Gonzalez-Sprinberg, I. Pan, On the monomial birational maps of the projective space. Anais da Academia Brasileira de Ciências **75**, 129–134 (2003)
6. P. Johnson, Inverses of monomial Cremona transformations. arXiv:1105.1188 [math.AG]
7. A. Simis, R. Villarreal, Constraints for the normality of monomial subrings and birationality. Proc. Am. Math. Soc. **131**, 2043–2048 (2003)

Chapter 9
Progress in the Theory of Nonlinear Diffusion: Asymptotics via Entropy Methods

Juan Luis Vázquez

Abstract We report on recent progress in the study of nonlinear diffusion equations in which the author has been involved. The main topic we discuss here is the use of entropy methods to obtain a precise description of the asymptotic behaviour of the solutions of evolution problems posed in the whole space. A detailed account is given of the analysis of the fast diffusion flow for low values of the equation exponent, which entails a delicate entropy analysis via weighted linearization. Connections and extensions are mentioned.

9.1 Introduction: Linear Models

In this article I will discuss an active topic in the theory of nonlinear diffusion where I have been involved in recent years. It may be considered as a continuation of the presentation at the International Congress of Mathematicians held in Madrid in 2006, [34]. The main subject is the same, nonlinear diffusion, with emphasis on porous medium and fast diffusion equations. The subject matter I will cover here is large-time asymptotics using self-similar solutions and entropy methods. These methods will be the focus of the presentation, and we will show how they work in practice, starting with the heat equation, and passing to the porous medium equation and then the fast diffusion equation. I will explain in some detail this last case since it involves some delicate mathematical analysis that we did between 2006 and 2010.

Linear Parabolic Equations. Linear heat flows. From Fourier's decisive contribution in 1822 until well into the twentieth century, the mathematical study of diffusion was almost exclusively centered on the heat equation, a part of the classical theory of PDEs. It has motivated a wealth of mathematics in the areas now

J.L. Vázquez (✉)
Departamento de Matemáticas, Universidad Autónoma de Madrid, 28049 Madrid, Spain
e-mail: juanluis.vazquez@uam.es

V. Ancona and E. Strickland (eds.), *Trends in Contemporary Mathematics*, Springer INdAM Series 8, DOI 10.1007/978-3-319-05254-0_9,

called Fourier Analysis, Functional Analysis, Spectral Theory, Potential Theory, Semigroup Theory, and it has had a deep influence on Mathematical Physics, Probability, and recently on Geometry. Let us also mention that the Fourier analysis of functions motivated Cantor in his work on the foundations of Set Theory.

In parallel with the heat equation, we have witnessed the development of a theory for other linear equations. The equations that can be included into this family are called parabolic equations. There is a well-established theory for the linear subfamily that is collected in books under the name of "Theory of Parabolic Equations". The typical form is

$$u_t = \sum_{i,j=1}^{n} a_{ij}\, \partial_i \partial_j u + \sum_{i=1}^{n} b_i\, \partial_i u + c\, u + f \tag{9.1}$$

where the coefficients a_{ij}, b_i, and c are more or less regular functions (constants in the simplest case), and the forcing term $f(x,t)$ accounts for external forces. The main restriction in all the theory is algebraic and consists of requiring the ellipticity of the $n \times n$ matrix $\mathtt{A} = (a_{ij})$, i.e.,

$$\lambda_1\, |\xi|^2 \le \sum_{i,j} a_{ij}\xi_i \xi_j \le \lambda_2\, |\xi|^2$$

for some positive $\lambda_1 \le \lambda_2$ and all vectors $\xi = (\xi_i)_i \in \mathbb{R}^N$. This is called in other areas uniform positivity of $\mathtt{A}$. The theory is displayed in a quite organized way in several levels of generality, each of them involving a number of subtle and fruitful new concepts and corresponding deep results, like Maximum Principles, Schauder estimates, Harnack inequalities, Calderón-Zygmund theory, and so on.

The probabilistic approach: Diffusion has another face. The development of the theory of Stochastic Processes led to the concept of Ito derivative and integral. It leads to the development of Stochastic Differential Equations, like

$$dX = bdt + \sigma dW. \tag{9.2}$$

Believe or not, this is another way of looking at the differential object described in Eq. (9.1). This connection is explained in books like [29, 30]. Probability motivated some of the concepts we use.

9.2 Nonlinear PDEs

In the last 50 years emphasis has shifted heavily towards the Nonlinear World. The motivation for this shift comes from the realization that many of the natural phenomena that we want to describe with this mathematical tool are essentially nonlinear and its more salient characteristics are not reflected by the linear theories that had been developed, notwithstanding the fact that such linear theories had been and continue to be very efficient for a huge number of applications.

The main obstacle to the systematic study of the Nonlinear PDE Theory was the perceived difficulty and lack of tools. This is reflected in a passage by the talented John Nash [28]. He said: "*The open problems in the area of nonlinear PDE are very relevant to applied mathematics and science as a whole, perhaps more so that the open problems in any other area of mathematics, and the field seems poised for rapid development. It seems clear, however, that fresh methods must be employed...*". This interest led him to his famous work on the regularity of heat flows that has influenced in such a tremendous way the research in nonlinear elliptic and parabolic equations, [28].

Once the tools were ready to start attacking Nonlinear PDEs in a rigorous way, it was discovered that resulting mathematics is quite difficult and complex, and more realistic than the linearized models in the applications to real-world phenomena. In the last decades we have been shown a multiplicity of new qualitative properties and surprising phenomena encapsulated in the nonlinear models supplied by the applied sciences, some of them very popular nowadays, like free boundaries, solitons and shock waves.

9.2.1 Nonlinear Heat Flows

In the case of parabolic equations, and the processes of heat propagation and diffusion, the Laplace operator has been often replaced by more general types of "elliptic operators with variable coefficients", and later by nonlinear differential operators; a huge body of theory is now available, both for the evolution equations [27] and for the stationary states, described by elliptic equations of different kinds [26]. The work on nonlinear parabolic equations can be summed up as the study of equations of the general form

$$u_t = \sum \partial_i \mathscr{A}_i(u, \nabla u) + \sum \mathscr{B}(x, u, \nabla u) \tag{9.3}$$

where $\mathscr{A}_i$ and $\mathscr{B}$ must satisfy so-called structure conditions, the main one is again the ellipticity condition on the function $A(x, u, z)$ as a function of the vector variable $z = (z_i)$. Notice that we are selecting the so-called divergence structure for the term representing diffusion; the parallel development involving non-divergence form leads to the theory of Fully Nonlinear Parabolic Equations that we will not touch here.

9.2.2 Classical Problems in Nonlinear Diffusion

The work on nonlinear parabolic equations in the mathematical research community to which I belong focussed attention on the analysis of a number of paradigmatic models involving the occurrence of free boundaries, for which new tools were developed and tested. Since the multiplicity of quite different phenomena precluded the elaboration of a general theory, a more successful strategy favored the detailed study of a number of important models. Some of the most famous are:

- **The Obstacle Problem.** Given a domain $\Omega \in \mathbb{R}^N$, an obstacle $\Phi(x)$ and functions $f(x), g(x)$, with the compatibility condition $\Phi \geq g$ on $\partial\Omega$, the problem is to find a continuous function u and a subdomain Ω_1 such that

$$\Delta u + f = 0 \text{ for } x \in \Omega_1, \quad u = \Phi(x) \text{ for } x \in \Omega \setminus \overline{\Omega_1},$$

  and $u = g$ on $\partial\Omega$. Of course we need suitable functional spaces in which the problem is set and then the methods of the Calculus of Variations produce a unique solution. See Caffarelli's presentation in [11].
- **The Stefan Problem.** (Lamé and Clapeyron, 1833; Stefan 1880)

$$SE : \begin{cases} u_t = k_1 \Delta u \text{ for } u > 0, \\ u_t = k_2 \Delta u \text{ for } u < 0. \end{cases} \quad TC : \begin{cases} u = 0, \\ \mathbf{v} = L(k_1 \nabla u_1 - k_2 \nabla u_2). \end{cases}$$

  Main feature: the free boundary or moving boundary is the boundary of the set where $u > 0$. TC= Transmission conditions at the free boundary.
- **The Hele-Shaw cell equation.** (Hele-Shaw, 1898; Saffman-Taylor, 1958)

$$u > 0, \ \Delta u = 0 \quad \text{in} \quad \Omega(t); \quad u = 0, \ \mathbf{v} = L\partial_n u \quad \text{on} \quad \partial\Omega(t).$$

  When posed in two space dimensions, this simplified version of the Stefan Problem has deep connections with complex variable theory.
- **The Porous Medium Equation.** This is maybe the simplest model of nonlinear heat equation

$$u_t = \Delta u^m, \quad m > 1.$$

  and was proposed in the twentieth century in a number of contexts: groundwater infiltration, flows in porous media, thermal propagation in plasmas, population dynamics, etc. A whole presentation of the mathematical study of the equation covering results until 2005 is given in [33]. A curious fact is that at first glance solving the typical Initial-Boundary Value problems for this equation is not a free boundary problem, and the methods of solution do not have to take this possibility into account. However, the solutions corresponding to compactly supported initial data exhibit a free boundary, i.e., it is an implicit free boundary problem. This is easier to see if we write the equation as

$$u_t = \nabla \cdot (c(u)\nabla u)$$

where $c(u)$ indicates the density-dependent diffusivity, in this case

$$c(u) = mu^{m-1}.$$

For $m = 1$ get the heat equation with constant c, but when If $m > 1$ the coefficient degenerates at $u = 0$, which on physical grounds means *slow diffusion* and on the mathematical side the equation is no more parabolic at $u = 0$. Another way of concentrating on the new features of the PME if to pass to the variable $v = u^{m-1}$, thus getting the equivalent equation

$$v_t = mv\Delta v + (m/(m-1))|\nabla v|^2$$

where we see that the equation is no more second order at $v = 0$ but first order, hence the finite propagation. The study of the free boundaries adds a difficult and interesting geometric flavor to an otherwise nonlinear parabolic equation. A detailed comment on this issue is contained in [34].

The properties are quite different if we take $m < 1$ in the above model so that $c(u) \to \infty$ as $u \to 0$. Hence, the equation receives a separate name, the Fast Diffusion Equation, it exhibits no free boundaries, but, as a kind of compensation. It has a very rich asymptotic behaviour that we will discuss below in some detail.

- **The p-Laplacian Equation.** A very related equation is the nonlinear diffusion equation with gradient-dependent diffusivity

$$u_t = \text{div}\,(|\nabla u|^{p-2}\nabla u)$$

where $c(u) = |\nabla u|^{p-2}$. This equation has solutions with free boundaries for $p > 2$, it becomes the heat equation for $p = 2$ and behaves like a fast diffusion equation for $1 < p < 2$. The limit cases $p = 1$ and $p = \infty$ have attracted lots of attention because of the mathematics and the applications.

There are a number of interesting variants of these models and also quite a few popular models of nonlinear diffusion with or without free boundaries that have been investigated in the last decades, like the chemotaxis system, thin film equations, different geometric flows,...

9.3 Asymptotics for the PME

Though the theory of the porous medium equation (PME) is very different from the Heat Equation due to the nonlinear and degenerate character, the asymptotic behaviour bears strong resemblance due to the explaining power of self-similarity and the tools of Functional Analysis.

For the sake of definiteness, we will consider the equation posed in the whole space $\mathbb{R}^N$. Fist of all, we have to replace the Gaussian fundamental solution of the Heat Equation with a hopefully similar solution for the PME. These are the Barenblatt solutions, also called source-type solutions, since they take initial data a point source, $u(x,t) \to M\,\delta(x)$ as $t \to 0$. There is fortunately an explicit formula of self-similar type (obtained in the period 1950, 1952)

$$\mathbf{B}(x,t;M) = t^{-\alpha}\mathbf{F}(x/t^{\beta}), \quad \mathbf{F}(\xi) = \left(C - k\xi^2\right)_+^{1/(m-1)}$$

where $k = (m-1)/2mN$, and

$$\alpha = \frac{N}{2+N(m-1)}, \qquad \beta = \frac{1}{2+N(m-1)}.$$

We see that the solution has maximum height $u = Ct^{-\alpha}$ and also has a free boundary at distance $|x| = ct^{\beta}$. Since $m > 1$ we have $\beta < 1/2$. These values differ from the typical values of the linear diffusion theory, in particular the scaling $\langle x^2 \rangle \approx ct$ of Brownian motion is no more valid. It falls therefore into the category of anomalous diffusion.

Nonlinear Central Limit Theorem. We will now use these explicit solutions to give a general answer to the question of long-time behaviour of a large class of solutions of the PME posed in the whole space. We need to make a choice of data, and this is $u_0(x) \in L^1(\mathbb{R}^N)$, that answers the usual physical requirements. We will write the equation in the general complete form

$$u_t = \Delta(|u|^{m-1}u) + f,$$

to accept both signs for the solutions, and a possible forcing term f. We assume that $f \in L^1_{x,t}$. Let $M = \int u_0(x)\,dx + \iint f\,dxdt$. The Asymptotic Theorem is

Theorem 9.1. *Let $B(x,t;M)$ be the Barenblatt with the asymptotic mass M; u converges to $\boldsymbol{B}$ in $L^1(\mathbb{R}^N)$:*

$$\|u(x,t) - \mathbf{B}(x,t)\|_1 \to 0.$$

Let $f = 0$ (or small for large times in L^p). For every $1 \le p \le \infty$ we have

$$\|u(t) - \mathbf{B}(t)\|_p = o(t^{-\alpha/p'}), \quad p' = p/(p-1).$$

Note: α *and* $\beta = \alpha/N = 1/(2+N(m-1))$ *are the zooming exponents as in* $\mathbf{B}(x,t)$.

The basic result is due to Friedman and Kamin [22] takes $u_0 \ge 0$, and $f = 0$ and proves uniform convergence on expanding sets. Full proof is done in [31]. Proof is done by rescaling method, and needs a good uniqueness theorem for fundamental solutions. For $M = 0$ the result we state is not precise and the next term in the asymptotics is studied in [25]. The proof was extended to the p-Laplacian equation by Kamin and Vázquez in [24].

9.4 Entropy Methods for Heat Equations

We are going to use energy functions of different types to study the evolution of dissipation equations. The basic equation here will be the classical heat equation. Our aim is not to establish the convergence of general solutions to the fundamental solution (which is well done by other methods), but a bit more: to find the speed of convergence. After change of variables (renormalization) this reads as rate of convergence to equilibrium and relies on important functional inequalities.

The methods will apply to more general linear parabolic equations that generate semigroups. The method also works for equations evolving on manifolds as a base space. The method has been applied to nonlinear diffusion equations since around year 2000, as we will see below.

First entropy method for the Heat Equation. Take the classical Heat Equation posed in the whole space $\mathbb{R}^N$ for $\tau > 0$:

$$u_\tau = (1/2)\Delta_y u\,, \tag{9.4}$$

with notation $u = u(y,\tau)$ that is convenient at this stage. We know the (self-similar) fundamental solution, i.e., the Gaussian function,

$$U(x,t) = C\,\tau^{-N/2}e^{-y^2/2\tau}\,,$$

that is an attractor of its basin consisting of all in the integrable functions with mass $\int u_0(y)\,dy = C$, $C \neq 0$. We now introduce a key transformation, the time-space rescaling

$$u(y,\tau) = v(x,t)\,(1+\tau)^{-N/2}, \quad y = x(1+\tau)^{1/2}, \quad t = \log(1+\tau),$$

(note the change to the logarithmic time scale). This gives the well-known Fokker-Plank equation for $v(x,t)$:

$$2v_t = \Delta_x v + \nabla\cdot(x\,v) \tag{9.5}$$

for which $G = C\,e^{-x^2/2}$ is the stationary state (stationary Gaussian kernel), corresponding in the transformation to U. If we now pass to the quotient $w = v/G$, we get the Ornstein-Uhlenbeck (OU) version

$$2w_t = G^{-1}\,\nabla\cdot\big(G\,\nabla w\big). \tag{9.6}$$

The equivalence of these three equations is a main tool in Linear Diffusion and Semigroup Theory.

Restricting ourselves for simplicity to nonnegative and nontrivial solutions, we may assume without lack of generality that $\int w\,d\mu = \int v\,dx = \int u\,dy = 1$, where $d\mu = G(x)dx$. We now make a crucial estimate on the time decay of the energy for the OU equation (9.6). If the relative entropy is defined as

$$\mathscr{F}(w(t)) = \int_{\mathbb{R}^N} |w-1|^2\, G\, dx, \tag{9.7}$$

then

$$\frac{d\mathscr{F}(w(t))}{dt} = -\int_{\mathbb{R}^N} |\nabla w|^2\, G\, dx = -\mathscr{D}(w(t)). \tag{9.8}$$

$\mathscr{D}$ is called the dissipation of $\mathscr{F}$. We can now use a result from abstract functional analysis: the Gaussian Poincaré inequality with measure $d\mu = G(x)\,dx$:

$$\int_{\mathbb{R}^N} w^2 d\mu - \left(\int_{\mathbb{R}^N} w\, d\mu\right)^2 \le \int_{\mathbb{R}^N} |\nabla w|^2\, d\mu$$

Then, the LHS is just $\mathscr{F}$ and the inequality implies after integration that:

$$\int_{\mathbb{R}^N} |w-1|^2\, \mathrm{d}\mu \le \mathrm{e}^{-t} \int_{\mathbb{R}^N} |w_0 - 1|^2\, \mathrm{d}\mu \quad \forall\, t \ge 0.$$

Then by regularity and interpolation we get $\|w-1\|_\infty \le \mathscr{K}\mathrm{e}^{-t}$, that means, once we go back to the original variables that:

$$(1-\mathscr{K}e^{-t})G(x) \le\; v(x,t)\; \le (1+\mathscr{K}e^{-t})\, G(x) \qquad (e^{-t} = \frac{1}{1+\tau}).$$

These are the well known Heat Kernel Estimates of solutions to the HE.

Second entropy method for the Heat Equation. There is another approach that starts the analysis from Boltzmann's ideas on entropy dissipation. We start from the Fokker-Planck equation $v_t = \Delta v + \nabla\cdot(xv)$ and consider the following **entropy**

$$\mathscr{E}(v) = \int_{\mathbb{R}^N} v\, \log(v/G)\, dx = \int_{\mathbb{R}^N} v\, \log(v) dx + \frac{1}{2}\int_{\mathbb{R}^N} x^2 v\, dx + C\,.$$

Differentiating along the flow (i.e., for a solution) leads to

$$\frac{d\mathscr{E}(v)}{dt} = -\frac{1}{2}\mathscr{I}(v), \quad \mathscr{I}(v) = \int_{\mathbb{R}^N} v\left|\frac{\nabla v}{v} + x\right|^2 dx = \int_{\mathbb{R}^N} v\,|\nabla \log(v/G)|^2\, dx\,.$$

Put now $v = Gf^2$ to find that

$$\mathscr{E}(v) = 2\int_{\mathbb{R}^N} f^2 \log(f)\, d\mu, \quad \mathscr{I}(v) = 4\int_{\mathbb{R}^N} |\nabla f|^2\, d\mu\,.$$

The famous logarithmic Sobolev inequality [23] says that (for all functions, not only solutions, with unit L^2 norm) $\mathscr{E} \le \frac{1}{2}\mathscr{I}$ and we obtain the decay $\mathscr{E}(t) \le \mathscr{E}(0)\, e^{-t}$. It is now a question of some work to translate this result into a good norm.

- Entropy has been introduced as a state function in thermodynamics by R. Clausius in 1865, in the framework of the second law of thermodynamics, in order to interpret the results of S. Carnot. A statistical physics approach appears with Boltzmann's formula (1877), the entropy of a system is defined in terms of a counting of the micro-states of a physical system. Boltzmann's equation is kinetic, different from our models: $\partial_t f + v\cdot\nabla_x f = Q(f,f)$. It describes the evolution of a gas of particles having binary collisions at the kinetic level; $f(t,x,v)$ is a time dependent distribution function (probability density) defined on the phase space $\mathbb{R}^N \times \mathbb{R}^N$.

The Boltzmann entropy $H[f] := \iint f\log(f)dxdv$ measures the irreversibility of the process. The H-Theorem (1872) says that

$$\frac{d}{dt}H[f] = \iint Q(f,f) log(f)\, dxdv \le 0\,.$$

9.5 Entropy Method for the PME

The next step of information after establishing plain convergence, see Theorem 9.1, is the calculation of the entropy rates. We want to know how large the error is between u and **B** in relative size. We explain here the well-known contribution of Carrillo and Toscani [14] who used a new entropy. As in the heat equation, we first rescale the function as

$$u(x,t) = R(t)^N \rho(y\, R(t), s)$$

where $R(t)$ is the Barenblatt radius at $t+1$, and "new time" is $s = \log(1+t)$. The equation becomes

$$\partial_s \rho = \operatorname{div}\left(\rho(\nabla\rho^{m-1} + \frac{c}{2}\nabla y^2)\right). \tag{9.9}$$

Then define a new entropy (not Boltzmann entropy, but a new type called Rényi entropy)

$$\mathscr{E}(\rho)(s) = \int (\frac{1}{m}\rho^m + \frac{c}{2}\rho y^2)\, dy\,. \tag{9.10}$$

The minimum of entropy is identified as the Barenblatt profile. We calculate to get

$$\frac{d\mathscr{E}}{ds} = -\int \rho|\nabla\rho^{m-1} + cy|^2\,dy = -\mathscr{D}\,. \tag{9.11}$$

Moreover, a calculation known as the Bakry-Emery method gives $d\mathscr{D}/ds = -\mathscr{R}$, with $\mathscr{R} \le \lambda\mathscr{D}$. We conclude exponential decay of $\mathscr{D}$ in the new time s, i.e., a power rate in real time t. It follows that $\mathscr{E}$ decays to a minimum $\mathscr{E}_\infty > 0$ and we then prove that this is the level of the Barenblatt solution, which attains the functional minimum. See whole details in [14, 16].

9.6 The Fast Diffusion Problem in $\mathbb{R}^N$

The application of the above machinery to the case $m < 1$ of the PME, the so-called Fast Diffusion Equation (FDE), is not very different when $m \approx 1$ but encounters serious difficulties when m goes below some critical value.

Preliminary analysis and self-similarity. Indeed, when we want to extend the formulas for the self-similar Barenblatt profiles to exponent m less than one, we succeed if $m > m_c$ with $m_c = (N-2)/N$ and we get the formulas:

$$\mathbf{B}(x,t;M) = t^{-\alpha}\mathbf{F}(x/t^{\beta}), \quad \mathbf{F}(\xi) = \frac{1}{(C + k\xi^2)^{1/(1-m)}}\,.$$

We have $k = k(m,N) > 0$ and the scaling exponents are

$$\alpha = N/(2 - N(1-m)), \qquad \beta = 1/(2 - N(1-m)),$$

and the latter is now larger than $1/2$. The algebra is the same as for $m > 1$ but the changes in some signs make for a completely different behaviour. Note that both α and $\beta \to \infty$ as m goes down to m_c. Thus, the new solutions with $m < 1$ do not have compact support. Much to the contrary, they exhibit polynomial decay as $x \to \infty$, which is known in probability and the applications as *fat tails*. They are another form of *anomalous distributions*.

The Nonlinear Central Limit Theorem proved for $m > 1$ can be adapted to the case $m < 1$ as long as $m > m_c$, i.e. as long as suitable Barenblatt solutions exist. The analysis of the questions of existence and qualitative behaviour for the subcritical case $m < m_c$ showed that remarkable phenomena occur like non-existence for very fast diffusion, non-uniqueness for very fast diffusion, extinction, universal estimates, lack of standard Harnack, cf. [9]. See the monograph [32], where the main results are described.

A physical property that plays a big role in the behaviour of the solutions is the conservation of the total mass:

(i) For $\infty > m > m_c$ *the mass* $\int_{\mathbb{R}^N} u(y,t)\,dy$ *is preserved in time if* $u_0 \in L^1(\mathbb{R}^N)$. Non-negative solutions are positive and smooth for all $x \in \mathbb{R}^N$ and $t > 0$.
(ii) On the other hand, if $m < m_c$ *mass is not preserved* and many solutions extinguish in finite time:

$$u_0 \in L^{p_c}(\mathbb{R}^N)\,, \quad p_c = \frac{N\,(1-m)}{2} \quad \Longrightarrow \exists\, T = T(u_0)\ : u(\tau,\cdot) \equiv 0 \forall\, t \geq T\,.$$

If we extend the range of exponents to $m < 0$ as we will do in a moment it can even happen that $T = 0$ (complete extinction of the solution since $t = 0^+$).

Setup of the Asymptotic Problem. A big problem that began to be investigated in the early 2000s was as follows: what happens for $m < (N-2)/N$ regarding long time behaviour? We will describe in the following subsections a unified approach based on entropies. We will consider the solutions $u \geq 0$ of the Cauchy Problem for the FDE, which is written in the convenient form

$$\partial_\tau u = \Delta\left(\frac{u^m}{m}\right) = \nabla\cdot\left(u^{m-1}\nabla u\right)\,, \qquad (\tau, y) \in (0,T)\times\mathbb{R}^N\,.$$

We take initial data

$$u(0,\cdot) = u_0\ , u_0 \in L^1_{\text{loc}}\big(\mathbb{R}^N\big).$$

We will consider non-negative initial data and solutions. Written in this way, we can also treat the cases $m \leq 0$; in particular $m = 0$ corresponds to *logarithmic diffusion*, very important case in the literature $\big($write then $\partial_\tau u = \Delta \log(u) = \nabla\cdot\big(\nabla u/u\big)\big)$.

Existence and uniqueness of weak solutions is known by a priori estimates and nonlinear semigroup theory. We have already noticed the presence of the critical exponent m_c in the existence of Barenblatt solutions.

Rates through entropies for Fast Diffusion. Let us summarize the increasing difficulties that we face:

(i) The nice properties of the PME entropy from the point of view of transport theory are lost soon, when $m = (N-1)/N$.
(ii) Finite entropy for the Barenblatt solutions is lost when the second moment is infinite, i.e. for $m = (N-1)/(N+1)$.
(iii) The Barenblatt solutions, which are the finite-mass, stable states after rescaling are lost when m reaches the value $m_c = (N-2)/N$.

On the other hand, there is an important fact that was known from the linear and porous medium theory: functional inequalities play a crucial role in the asymptotic analysis, there are so to say "equivalent" to the decay rates that we want to obtain. A large effort was made to implement the entropy method for fast diffusion. There is work by many authors. Preliminary works by Carrillo, Dolbeault, Del Pino, Markowich, the author, and others, did not break the barrier $m = m_c$. The solution of the subcritical range that I will report here is the result of collaboration with Blanchet, Bonforte, Dolbeault and Grillo [2, 3, 6, 7]. I must also mention works on this topic by Daskalopoulos and Sesum [17] and Denzler, McCann [20]. All these papers contain references to a wide literature.

- **Barenblatt and Pseudo-Barenblatt Solutions.** A key point in the joint analysis of the ranges $m > m_c$ and $m \le m_c$ is to find a continuation for the Barenblatt solutions in a form that shares certain algebraic structure. Actually, it was found that there is a self-similar form valid for every $m \in \mathbb{R}$:

$$U_{D,T}(\tau, y) := \frac{1}{R(\tau)^N}\left(D + \frac{1-m}{2m}\left|\frac{y}{R(\tau)}\right|^2\right)_+^{-\frac{1}{1-m}}.$$

The time scaling is delicate, since it has to include cases with global existence and cases with finite extinction. It is as follows:

$$\begin{cases} R(\tau) : = \left[N\,(m-m_c)\,(T+\tau)\right]^{\frac{1}{N\,(m-m_c)}} & \text{if } m_c < m < 1, \text{ Super-Critical Range} \\ R(\tau) := e^{T+\tau} & \text{if } m_c = m\,, \quad \text{First Critical Exp.} \\ R(\tau) := \left[N\,(m_c-m)\,(T-\tau)\right]^{-\frac{1}{N\,(m_c-m)}} & \text{if } m < m_c, \quad \text{Sub-Critical Range} \end{cases}$$

T is a free parameter to be suitably chosen later. It is only important in the last case, where we have extinction in finite time. In this case we have called the solutions pseudo-Barenblatt, since they have infinite mass. Let us point out that they are not source-type solutions, i.e., their initial data is never a Dirac mass or an isolated singularity, just a plain non-integrable distribution of mass.
- **Assumptions on the data. Reduced basin of attraction.** For $m < m_c$ the generality of the data is reduced, the pseudo-Barenblatt are not so attractive, and we will need the initial data to be finite perturbations of the old or new Barenblatt profiles in the following sense:

(H1) $u_0 \in L^1_{\rm loc}(\mathbb{R}^N)$, non-negative, and there exist positive constants T and $D_0 > D_1$ such that

$$U_{D_0,T}(0, y) \le u_0(y) \le U_{D_1,T}(0, y) \quad \forall\ y \in \mathbb{R}^N .$$

(H2) There exist $D_* \in [D_1, D_0]$ and $f \in L^1(\mathbb{R}^N)$ such that

$$u_0(y) = U_{D_*,T}(0, y) + f(y) \quad \forall\ y \in \mathbb{R}^N .$$

$(H1)$ implies $(H2)$ when $m_c < m < 1$, even for $m > m_*$ where $m_* = (N-4)/(N-2)$ will play a role below. On the other hand, when $m < m_c$, by Comparison Principle, (H1) implies that the extinction of $u(t,\cdot)$ occurs exactly at time T. This is a strong restriction on the data, we are fixing the extinction time T.

Conservation of Relative Mass. A special role in the formulation of the result and also in the proof is played by the weak form of mass conservation, stated as follows

Proposition 9.1. *Let $m < 1$. Consider a solution u of the FDE with initial data u_0 satisfying* (H1)-(H2). *If for some $D > 0$, $\int_{\mathbb{R}^N} \left[u_0 - U_{D,T}(0,\cdot)\right] dy$ is finite, then*

$$\int_{\mathbb{R}^N} \left[u(\tau,y) - U_{D,T}(\tau,y)\right] \mathrm{d}y = \int_{\mathbb{R}^N} \left[u_0(y) - U_{D,T}(0,y)\right] \mathrm{d}x \quad , \forall\, \tau \in (0,T)\,.$$

We summarize the fact that $(\mathrm{d}/\,\mathrm{d}t) \int_{\mathbb{R}^N} \left[u_0 - U_{D_*,T}(0,\cdot)\right] dy = 0$ by saying that the *relative mass is conserved.* The map $D \mapsto \int_{\mathbb{R}^N} (v_0 - V_D)\, dx$ is continuous and monotone increasing. We can define a unique $D_* \in [D_1, D_0]$ such that if $m > m_*$, then

$$\int_{\mathbb{R}^N} \left[u(\tau,y) - U_{D_*,T}(\tau,y)\right] \mathrm{d}y = 0 \quad \forall\, t > 0\,.$$

New exponent. Note next that there is a lower exponent $m_* = (N-4)/(N-2) < m_c$ such that the difference of two Barenblatt solutions $U_{D_1,T} - U_{D_2,T}$ has finite relative mass only if $m > m_*$. We conclude that when $m \le m_*$, integrals in the relative mass expression are infinite unless $D = D_*$ and then,

$$\int_{\mathbb{R}^N} \left[u_0 - U_{D_*,T}(0,\cdot)\right] \mathrm{d}y = \int_{\mathbb{R}^N} f\, dx \quad \forall\, t > 0\,.$$

In this case the perturbation $f \in L^1(\mathbb{R}^N)$ of $U_{D_*,T}$, has in general nonzero mass.

We are ready to state the main result.

Theorem 9.2. *Let $N \ge 3$, $m < 1$, $m \ne m_*$. Consider a solution u of the FDE, with initial data satisfying* (H1)-(H2). *For τ large enough, for any $q \ge 1$, $q \in (q_*, \infty]$, there exists a positive constant C such that*

$$\|u(\tau) - U_{D_*}(\tau)\|_q \le C\, R(\tau)^{-\alpha}$$

where the optimal rate is given by

$$\alpha = \Lambda_{m,N} + N\,(q-1)/q$$

and $\Lambda_{m,N}$ is the inverse of the Hardy-Poincaré constant $\mathscr{C}_{m,N} = \Lambda_{m,N}^{-1}$.

Large means in the real time that $\tau \to T$ if $m < m_c$, and $\tau \to \infty$, if $m \ge m_c$. We have a formula for $q_* := \dfrac{2\,N\,(1-m)}{2\,(2-m) + N\,(1-m)}$.

Hardy-Poincaré constant and inequalities. We put $V_D = (D + |x|^2)^{-\frac{1}{1-m}}$ for the self-similar profile of the Barenblatt solutions. Here is the basic functional result we need in the stage of asymptotic entropy estimate: the Hardy-Poincaré Inequalities, that imply the existence of a spectral gap if $m \ne m_* = (N-4)/(N-2)$.

Theorem 9.3. *Let $N \geq 1$ and $D > 0$. There exists $\Lambda_{m,N}$, not depending on D, such that*

(i) POINCARÉ CASE. *If $m \in (0,1)$ and $1 \leq N \leq 4$, or $m \in (m_*, 1)$ and $N \geq 5$, then*

$$\Lambda_{m,N} \int_{\mathbb{R}^N} |g - \overline{g}|^2 \; V_D^{2-m} \, dx \leq \int_{\mathbb{R}^N} |\nabla g|^2 \; V_D \, dx \,, \qquad \overline{g} = \frac{\int_{\mathbb{R}^N} g \; V_D^{2-m} \, dx}{\int_{\mathbb{R}^N} V_D^{2-m} \, dx} \,.$$

(ii) HARDY CASE. *In case $N \geq 3$ and $m < m_*$, we have*

$$\Lambda_{m,N} \int_{\mathbb{R}^N} g^2 \; V_D^{2-m} \, dx \leq \int_{\mathbb{R}^N} |\nabla g|^2 \; V_D \, dx \,.$$

The optimal constant is

$$\Lambda_{m,N} = \begin{cases} \dfrac{2}{1-m} & \text{if} \quad \dfrac{N-1}{N} < m < 1 \,, \\ 2\dfrac{2-N(1-m)}{1-m} & \text{if} \quad \dfrac{N}{N+2} < m < \dfrac{N-1}{N} \,, \\ \dfrac{[(N-2)\,(m-m_*)]^2}{4\,(1-m)^2} & \text{if} \quad m < \dfrac{N}{N+2}, \quad m \neq m_* \,. \end{cases}$$

Remarks. (1) when $m = m_*$ we have $\Lambda_{m,N} = 0$ and there is no spectral gap! An extra functional analysis is needed. It was done in [7], and it is a different story that leads to a flow on a Riemannian manifold which needs further tools from geometry.

(2) We observe that the weight is a power of the Barenblatt and has a simple "fat tail"

$$V_D^{2-m} \sim V_D/|x|^2 \,, \qquad \text{as } |x| \to \infty$$

- If $m < m_*$, we get the *Hardy-type*: the weight V_D^{2-m} is *not integrable*, the formula includes no average, the infimum of the spectrum is positive, and $\mathscr{C}_{m,N}$ is the best constant.
- If $m_* < m < 1$, we are in the *Poincaré-type*: the weight V_D^{2-m} is integrable, and the spectral gap inequality involves the average as in the classical Poincarè inequality, but now with *weights.*

(3) The optimal rate of convergence has been calculated by Del Pino-Dolbeault [18] when $m > (N-1)/N$, by Carrillo-Vázquez [15] when $m > m_c$, while no other results where known for $m \leq m_c$.

9.6.1 *Idea of the Proof When $m \neq m_*$*

Rescaling into the FP equation. We first pass to self-similar variables to obtain a Nonlinear Fokker-Plank Equation:

- When $m < m_c$, assume that u extinguishes in finite time T.
- When $m_c < m < 1$, T is a free parameter to be suitably chosen later.

Let $a = (1-m)/2[N(1-m)-2]$. Define the rescaled function v by

$$v(t,x) := R^N(\tau)\, u(\tau, y)\,, \qquad t := a \log\left(\frac{R(\tau)}{R(0)}\right), \qquad x := \sqrt{a}\frac{y}{R(\tau)}\,.$$

The function v is solution to the *non-linear Fokker-Planck equation* (NLFP):

$$\begin{cases} \dfrac{\partial v}{\partial t} = \Delta(v^m) + \dfrac{2}{1-m}\nabla(x\, v) = \nabla\cdot\left[v\nabla\left(\dfrac{v^{m-1}-V_D^{m-1}}{m-1}\right)\right] & \text{in } (0,+\infty)\times\mathbb{R}^N, \\ v(0,\cdot) = v_0 = R(0)^d\, u_0(\cdot\, R(0)) & \text{in } \mathbb{R}^N\,. \end{cases}$$

The time T has disappeared from the equation but is still part of the change of variable. Notice that the stationary solution of the NLFP equation is the (pseudo)-Barenblatt solution:

$$V_D(x) := \left(D + |x|^2\right)^{-\frac{1}{1-m}}\,, \qquad \text{we leave } D \text{ as a free “mass” parameter.}$$

- Let us translate the assumptions on the data to this setting:

  (H1) $u_0 \in L^1_{loc}(\mathbb{R}^N)$, non-negative and there exist positive constants $D_0 > D_1$ such that $V_{D_0}(x) \leq v_0(x) \leq V_{D_1}(x)$ for all $x \in \mathbb{R}^N$.

  (H2) There exist $D_* \in [D_1, D_0]$ and $f \in L^1(\mathbb{R}^N)$ such that $v_0(x) = V_{D_*}(x) + f(x)$ for all $x \in \mathbb{R}^N$. The center of mass of the initial datum is not fixed. Fixing the center of mass improves the rate in the range $m_c < m < 1$. Remember that the “mass” parameter D_* is fixed by *conservation of relative mass.*

- We state now the main result about convergence with rate. Since we have now a stationary state to tend to, the result is much more evident.

Theorem 9.4. *Under the assumptions of Theorem 1, if $m \neq m_*$, there exists $t_0 \geq 0$, $C > 0$ such that, for all $q > q_*$ with q_* given above, one has*

$$\|v(t) - V_{D_*}\|_{L^q(\mathbb{R}^N)} \leq C_q\, e^{-\Lambda_{m,N}\, t} \quad \forall\, t \geq t_0\,.$$

where $\Lambda_{m,N}$ is the eigenvalue in the Hardy-Poincaré inequality. Moreover, for all $p \ge N/2$ one has convergence in relative error, namely

$$\left\| \frac{v(t)}{V_{D*}} - 1 \right\|_{L^p(\mathbb{R}^N)} \le C_p\, \mathrm{e}^{-\lambda(p)\,t} \quad \forall\, t \ge t_0$$

for some $\lambda(p) > 0$. Finally, uniform convergence of all derivatives also hold with an exponential rate.

Relative entropy. We choose D_* by relative conservation of mass. The function $w = v/V_{D*}$ satisfies the NonLinear Ornstein-Uhlenbeck equation

$$\text{(NLOU)} \qquad w_t = \frac{1}{V_{D*}}\,\nabla\cdot\left[w\,V_{D*}\nabla\left(\frac{w^{m-1}-1}{m-1}\;V_{D*}^{m-1}\right)\right] \qquad \text{in } (0,+\infty)\times\mathbb{R}^N$$

whenever v satisfies (NLFP).

Relative entropy/entropy production. Define the nonlinear *relative entropy*

$$\mathscr{F}[w] := \int_{\mathbb{R}^N}\left[\frac{1}{m-1}(w^m-1) - \frac{m}{m-1}\,(w-1)\right]\,V_{D*}^m\,dx$$

and the nonlinear *relative entropy production* functional (or *Fisher information*)

$$\mathscr{I}[w] := \int_{\mathbb{R}^N}\left|\nabla\left[\left(\frac{w^{m-1}-1}{m-1}\right)\,V_{D*}^{m-1}\right]\right|^2\, w\,V_{D*}\,dx\,.$$

If v is solution to (NLFP) or, equivalently, if $w = v/V_{D*}$ satisfies (NLOU) then

$$\frac{\mathrm{d}}{\mathrm{d}t}\mathscr{F}[w] = -\mathscr{I}[w]\,.$$

A weighted Linearization. This is one of trickiest points. We define the function g by

$$w(t,x) = 1 + \varepsilon\,\frac{g(t,x)}{V_{D*}^{m-1}(x)} \quad \forall\, t>0\,, \qquad \forall\, x\in\mathbb{R}^N\,.$$

Letting $\varepsilon \to 0$ we **formally get a linear evolution equation** for g, namely

$$g_t = V_{D*}^{m-2}\,\nabla\cdot\left[V_{D*}\,\nabla g\right]\,.$$

Logically, we now define the "linearized entropy" functional

$$\mathsf{F}[g] := \frac{1}{2}\int_{\mathbb{R}^N}|g|^2\,V_{D*}^{2-m}\,dx$$

and notice that its time derivative (along linear flow) is

$$\frac{d}{dt}\mathsf{F}[g] = -\mathsf{I}[g] := -\int_{\mathbb{R}^N} |\nabla g|^2 \, V_{D_*} \, dx \,.$$

We use the spectral gap to obtain the convergence with rate for the linearized flow thanks to the Hardy-Poincaré inequality that we have proved

$$2\,\mathsf{F}[g(t)] \le \frac{1}{\Lambda_{m,N}}\mathsf{I}[g] \quad \Longrightarrow \quad \mathsf{F}[g(t)] \le \mathsf{F}[g(0)]\,\mathrm{e}^{-2\,\Lambda_{m,N}\,t} \quad \forall\, t \ge 0\,.$$

Comparing linear and nonlinear quantities. This step uses the knowledge of the linear analysis to wrap up the argument. We define

$$h = h(t) = \max\left\{ \sup_{x\in\mathbb{R}^N} w(t,x)\,,\, \left[\inf_{x\in\mathbb{R}^N} w(t,x)\right]^{-1}\right\}\,.$$

If t is sufficiently large, then

$$h^{m-2}\,\mathsf{F}[g] \le 2\mathscr{F}[w] \le h^{2-m}\,\mathsf{F}[g]\,,$$

$$\mathsf{I}[g] \le \big[1 + X(h)\big]\,\mathscr{I}[w] + Y(h)\,\mathsf{F}[g]$$

with $g := (w-1)\,V_{D_*}^{m-1}$. Notice that $h(t) \to 1$ as $t \to \infty$, and that

$$0 < X(h) + 1 = h^{5-2m} \to 1 \quad \text{as } t \to +\infty$$

$$0 < Y(h) = N(1-m)\big[h^{4(2-m)} - 1\big] \to 0 \quad \text{as } t \to +\infty.$$

This is a consequence of convergence without rate, proved before.

Nonlinear entropy method. By the Hardy-Poincaré inequality

$$\mathsf{F}[g] \le \frac{1}{\Lambda_{m,N}}\,\mathsf{I}[g] \le \frac{1}{\Lambda_{m,N}}\Big[(1 + X(h))\,\mathscr{I}[w] + Y(h)\,\mathsf{F}[g]\Big]\,,$$

so we deduce that

$$\mathscr{F}[w] \le \frac{h^{2-m}}{2}\,\mathsf{F}[g] \le \frac{h^{2-m}\big[1 + X(h)\big]}{2\big[\Lambda_{m,N} - Y(h)\big]}\mathscr{I}[w]$$

as soon as $0 < h < h_* := \min\big\{h > 0 : \Lambda_{m,N} - Y(h) \ge 0\big\}$. Moreover,

$$0 \le h - 1 \le C\,\mathscr{F}[w]^{\frac{1-m}{d+2-(d+1)m}}$$

for a suitable constant $C > 0$. Recall that $h \to 1$, $X(h), Y(h) \to 0$ as $t \to \infty$. When t is large, there exists a suitable $\gamma > 0$:

$$\gamma\, \mathscr{F}[w] \le \mathscr{I}[w] = -\frac{\mathrm{d}\mathscr{F}[w]}{\mathrm{d}t} \qquad \Longrightarrow \qquad \mathscr{F}[w(t)] \le \mathscr{F}[w_0]\, \mathrm{e}^{-\gamma t} .$$

That is, for the L^2-norm:

$$\|v - V_{D_*}\|_{L^2}^2 \le \left\| V_{D_*}^{2-m} \right\|_{L^\infty} \int |v - V_{D_*}|^2\, V_{D_*}^{m-2}\, dx = C\, \mathsf{F}[g] \le C \frac{1}{C_0} \mathscr{F}[w] \le \tilde{C}\, \mathrm{e}^{-\gamma t} .$$

Improvement of convergence: First we prove uniform convergence of w to 1 by an interpolation lemma. Letting then $h(t) = 1 + C\,\mathrm{e}^{-\gamma t}$ in the above estimates, we conclude that γ can be improved up to $\Lambda_{m,N}$. This ends the proof. Whole details of these calculations are given in [3, 6]. Notice that the asymptotic calculations are performed at the level of the entropy, its dissipation and their linearizations, all of them integrals of the function and its derivatives, thus avoiding the more difficult analysis of the actual differential equation.

Remark. Asymptotic estimates can be obtained by other methods, see [32], Chapter 7, and [21] for two applications to this fast diffusion model. The class of data to which they apply is different.

9.7 Extensions and Current Work

The application of the entropy method described above to study the asymptotic behaviour of the solutions of other nonlinear evolution problems has been considered by different authors. Progress has been done but the results are less conclusive, and much work is being done at this moment.

Thus, the method has been applied to study the p-Laplacian evolution equation by Del Pino and Dolbeault [19] and then by other authors, but the most difficult range of values of p near 1 are still not researched. Doubly nonlinear equations are treated by Agueh et al. in [1], also in a restricted range of parameters.

There is also partial work on the chemotaxis model system by Carrillo et al. [4].

A new field of application concerns the area of nonlinear diffusion with nonlocal operators, in particular, fractional Laplacian operators. The author has been involved in the application of the entropy method in the model equation

$$u_t + \nabla \cdot (u\, \nabla (-\Delta u)^{-s} u) = 0.$$

where $(-\Delta u)^{-s}$ is the inverse fractional Laplacian operator of index $0 < s < 1$, given by convolution with a kernel of the form $K(x, y) = c|x - y|^{2s-N}$, $N \ge 2$, see [12]. The analysis has been performed in [13], and the whole topic has been surveyed in detail in [35].

Finally, the method does not seem to be so well suited to the study of problems in bounded domains with Dirichlet data, but see [5, 8].

Acknowledgements Work partially supported by Spanish Project MTM2011-24696 (Spain).

References

1. M. Agueh, A. Blanchet, J.A. Carrillo, Large time asymptotics of the doubly nonlinear equation in the non-displacement convexity regime. J. Evol. Equ. **10**(1), 59–84 (2010)
2. A. Blanchet, M. Bonforte, J. Dolbeault, G. Grillo, J.L. Vázquez, Hardy-Poincaré inequalities and applications to nonlinear diffusions. C. R. Math. Acad. Sci. Paris **344**, 431–436 (2007)
3. A. Blanchet, M. Bonforte, J. Dolbeault, G. Grillo, J.L. Vázquez, Asymptotics of the fast diffusion equation via entropy estimates. Arch. Ration. Mech. Anal. **191**, 347–385 (2009)
4. A. Blanchet, E.A. Carlen, J.A. Carrillo, Functional inequalities, thick tails and asymptotics for the critical mass Patlak-Keller-Segel model. J. Funct. Anal. **262**(5), 2142–2230 (2012)
5. T. Bodineau, J.L. Lebowitz, C. Mouhot, C. Villani, Lyapunov functionals for boundary-driven nonlinear drift-diffusions. Preprint arXiv:1305.7405 [math.AP]
6. M. Bonforte, J. Dolbeault, G. Grillo, J.L. Vázquez, Sharp rates of decay of solutions to the nonlinear fast diffusion equation via functional inequalities. Proc. Natl. Acad. Sci. **107**(38), 16459–16464 (2010)
7. M. Bonforte, G. Grillo, J.L. Vázquez, Special fast diffusion with slow asymptotics. Entropy method and flow on a Riemannian manifold. Arch. Ration. Mech. Anal. **196**, 631–680 (2010)
8. M. Bonforte, G. Grillo, J.L. Vázquez, Behaviour near extinction for the fast diffusion equation on bounded domains. J. Math. Pures Appl. **97**, 1–38 (2012)
9. M. Bonforte, J.L. Vázquez, Global positivity estimates and Harnack inequalities for the fast diffusion equation. J. Funct. Anal. **240**, 399–428 (2006)
10. M. Bonforte, J.L. Vázquez, Positivity, local smoothing, and Harnack inequalities for very fast diffusion equations. Adv. Math. **223**, 529–578 (2010)
11. L.A. Caffarelli, *The Obstacle Problem*. Lezioni Fermiane [Fermi Lectures] (Accademia Nazionale dei Lincei/Scuola Normale Superiore, Rome/Pisa, 1998)
12. L.A. Caffarelli, J.L. Vázquez, Nonlinear porous medium flow with fractional potential pressure. Arch. Ration. Mech. Anal. **202**, 537–565 (2011)
13. L.A. Caffarelli, J.L. Vázquez, Asymptotic behaviour of a porous medium equation with fractional diffusion. Discret. Contin. Dyn. Syst. A **29**(4), 1393–1404 (2011)
14. J.A. Carrillo, G. Toscani, Asymptotic L^1-decay of solutions of the porous medium equation to self-similarity. Indiana Univ. Math. J. **49**, 113–141 (2000)
15. J.A. Carrillo, J.L. Vázquez, Fine asymptotics for fast diffusion equations. Commun. Partial Differ. Equ. **28**(5–6), 1023–1056 (2003)
16. J.A. Carrillo, A. Jüngel, P.A. Markowich, G. Toscani, A. Unterreiter, Entropy dissipation methods for degenerate parabolic problems and generalized Sobolev inequalities. Monatsh. Math. **133**(1), 1–82 (2001)
17. P. Daskalopoulos, N. Sesum, On the extinction profile of solutions to fast diffusion. J. Reine Angew. Math. **622**, 95–119 (2008)
18. M. Del Pino, J. Dolbeault, Best constants for Gagliardo-Nirenberg inequalities and applications to nonlinear diffusions. J. Math. Pures Appl. (9) **81**(9), 847–875 (2002)
19. M. Del Pino, J. Dolbeault, Asymptotic behavior of nonlinear diffusions. Math. Res. Lett. **10**(4), 551–557 (2003)
20. J. Denzler, R.J. McCann, Fast diffusion to self-similarity: complete spectrum, long-time asymptotics, and numerology. Arch. Ration. Mech. Anal. **175**(3), 301–342 (2005)

21. M. Fila, J.L. Vázquez, M. Winkler, E. Yanagida, Rate of convergence to Barenblatt profiles for the fast diffusion equation. Arch. Ration. Mech. Anal. **204**(2), 599–625 (2012)
22. A. Friedman, S. Kamin, The asymptotic behavior of gas in an N-dimensional porous medium. Trans. Am. Math. Soc. **262**, 551–563 (1980)
23. L. Gross, Logarithmic Sobolev inequalities. Am. J. Math. **97**(4), 1061–1083 (1975)
24. S. Kamin, J.L. Vázquez, Fundamental solutions and asymptotic behaviour for the p-Laplacian equation. Rev. Mat. Iberoam. **4**(2), 339–354 (1988)
25. S. Kamin, J.L. Vázquez, Asymptotic behaviour of solutions of the porous medium equation with changing sign. SIAM J. Math. Anal. **22**(1), 34–45 (1991)
26. O.A. Ladyzhenskaya, N.N. Uraltseva, *Linear and Quasilinear Equations of Elliptic Type*, Moscow (1964) [in Russian] (Academic, New York, 1968). MR 0244627 (39:5941)
27. O.A. Ladyzhenskaya, V.A. Solonnikov, N.N. Ural'tseva, *Linear and Quasilinear Equations of Parabolic Type*. Translations of Mathematical Monographs, vol. 23 (American Mathematical Society, Providence, 1968)
28. J. Nash, Continuity of solutions of elliptic and parabolic equations. Am. J. Math. **80**(4), 931–954 (1958)
29. L.C.G. Rogers, D. Williams, *Diffusions, Markov Processes, and Martingales. Vol.1. Foundations and Vol. 2. Ito Calculus* (Cambridge University Press, Cambridge, 2000). Reprint of the second (1994) edition
30. S.R.S. Varadhan, *Lectures on Diffusion Problems and Partial Differential Equations*. Tata Institute of Fundamental Research Lectures on Mathematics and Physics, vol. 64 (Tata Institute of Fundamental Research, Bombay, 1980). MR0607678 (83j:60087)
31. J.L. Vázquez, Asymptotic behaviour for the porous medium equation posed in the whole space. J. Evol. Equ. **3**, 67–118 (2003)
32. J.L. Vázquez, *Smoothing and Decay Estimates for Nonlinear Parabolic Equations. Equations of Porous Medium Type*. Oxford Lecture Series in Mathematics and Its Applications, vol. 33 (Oxford University Press, Oxford, 2006)
33. J.L. Vázquez, *The Porous Medium Equation. Mathematical Theory*. Oxford Mathematical Monographs (Oxford University Press, Oxford, 2007)
34. J.L. Vázquez, *Perspectives in Nonlinear Diffusion: Between Analysis, Physics and Geometry*. International Congress of Mathematicians, vol. I (European Mathematical Society, Zürich, 2007), pp. 609–634
35. J.L. Vázquez, Nonlinear diffusion with fractional Laplacian operators, in *Nonlinear Partial Differential Equations: The Abel Symposium 2010*, ed. by H. Holden, K.H. Karlsen (Springer, Berlin/Heidelberg, 2012), pp. 271–298

Chapter 10
Challenges in Geometric Numerical Integration

Ernst Hairer

Abstract Geometric Numerical Integration is a subfield of the numerical treatment of differential equations. It deals with the design and analysis of algorithms that preserve the structure of the analytic flow. The present review discusses numerical integrators, which nearly preserve the energy of Hamiltonian systems over long times. Backward error analysis gives important insight in the situation, where the product of the step size with the highest frequency is small. Modulated Fourier expansions permit to treat nonlinearly perturbed fast oscillators. A big challenge that remains is to get insight into the long-time behavior of numerical integrators for fully nonlinear oscillatory problems, where the product of the step size with the highest frequency is not small.

10.1 Geometric Numerical Integration

Ordinary differential equations arise everywhere in science and their numerical treatment is of great importance. The development took place in three periods: the numerical solution of non-stiff differential equations started in the end of the nineteenth century, whereas that of stiff problems began in the middle of the twentieth century and was motivated by space discretizations of parabolic differential equations and by simulations of chemical reactions. With the interest in computations over long time intervals one discovered that certain methods reproduce the qualitative behavior of the exact flow much better than others. In the late 1980s numerical analysts started to design and study (we quote from the preface of the monograph [9])

E. Hairer (✉)
Sect. de mathématiques, Univ. de Genève, 2-4 rue du Lièvre, CH-1211 Genève 4, Switzerland
e-mail: Ernst.Hairer@unige.ch

V. Ancona and E. Strickland (eds.), *Trends in Contemporary Mathematics*, Springer INdAM Series 8, DOI 10.1007/978-3-319-05254-0_10,

> ... numerical methods that preserve geometric properties of the flow of a differential equation: symplectic integrators for Hamiltonian systems, symmetric integrators for reversible systems, ... and methods for problems with highly oscillatory solutions.

This period is called "Geometric Numerical Integration" and much research has been devoted to this topic during the last decades. Special attention has been paid to the long-time integration of Hamiltonian systems and, in particular, to simulations in astronomy (planetary motion) and in molecular dynamics. The present work focuses on algorithms that nearly preserve energies (total and oscillatory) over long time intervals. Our main interest is in getting theoretical insight into their long-time behavior. We distinguish between three degrees of difficulty:

- *Non oscillatory Hamiltonian systems:* the term 'non oscillatory' means that the product of the highest frequency in the system with the time step size of the numerical integrator is small. In this situation *backward error analysis* gives much insight into the long-time behavior of numerical solutions.
- *Nonlinearly perturbed fast oscillators:* if several high frequency harmonic oscillators are nonlinearly coupled, then the technique of *modulated Fourier expansions* yields much information on the numerical preservation of total and oscillatory energies over long times.
- *Fully nonlinear, highly oscillatory Hamiltonian problems:* this is the situation where high frequencies stem from a nonlinear part in the problem, and the numerical integrator is applied, such that the product of the time step size with the highest frequency is not small. A good understanding of the long-time behavior (e.g., near energy preservation) of numerical solutions is still missing.

In the following sections each of these types of problems is treated individually.

10.2 Hamiltonian Systems: Backward Error Analysis

We start by considering Hamiltonian systems

$$\dot{p} = -\nabla_q H(p,q), \qquad \dot{q} = \nabla_p H(p,q),$$

where $H(p,q)$ is a smooth scalar function (called total energy) of position variables $q \in \mathbb{R}^d$ and momenta $p \in \mathbb{R}^d$. A characteristic property of such systems is that the exact flow is *symplectic*, i.e., the derivative of the flow map φ_t with respect to initial values satisfies $(\varphi_t')^{\mathsf{T}} J \varphi_t' = J$ for the canonical structure matrix J (see [9, Chapter VI] for more details). Moreover, the total energy $H(p,q)$ is preserved along the exact flow of the system. Here, we are interested to which extent the total energy can also be preserved by a numerical solution.

Numerical Experiment. As a representative example of numerical integrators we consider the *explicit Euler method*

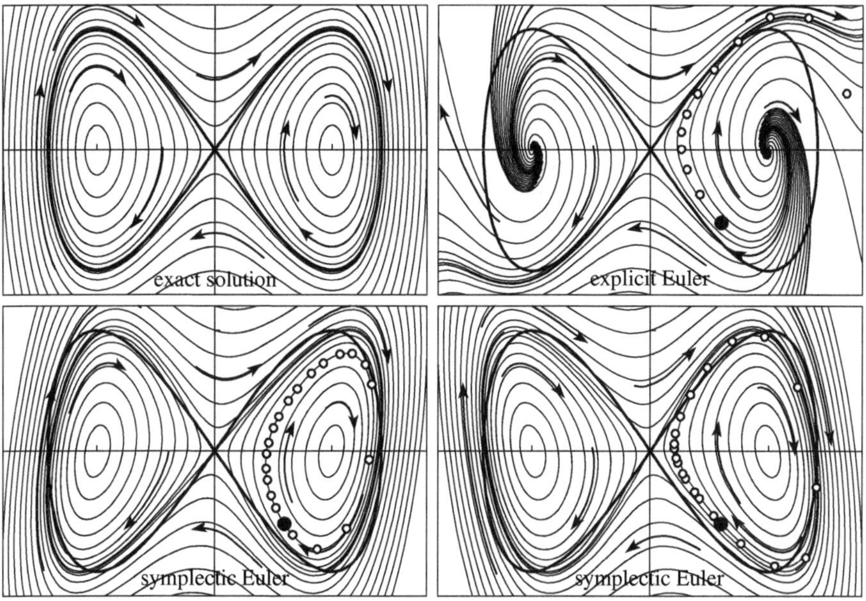

Fig. 10.1 Illustration of backward error analysis for the Hamiltonian $H(p,q) = \frac{1}{2}p^2 + U(q)$ with the double well potential $U(q) = \frac{1}{2}(q^2 - 1)^2$. The numerical solution for an initial value, indicated by a *big bullet*, and solutions of the modified differential equation are shown in the (q, p) phase space

$$\begin{aligned} p_{n+1} &= p_n - h\nabla_q H(p_n, q_n) \\ q_{n+1} &= q_n + h\nabla_p H(p_n, q_n). \end{aligned}$$

It provides approximations (p_n, q_n) to the solution of the Hamiltonian system at time $t = nh$. This method and most of the classical Runge–Kutta and multistep methods are not suited for the long-time integration of Hamiltonian systems. Even for very simple problems a linear drift in the energy $H(p_n, q_n)$ along their numerical solution can be observed.

Already very early de Vogelaere [5] noticed that each of the so-called *symplectic Euler methods*

$$\begin{aligned} p_{n+1} &= p_n - h\nabla_q H(p_{n+1}, q_n) \qquad & p_{n+1} &= p_n - h\nabla_q H(p_n, q_{n+1}) \\ q_{n+1} &= q_n + h\nabla_p H(p_{n+1}, q_n) \qquad & q_{n+1} &= q_n + h\nabla_p H(p_n, q_{n+1}) \end{aligned}$$

has a much better long-time behaviour. This can be seen in the experiment of Fig. 10.1, where the explicit and both symplectic Euler methods are applied to a simple Hamiltonian system (with step size $h = 0.3$). Whereas the numerical solution of the explicit Euler method spirals outwards and gives a qualitatively wrong approximation to the exact flow, those of the symplectic Euler methods remain apparently on a closed curve.

Backward Error Analysis for the Example of Fig. 10.1. The idea of backward error analysis is the following: for a given numerical integrator search a modified differential equation, such that the exact solution of this modified equation approximates very well the numerical solution. An analysis of the modified differential equation then gives much insight into the numerical flow. The construction of the modified differential equation is straight-forward. One makes an ansatz as a formal series in powers of the step size h, inserts its solution into the numerical method, and compares like powers of h. For a problem of the form $\dot q = p,\ \dot p = -U'(q)$ the explicit Euler method yields

$$\begin{pmatrix}\dot q\\ \dot p\end{pmatrix} = \begin{pmatrix}p\\ -U'(q)\end{pmatrix} + \frac{h}{2}\begin{pmatrix}U'(q)\\ U''(q)p\end{pmatrix} + \frac{h^2}{4}\begin{pmatrix}-2\,U''(q)p\\ 2\,U'(q)U''(q) - U'''(q)p^2\end{pmatrix} + \dots$$

and the symplectic Euler method (explicit in q, implicit in p) gives

$$\begin{pmatrix}\dot q\\ \dot p\end{pmatrix} = \begin{pmatrix}p\\ -U'(q)\end{pmatrix} + \frac{h}{2}\begin{pmatrix}-U'(q)\\ U''(q)p\end{pmatrix} + \frac{h^2}{12}\begin{pmatrix}2\,U''(q)p\\ -2\,U'(q)U''(q) - U'''(q)p^2\end{pmatrix} + \dots .$$

In Fig. 10.1, solutions of the truncated modified differential equation (with $h = 0.3$) are included. The accordance with the numerical solution is striking. For the double well potential $U(q) = \frac{1}{2}(q^2-1)^2$ the stationary points $(q,p) = (\pm 1, 0)$ turn into a spiral source for the modified equation of the explicit Euler method, whereas they remain stable centers for the symplectic Euler method. The modified equation is a Hamiltonian system only for the symplectic Euler method, in which case the Hamiltonian is given by $H_h(p,q) = \frac{1}{2}p^2 + U(q) - \frac{h}{2}U'(q)p + \frac{h^2}{12}\big(U'(q)^2 + U''(q)p^2\big) + \dots$.

Backward Error Analysis – General Situation. Consider an ordinary differential equation $\dot y = f(y)$ and a numerical integrator $y_{n+1} = \Phi_h(y_n)$. If both, the vector field f and the discrete flow Φ_h are sufficiently differentiable, one can find a (truncated) modified equation

$$\dot y = f(y) + hf_2(y) + h^2 f_3(y) + \dots + h^{N-1} f_N(y),$$

such that

$$\|\Phi_h(y) - \varphi_{N,h}(y)\| \le C(y)\, h^{N+1},$$

where $\varphi_{N,t}$ is the exact flow of the truncated modified equation. This means that the numerical solution after one step, $y_1 = \Phi_h(y_0)$, is very close to the exact solution of the modified equation at time $t = h$ corresponding to the initial value y_0. The constant $C(y)$ depends on the truncation index and on bounds for derivatives of the vector field, but it is independent of h. If the vector field is real analytic and the integrator falls into the class of (partitioned) Runge–Kutta methods then, by choosing N proportional to h^{-1}, one can prove the estimate (see [3] and [7])

$$\|\Phi_h(y) - \varphi_{N,h}(y)\| \leq h\,\alpha\,\mathrm{e}^{-\gamma/(\omega h)}$$

with constants that are independent of N and h. Here, $\gamma > 0$ only depends on the numerical integrator and $\omega > 0$ is related to the highest frequency present in the solution of the differential equation.

One of the most important applications of backward error analysis is the long-time energy preservation of symplectic integrators (see [9, Chapter IX]). In fact, if the vector field is Hamiltonian (i.e., $f(y) = J^{-1}\nabla H(y)$) and if the discrete flow Φ_h is a symplectic transformation, then the modified differential equation is also Hamiltonian with

$$H_h(y) = H(y) + hH_2(y) + h^2 H_3(y) + \ldots + h^{N-1} H_N(y).$$

This implies that $H_h(y)$ is exactly preserved along the flow $\varphi_{N,t}$ of the modified equation, and consequently $\|H_h(\Phi_h(y)) - H_h(y)\| \leq c\,h\,\mathrm{e}^{-\gamma/(\omega h)}$. Telescoping summation gives $\|H_h(y_n) - H_h(y_0)\| \leq c\,nh\,\mathrm{e}^{-\gamma/(\omega h)}$ and, since for a method of order r we have $H_j(y) = 0$ for $j = 2, \ldots, r$, this implies that

$$\|H(y_n) - H(y_0)\| \leq C\,h^r \qquad \text{for} \qquad nh \leq \mathrm{e}^{\gamma/(2\omega h)}.$$

Consequently, for symplectic (Runge–Kutta) methods the Hamiltonian is preserved up to an error of size $\mathcal{O}(h^r)$ on exponentially long time intervals.

10.3 Perturbed Fast Oscillators: Modulated Fourier Expansions

Whenever applicable, backward error analysis is an excellent tool for getting insight into the long-time behaviour of numerical solutions. The disadvantage is that for situations, where the product of the step size h with the highest frequency ω is not small, it does not give any information. In this section we consider nonlinearly perturbed harmonic oscillators of the form

$$\ddot{q}_j + \omega_j^2 q_j = -\nabla_j U(\mathbf{q}), \qquad j = 0, 1, \ldots, m,$$

where $\mathbf{q} = (q_0, q_1, \ldots, q_m)$ with $q_j \in \mathbb{R}^{d_j}$, and ∇_j denotes the partial derivative with respect to q_j. We assume $\omega_0 = 0$ and

$$\omega_j \geq \varepsilon^{-1}, \qquad 0 < \varepsilon \ll 1, \qquad j = 1, \ldots, m.$$

This system is Hamiltonian with energy

$$H(\mathbf{q}, \dot{\mathbf{q}}) = \frac{1}{2} \sum_{j=0}^{m} \left(\dot{q}_j^{\mathsf{T}} \dot{q}_j + \omega_j^2 q_j^{\mathsf{T}} q_j \right) + U(\mathbf{q}).$$

With the notation Ω for the diagonal matrix with entries ω_j, and ∇U for the vector that collects all $\nabla_j U$, the differential equation can be written as $\ddot{\mathbf{q}} = -\Omega^2 \mathbf{q} - \nabla U(\mathbf{q})$.

It turns out that the study of the near energy preservation of numerical integrators requires the consideration of the *oscillatory energy*

$$H_\omega(\mathbf{q}, \dot{\mathbf{q}}) = \frac{1}{2} \sum_{j=1}^{m} \left(\dot{q}_j^{\mathsf{T}} \dot{q}_j + \omega_j^2 q_j^{\mathsf{T}} q_j \right),$$

and it is essential to assume that $H_\omega(\mathbf{q}(0), \dot{\mathbf{q}}(0)) \leq E$ is bounded independently of ε. The oscillatory energy is then nearly preserved along the analytic solution of the differential equation over long times [6].

Numerical Experiment with the Störmer–Verlet Method. We consider a chain with alternating soft nonlinear and stiff linear springs as described in [9, Section I.5]. It is of the above form with $m = 1$, $d_0 = d_1 = 3$, and has a quartic potential U. For our experiment we choose $\omega_1 = \omega = 50$. As numerical integrator we consider the *Störmer–Verlet method*

$$\begin{aligned} \mathbf{q}_{n+1} - 2\mathbf{q}_n + \mathbf{q}_{n-1} &= h^2 \big(- \Omega^2 \mathbf{q}_n - \nabla U(\mathbf{q}_n) \big) \\ 2h\, \dot{\mathbf{q}}_n &= \mathbf{q}_{n+1} - \mathbf{q}_{n-1}, \end{aligned}$$

which is frequently used in molecular dynamics simulations. Considered as a mapping $(\mathbf{q}_n, \dot{\mathbf{q}}_n) \mapsto (\mathbf{q}_{n+1}, \dot{\mathbf{q}}_{n+1})$ it is symplectic, symmetric, and of order 2, and it is perfectly suited for computations requiring low accuracy. Note that stable numerical solutions (for the harmonic oscillator $\ddot{q} + \omega^2 q = 0$) are obtained only under the step size restriction $h\omega < 2$.

Figure 10.2 shows the error in the energy of the Störmer–Verlet scheme applied to the above mentioned problem of alternating soft and stiff springs. Even for a very large step size $h = 1.95/\omega$ the error (although very large) remains bounded without any drift. For more reasonable step sizes $h\omega \in \{1, 0.5, 0.25\}$ the error in the energy behaves as expected. There is no drift and the bound on the error decreases when the step size becomes smaller. This excellent long-time behaviour cannot be explained with the techniques of the previous section.

A Theorem on the Numerical Energy Preservation. To simplify the notation we present a result for the case of only one high frequency ($m = 1$), and we denote $\omega = \omega_1$. For the numerical solution, obtained by the Störmer–Verlet method, we assume that the step size satisfies $0 < c_0 \leq h\omega \leq c_1 < 2$ and that the numerical non-resonance condition

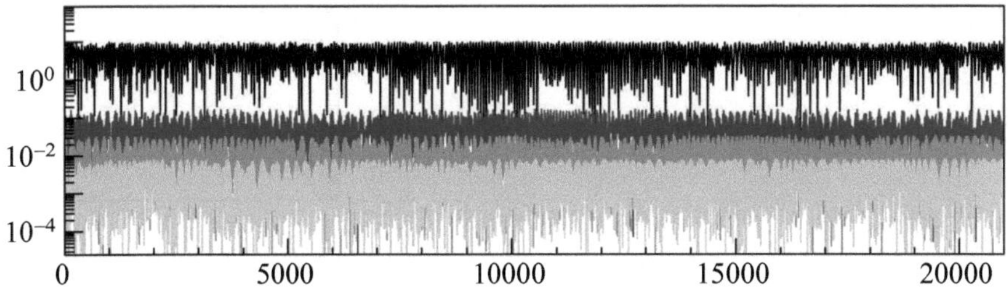

Fig. 10.2 Numerical energy error of the Störmer–Verlet method applied to a nonlinearly perturbed fast harmonic oscillator as function of time. The step sizes are such that $h\omega = 1.95$ (*black*) and $h\omega = 1, 0.5, 0.25$ (different grades of *gray*)

$$\left|\sin(\tfrac{1}{2}kh\tilde{\omega})\right| \geq c\sqrt{h}, \qquad k = 1, \ldots, N$$

holds for some $N \geq 2$ and $c > 0$, where $h\tilde{\omega}$ is defined by the relation $\sin(\frac{1}{2}h\tilde{\omega}) = \frac{1}{2}h\omega$. The latter condition is in fact a definition of N. We further suppose that the numerical solution stays in a region on which all derivatives of U are bounded. With the modified energies

$$\begin{aligned} H^*(\mathbf{q}, \dot{\mathbf{q}}) &= H(\mathbf{q}, \dot{\mathbf{q}}) + \tfrac{1}{2}\gamma(h\omega)\|\dot{q}_1\|^2 \\ H^*_\omega(\mathbf{q}, \dot{\mathbf{q}}) &= H_\omega(\mathbf{q}, \dot{\mathbf{q}}) + \tfrac{1}{2}\gamma(h\omega)\|\dot{q}_1\|^2, \end{aligned}$$

where $\gamma(h\omega)$ is given in Fig. 10.3, it then holds that

$$\left.\begin{aligned} H^*(\mathbf{q}_n, \dot{\mathbf{q}}_n) &= H^*(\mathbf{q}_0, \dot{\mathbf{q}}_0) + \mathcal{O}(h) \\ H^*_\omega(\mathbf{q}_n, \dot{\mathbf{q}}_n) &= H^*_\omega(\mathbf{q}_0, \dot{\mathbf{q}}_0) + \mathcal{O}(h) \end{aligned}\right. \qquad \text{for} \qquad 0 \leq nh \leq h^{-N+1}.$$

The constants symbolised by $\mathcal{O}$ are independent of n, h, ω under the above conditions. Along the numerical solution the expression $\|\dot{q}_1\|^2$ is highly oscillatory and of size $\mathcal{O}(1)$. Moreover, its time average over intervals of length T is nearly constant:

$$\frac{h}{T}\sum_{|jh|\leq T} \|\dot{q}_{n+j,1}\|^2 = \frac{1}{1+\gamma(h\omega)} H^*_\omega(\mathbf{q}_0, \dot{\mathbf{q}}_0) + \mathcal{O}(h).$$

This result is taken from [9, Chapter XIII.8]. The long-time behaviour of Fig. 10.2 can now be explained. Due to the near preservation of the modified energies, the dominant error term comes from $\frac{1}{2}\gamma(h\omega)\|\dot{q}_1\|^2$. For $h\omega = 1.95$ we have $\frac{1}{2}\gamma(h\omega) \approx 10$, which explains the large (but bounded) energy error for this particular step size. For the step sizes, for which $h\omega \in \{1, 0.5, 0.25\}$, this expression is much smaller (see Fig. 10.3) and precisely corresponds to the observation of Fig. 10.2.

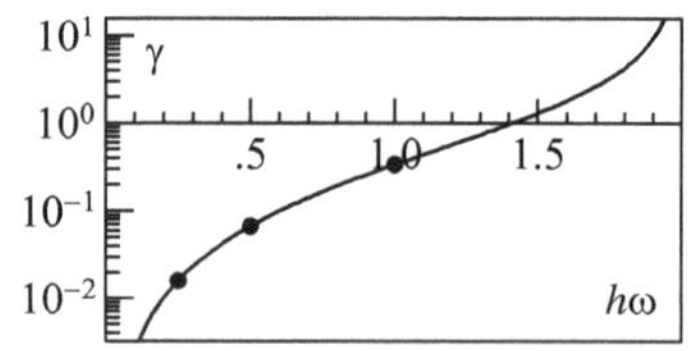

Fig. 10.3 The function $\gamma(h\omega)$ appearing in the modified energies:

$$\gamma(h\omega) = \frac{\frac{1}{4}(h\omega)^2}{1-\frac{1}{4}(h\omega)^2}$$

Idea of the Proof. A detailed proof of the above result can be found in [9, Chapter XIII.8]. An extension to the multi-frequency case is presented in [4]. We shortly mention here the two main ingredients (for the multi-frequency case):

- *Modifying the frequencies.* For vanishing potential U the Störmer–Verlet discretization reduces to a linear three-term recursion with exact solution $q_{n,j} = c_{j,1}\,\mathrm{e}^{\mathrm{i}\tilde{\omega}_j nh} + c_{j,2}\,\mathrm{e}^{-\mathrm{i}\tilde{\omega}_j nh}$, where the modified frequencies $\tilde{\omega}_j$ are given by

$$1 - \frac{(h\omega)^2}{2} = \cos\big(h\tilde{\omega}\big) \qquad \text{or, equivalently,} \qquad \sin\Big(\frac{h\tilde{\omega}}{2}\Big) = \frac{h\omega}{2}.$$

  The Störmer–Verlet scheme thus becomes the trigonometric integrator

$$\mathbf{q}_{n+1} - 2\cos\big(h\tilde{\Omega}\big)\mathbf{q}_n + \mathbf{q}_{n-1} = -h^2\nabla U(\mathbf{q}_n).$$

  which is easier to analyse, because the linear part exactly integrates a harmonic oscillator with modified frequencies.
- *Modulated Fourier expansion.* For the problem of this section the Störmer–Verlet method is a nonlinear perturbation of a three-term relation, for which the solution is a linear combination of exponentials $\mathrm{e}^{\pm\mathrm{i}\tilde{\omega}_j nh}$. It is therefore natural to approximate the numerical solution $\mathbf{q}_n$ of the complete discretisation as a linear combination of products of $\mathrm{e}^{\pm\mathrm{i}\tilde{\omega}_j nh}$ (called *modulated Fourier expansion*)

$$\mathbf{q}_n = \mathbf{y}(t) + \sum_{\mathbf{k}\in\mathcal{K}} \mathbf{z}^{\mathbf{k}}(t)\mathrm{e}^{\mathrm{i}(\mathbf{k}\cdot\tilde{\boldsymbol{\omega}})t} \qquad \text{with} \qquad t = nh.$$

  Here, $\mathbf{k} = (k_1, \ldots, k_n)$ is a multi-index, $\tilde{\boldsymbol{\omega}} = (\tilde{\omega}_1, \ldots, \tilde{\omega}_n)$ is the vector of modified high frequencies, $\mathbf{k}\cdot\tilde{\boldsymbol{\omega}} = k_1\tilde{\omega}_1 + \ldots + k_n\tilde{\omega}_n$, and $\mathcal{K}$ is a suitable finite index set. The coefficient functions $\mathbf{y}(t)$ and $\mathbf{z}^{\mathbf{k}}(t)$ are vector-valued with the same dimension and partitioning as $\mathbf{q}_n$, and they are assumed to be smooth. This means that together with all their derivatives they are bounded independently of ε for $0 < \varepsilon \le \varepsilon_0$ so that with this ansatz high oscillations are well separated from the slow motion. The proof is typically in three steps:

  - Construction of the modulation functions $\mathbf{y}(t)$ and $\mathbf{z}^{\mathbf{k}}(t)$ as the solution of a differential-algebraic system (on short intervals of length $\mathcal{O}(1)$).
  - Proof of the existence of formal invariants of the differential-algebraic system, which are close to the total and oscillatory energies (on short intervals).

– Concatenation of estimates on short intervals to get the near energy preservation of long time intervals.

Modulated Fourier expansions for the long-term analysis of (analytical and numerical) solutions of highly oscillatory differential equations have been introduced in [8] for the case of a single high frequency ω. The case of several high frequencies satisfying a non-resonance condition is studied in [4]. They are extensively treated in Chapters XIII and XIV of the monograph [9]. Related results for the analytic solution have been obtained in [1, 2] with canonical transformation techniques of Hamiltonian perturbation theory.

10.4 Fully Nonlinear, Highly Oscillatory Hamiltonian Problems

What happens, when neither backward error analysis can be applied nor the problem can be cast into the form of a perturbed fast oscillator? Let us consider a molecular dynamics model, which consists of an N-body problem

$$\ddot{\mathbf{q}} = -\nabla U(\mathbf{q}), \qquad U(\mathbf{q}) = \sum_{i=2}^{N} \sum_{j=1}^{i-1} V(\|q_i - q_j\|)$$

interacting with the Lennard–Jones potential $V(r) = r^{-12} - 2r^{-6}$. We assume particles $q_i \in \mathbb{R}^2$ in a plane, and an initial configuration consisting of $N = 100$ particles which are at randomly perturbed points of the lattice $\{(l,m)\,;\, l,m = 1,\ldots,10\}$. Initial velocities are taken to be zero.

Figure 10.4 shows the error in the energy of the Störmer–Verlet method applied to the N-body problem. For the step size $h = 0.06$ the error increases and is soon out of scale. However, for $h \leq 0.04$ the energy is well preserved, and the error decreases as expected, when the step size becomes smaller.

To check whether this behaviour can be explained either by the backward error analysis of Sect. 10.2 or by the technique of modulated Fourier expansions of Sect. 10.3 we have to compute the highest frequencies of the solutions in the system. Along the numerical solution we have computed the Hessian matrix $\nabla^2 U(\mathbf{q})$ and its eigenvalues. The dominant eigenvalues are negative and they represent $-\omega^2$, where ω corresponds to the frequencies in the system. In Fig. 10.5 we present the five largest frequencies obtained in this way. We see that they are slightly smaller than the value $\omega = 25$.

– Backward error analysis does not give any information on the long-time behaviour. The reason is that for the step sizes used in the experiment of Fig. 10.4 we have $h\omega = 1.5$ (for $h = 0.06$), and $h\omega = 1$ (for $h = 0.04$), which are not small.

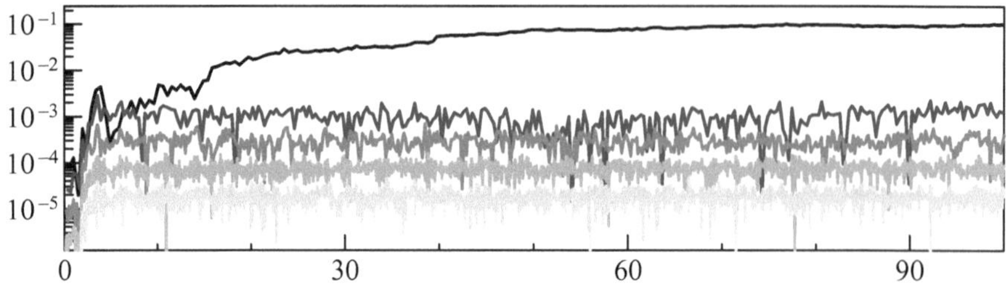

Fig. 10.4 Numerical energy error of the Störmer–Verlet method applied to the molecular dynamics model of Sect. 10.4. The step sizes are $h = 0.06$ (*black*) and $h = 0.04, 0.02, 0.01, 0.005$ (different grades of *gray*)

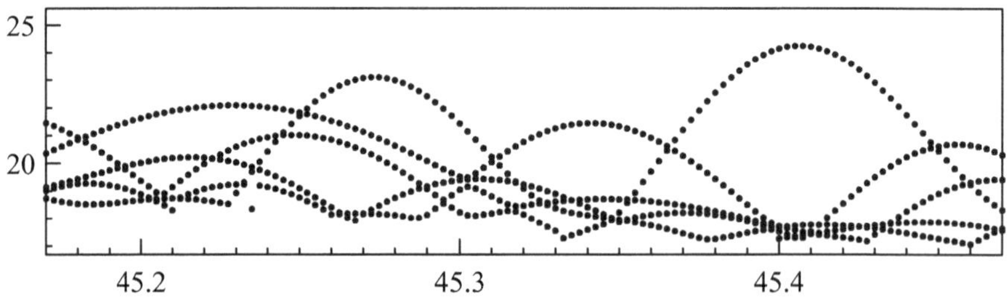

Fig. 10.5 The five largest frequencies corresponding to the solution of the N-body problem of Sect. 10.4 as a function of time

– The technique of modulated Fourier expansions requires that the differential equation is a perturbation of a set of harmonic oscillators. This means that the large frequencies of the system have to be nearly constant. Figure 10.5 shows that this is by far not the case.

An explanation of the good energy preservation observed in Fig. 10.4 is still missing.

Acknowledgements This review is an update of a talk given at the "giornata INdAM" in June 2007 (Pisa). We acknowledge the support over many years of the Fonds National Suisse, Project No. 200020-144313/1.

References

1. G. Benettin, L. Galgani, A. Giorgilli, Realization of holonomic constraints and freezing of high frequency degrees of freedom in the light of classical perturbation theory. Part I. Commun. Math. Phys. **113**, 87–103 (1987)
2. G. Benettin, L. Galgani, A. Giorgilli, Realization of holonomic constraints and freezing of high frequency degrees of freedom in the light of classical perturbation theory. Part II. Commun. Math. Phys. **121**, 557–601 (1989)
3. G. Benettin, A. Giorgilli, On the Hamiltonian interpolation of near to the identity symplectic mappings with application to symplectic integration algorithms. J. Stat. Phys. **74**, 1117–1143 (1994)

4. D. Cohen, E. Hairer, C. Lubich, Numerical energy conservation for multi-frequency oscillatory differential equations. BIT **45**, 287–305 (2005)
5. R. de Vogelaere, Methods of integration which preserve the contact transformation property of the Hamiltonian equations. Technical report, Department of Mathematics, University of Notre Dame, Notre Dame, 1956
6. L. Gauckler, E. Hairer, C. Lubich, Energy separation in oscillatory Hamiltonian systems without any non-resonance condition. Commun. Math. Phys. **321**, 803–815 (2013)
7. E. Hairer, C. Lubich, The life-span of backward error analysis for numerical integrators. Numer. Math. **76**, 441–462 (1997). Erratum: http://www.unige.ch/math/folks/hairer/
8. E. Hairer, C. Lubich, Long-time energy conservation of numerical methods for oscillatory differential equations. SIAM J. Numer. Anal. **38**, 414–441 (2000)
9. E. Hairer, C. Lubich, G. Wanner, *Geometric Numerical Integration. Structure-Preserving Algorithms for Ordinary Differential Equations*. Springer Series in Computational Mathematics, vol. 31, 2nd edn. (Springer, Berlin, 2006)

Chapter 11
Integral Hodge Classes, Decompositions of the Diagonal, and Rationality Questions

Claire Voisin

Abstract This paper is a short survey of classical and recent results on rationality or stable rationality for smooth projective varieties. We describe (or reinterpret in a geometric form) cohomological or Chow theoretic criteria for stable rationality, like the triviality of unramified cohomology or the universal triviality of the Chow group of zero cycles, and show that they are effective on examples.

11.1 Introduction

We consider smooth projective varieties X over $\mathbb{C}$. The simplest such variety is of course the projective space $\mathbb{P}^n$ itself. However one can consider the following three variants, the first of which was introduced in [18]:

Definition 11.1. A smooth projective variety X over $\mathbb{C}$ is rationally connected if for any two points $x, y \in X$, there is a curve $C \cong \mathbb{P}^1 \subset X$ passing through x and y.

The following definitions are much more classical:

Definition 11.2. A variety X as above is unirational if there is a dominant rational map

$$\phi : \mathbb{P}^n \dashrightarrow X.$$

Dominant means that after resolution of indeterminacies, the proper morphism $\tilde{\phi} : \widetilde{\mathbb{P}^n} \to X$ one gets is surjective. Note that by replacing $\mathbb{P}^n$ by a $\mathbb{P}^k \subset \mathbb{P}^n$ in

C. Voisin (✉)
CNRS, Centre de mathématiques Laurent Schwartz, École Polytechnique,
91128 Palaiseau Cédex, France
e-mail: voisin@math.polytechnique.fr

V. Ancona and E. Strickland (eds.), *Trends in Contemporary Mathematics*,
Springer INdAM Series 8, DOI 10.1007/978-3-319-05254-0_11,

general position, where $k = \dim X$, one can assume $n = \dim X$ in the definition above.

Definition 11.3. A variety X as above is rational if there is a birational map

$$\phi : \mathbb{P}^n \dashrightarrow X,$$

which means that after resolving the indeterminacies of ϕ, the morphism $\tilde{\phi}$ one gets has degree 1.

Obviously

$$\text{rational} \Rightarrow \text{unirational} \Rightarrow \text{rationally connected}.$$

In dimension ≤ 2, it is known that the three notions are equivalent. In higher dimension, a completely open problem is whether rationally connected varieties are necessarily unirational (the general belief is that the answer to this question is negative). It is known since the 1970s that starting from dimension 3, there exist unirational varieties which are not rational. A number of different criteria have been found which allowed to prove irrationality of smooth cubic threefolds in $\mathbb{P}^4$ [8], of certain desingularized nodal quartic double solids [1], and of quartic threefolds in $\mathbb{P}^4$ and more generally, hypersurfaces of degree n in $\mathbb{P}^n$ (see [11,17,21]). The Clemens-Griffiths criterion works only in dimension 3 but it applies to many classes of rationally connected threefolds, showing their irrationality. It involves the geometry of their intermediate Jacobian, which is a principally polarized abelian variety. The Iskovskikh-Manin method is based on birational geometry: They prove that the considered variety has a very small birational automorphisms group, which of course prevents rationality. The Artin-Mumford criterion does not apply to many rationally connected threefolds, but it has the advantage that unlike the other methods, it allows to detect as well varieties which are not stably rational, the definition being as follows:

Definition 11.4. A variety X as above is stably rational if there is a birational map

$$\phi : \mathbb{P}^n \dashrightarrow X \times \mathbb{P}^k.$$

It was proved in [4] that there exist irrational stably rational threefolds, so the last definition is again weaker than rationality, although it is stronger than unirationality.

These beautiful results can be considered as closing the subject of rationality. However, there is one major unsolved issue, which justifies the search for further birational invariants:

Question 11.1. *Are rationality or unirationality deformation invariant properties?*

Note that there are two aspects to this question, namely whether the central fiber of a family with rational general member should be rational (see [12] for the 3-dimensional case) and secondly whether the general member should be rational, assuming the central fiber (supposed smooth) is. A challenging example is the case of cubic fourfolds; some of them are rational but it is conjectured that the general one is not (see [15, 16]).

The purpose of this note is to describe recent results obtained in [10, 25, 28], and [29], translating into geometric terms birational invariants called unramified cohomology groups, introduced and developed from a K-theoretic point of view in [9], which allows in some cases their computations. Note that the first of these invariants was the Artin-Mumford invariant [1], which is purely topological. In higher degree, the invariants one gets should depend on the complex structure, that is, should not be deformation invariant, although no rationally connected example of this phenomenon has been exhibited yet. More precisely, we will see that they can be computed in some cases via Chow groups and cycle class map. Here we use the following definition:

Definition 11.5. Let X be a smooth variety defined over a field K. We define $\mathrm{CH}^k(X) = \mathrm{CH}_{n-k}(X)$, $n = \dim X$ as the group $\mathcal{Z}^k(X)/\mathcal{Z}^k(X)_{rat}$ of codimension k cycles of X defined over K, modulo the subgroup of cycles rationally equivalent to 0, which is generated by the cycles $\operatorname{div}\phi$, where ϕ is a nonzero rational function on an irreducible subvariety $W \subset X$ of of codimension k, everything being defined over K.

When $K = \mathbb{C}$, we have the cycle class $cl : \mathrm{CH}^k(X) \to H^{2k}(X, \mathbb{Z})$ and its Deligne refinement which takes into account the Abel-Jacobi invariant. The Chow groups depend on the complex structure, as one can see by considering the cycle class map: indeed, it takes value in the set of Hodge classes on X, (conjecturally, after passing to $\mathbb{Q}$ coefficients, its image is equal to the set of Hodge classes). On the other hand, the set of Hodge classes is not deformation invariant as there is in general some variation of the Hodge structure on the cohomology of the considered variety when we deform it. A typical example is that of a cubic fourfold in $\mathbb{P}^5$. The very general one has no Hodge class of degree 4 except for the multiples of the class h^2, $h = c_1(\mathcal{O}_X(1))$, but some special cubic fourfolds have extra Hodge classes (for example cubic fourfolds containing a plane, see [16] for more sophisticated examples).

Some of the results presented here (see Sect. 11.2) show the triviality of these birational invariants of cohomological or Chow theoretic type for certain rationally connected varieties. For example, we show in Sect. 11.2 show that for a smooth cubic fourfold in $\mathbb{P}^5$, the unramified cohomology groups $H^3_{nr}(X, \mathbb{Q}/\mathbb{Z})$ and $H^4_{nr}(X, \mathbb{Q}/\mathbb{Z})$ vanish. At the opposite, we also show the nontriviality of the *universal* CH_0 group and of the *universal* unramified cohomology of the general quartic double solid with 7 nodes. The "universal" variants of the invariants which appear here are introduced in [2] and we study them in [29] (see Sect. 11.4) under a geometric form which was already considered in [28]. The general idea is very nice: it is clear that a rationally connected variety defined over $\mathbb{C}$ has trivial Chow group $\mathrm{CH}_0(X)$ (all points are rationally equivalent). But there is a priori no canonical rational equivalence relation between two points, except for projective space, where we just have to consider the line passing through them. The defect of the universal triviality of the CH_0 group measures the impossibility of writing the equivalence relations $x \equiv_{rat} x_0$, $x \in X$ in family over X.

The geometric study of these notions, via the study of the diagonal, allows us to conclude that the very general desingularized quartic double solid with 7 nodes has a non trivial universal third unramified cohomology group, and a nontrivial universal CH_0 group. Thus it is not stably rational, although it satisfies the Clemens-Griffiths and the Artin-Mumford criterion.

11.2 Geometric Interpretation of Some Unramified Cohomology Groups

11.2.1 Unramified Cohomology

Let X be a smooth projective complex variety. We will denote X_{cl} the set $X(\mathbb{C})$ endowed with its classical (or Euclidean) topology, and X_{Zar} the set $X(\mathbb{C})$ endowed with its Zariski topology.

Let

$$\pi : X_{cl} \to X_{Zar}$$

be the identity of $X(\mathbb{C})$. This is obviously a continuous map, and Bloch-Ogus theory [6] is the study of the Leray spectral sequence associated to this map and any constant sheaf with stalk A on X_{cl}. In applications, A will be one of the following groups: $\mathbb{Z}$, $\mathbb{Q}$, $\mathbb{Q}/\mathbb{Z}$.

We are thus led to introduce the sheaves on X_{Zar}

$$\mathcal{H}^i(A) := R^i\pi_*A.$$

The Leray spectral sequence for π and A has terms

$$E_2^{p,q} = H^p(X_{Zar}, \mathcal{H}^q(A)).$$

Definition 11.6. Unramified cohomology of X with value in A is defined by the formula (cf. [9])

$$H^i_{nr}(X, A) = H^0(X_{Zar}, \mathcal{H}^i_X(A)).$$

The main result of the paper by Bloch and Ogus [6] is the following Gersten-Quillen resolution for the sheaves $\mathcal{H}^i_X(A)$. For any closed subvariety $D \subset X$, let $i_D : D \to X$ be the inclusion map and $H^i(\mathbb{C}(D), A)$ the constant sheaf on D with stalk

$$\lim_{\substack{\to \\ U \subset D \\ \text{nonempty Zariski open}}} H^i(U, A)$$

at any point of D. When $D' \subset D$ has codimension 1, there is a map induced by the topological residue (on the normalization of D):

$$Res_{D,D'} : H^i(\mathbb{C}(D), A) \to H^{i-1}(\mathbb{C}(D'), A).$$

For $r \geq 0$, let $X^{(r)}$ be the set of irreducible closed algebraic subsets of codimension r in X.

Theorem 11.1 ([6], Theorem 4.2). *For any A, and any integer $i \geq 1$, there is an exact sequence of sheaves on X_{Zar}*

$$0 \to \mathcal{H}^i_X(A) \to i_{X*}H^i(\mathbb{C}(X), A) \stackrel{\partial}{\to} \bigoplus_{D \in X^{(1)}} i_{D*}H^{i-1}(\mathbb{C}(D), A) \stackrel{\partial}{\to} \ldots$$

$$\stackrel{\partial}{\to} \bigoplus_{D \in X^{(i)}} i_{D*}A_D \to 0.$$

Here the components of the maps ∂ are induced by the maps $Res_{D,D'}$ when $D' \subset D$ (and are 0 otherwise). The sheaf A_D on D_{Zar} identifies of course to the constant sheaf with stalk $H^0(\mathbb{C}(D), A)$. A very important consequence established in [6] is the Bloch-Ogus formula

$$H^k(X_{Zar}, \mathcal{H}^k(\mathbb{Z})) = \mathrm{CH}^k(X)/\sim alg, \tag{11.1}$$

where $\sim$ *alg* is algebraic equivalence (the subgroup of cycles algebraically equivalent to 0 is generated by the cycles $\mathcal{Z}_t - \mathcal{Z}_{t'}$, where $\mathcal{Z}$ is a codimension k cycle in $C \times X$, for some connected smooth curve C, and t, t' are two points of C).

An other immediate corollary is the following:

Corollary 11.1. *Unramified cohomology groups are birationally invariant, hence trivial in degree >0 for rational varieties X.*

Indeed, they are computed as

$$\mathrm{Ker}\,(H^i(\mathbb{C}(X), A) \stackrel{\partial}{\to} \bigoplus_{D \in X^{(1)}} H^{i-1}(\mathbb{C}(D), A)),$$

where the maps are residue maps.

Concerning the structure of the sheaves $\mathcal{H}^i(\mathbb{Z})$, we have the following result, which is a consequence of the Bloch-Kato conjecture recently proved by Rost and Voevodsky [24] (we refer to [3,7,10] for more explanations concerning the way the very important result below is deduced from the Bloch-Kato conjecture).

Theorem 11.2. *The sheaves $\mathcal{H}^i(\mathbb{Z})$ of $\mathbb{Z}$-modules over X_{Zar} have no torsion. In particular, unramified cohomology groups $H^i_{nr}(X,\mathbb{Z})$ have no torsion.*

Note finally that unramified cohomology has good functoriality properties under correspondences (see [10, Appendix]). The Bloch-Srinivas decomposition of the diagonal [7] then implies that for a smooth complex projective variety X with $\mathrm{CH}_0(X) = \mathbb{Z}$, (which includes rationally connected varieties,) all the unramified cohomology groups $H^i_{nr}(X,A)$ are torsion for $i > 0$. For $A = \mathbb{Z}$, this implies by Theorem 11.2 that $H^i_{nr}(X,\mathbb{Z}) = 0$ for $i > 0$. We will thus focus in the sequel on unramified cohomology with finite or torsion coefficients.

11.2.2 Geometric Interpretation

The first unramified cohomology group with torsion coefficients which is interesting for the Lüroth problem, that is, which can be nontrivial for unirational or rationally connected varieties, is the group $H^2_{nr}(X,\mathbb{Q}/\mathbb{Z})$. In fact we have (see [30, Prop. 6.17]):

Proposition 11.1. *If X is a rationally connected variety, $H^2_{nr}(X,\mathbb{Q}/\mathbb{Z})$ is isomorphic to the torsion subgroup of $H^3(X,\mathbb{Z})$.*

Note that the torsion subgroup of $H^3(X,\mathbb{Z})$ is the so-called Artin-Mumford invariant, which was shown by Artin-Mumford to be nonzero for certain desingularized quartic double solids.

Higher degree unramified cohomology groups are interpreted as follows. First of all, it is observed in [22] that if one defines, for any smooth projective variety X, the group $Z^4(X)$ as the quotient of the group $\mathrm{Hdg}^4(X,\mathbb{Z})$ of integral Hodge classes by the subgroup of classes of codimension 2 cycles, $Z^4(X)$ is a birational invariant of X (hence trivial for rational or stably rational X).

Note also that under the assumption that $CH_0(X)$ is supported on a surface, Bloch and Srinivas prove that X satisfies the Hodge conjecture in degree 4, which can be stated as saying that $Z^4(X)$ is of torsion. The following result is proved in [10], using the Bloch-Ogus resolution, the Bloch-Ogus formula (11.1) and Theorem 11.2:

Theorem 11.3. *For any smooth projective variety X, there is an exact sequence*

$$0 \to H^3_{nr}(X,\mathbb{Z}) \otimes \mathbb{Q}/\mathbb{Z} \to H^3_{nr}(X,\mathbb{Q}/\mathbb{Z}) \to \mathrm{Tors}(Z^4(X)) \to 0.$$

Corollary 11.2. *If $\mathrm{CH}_0(X)$ is supported on a surface (for example, X is rationally connected), one has*

$$H^3_{nr}(X,\mathbb{Q}/\mathbb{Z}) = Z^4(X).$$

The paper [9] exhibits unirational sixfolds which have a nontrivial unramified cohomology group $H^3_{nr}(X, \mathbb{Q}/\mathbb{Z})$. These varieties X thus carry by Corollary 11.2 an integral Hodge class of degree 4 which is nonalgebraic. It would be very interesting to make this class more explicit, and to study whether it is stable or not under deformations of X.

In the other direction, there are vanishing results for the group $Z^4(X)$:

Theorem 11.4 ([25]). *Let X be an uniruled or Calabi-Yau threefold (that is K_X trivial, $H^1(X, \mathcal{O}_X) = 0$). Then $Z^4(X) = 0$.*

An uniruled variety is a variety which is swept-out by rational curves. It is clear that if one chooses an ample hypersurface $Y \subset X$, any point of X is rationally equivalent in X to a point on Y, since it lies on a rational curve which intersects Y. Together with Corollary 11.2, this shows that $H^3_{nr}(X, \mathbb{Q}/\mathbb{Z}) = 0$, for an uniruled threefold.

Another vanishing result is as follows (it had been obtained by a similar method in [26] for cubic fourfolds). It builds on the method of Zucker [31] which uses normal functions and on [14].

Theorem 11.5 ([28]). *Let X be a smooth projective fourfold which is fibered over a curve C. Assume that the smooth fibers are cubic threefolds and that the singular fibers have at worst ordinary quadratic singularities. Then $Z^4(X) = 0$, hence $H^3_{nr}(X, \mathbb{Q}/\mathbb{Z}) = 0$.*

This applies to cubic fourfolds blown-up along a general codimension 2 linear section and to the associated Lefschetz pencil of hyperplane sections.

Let us finish with some results on $H^4_{nr}(X, \mathbb{Q}/\mathbb{Z})$ (a group which has been used by Peyre [20] to prove irrationality of some quotient varieties). In the paper [27], we observe that one can construct a birational invariant from codimension 3-cycles modulo rational or algebraic equivalence. Precisely, if one considers the torsion of the subgroup of $\mathrm{CH}^3(X)$ consisting of cycles annihilated by the Deligne cycle class, this is a birational invariant and so is its image $T^3(X)$ in the group of codimension 3-cycles modulo algebraic equivalence. This is because for codimension 2-cycles, the Deligne cycle class is injective on torsion (see [19]). The following result is proved in [27]:

Theorem 11.6. *Let X be a smooth complex projective variety such that $H^5(X, \mathbb{Z}) = 0$. Then there is an exact sequence*

$$0 \to H^4_{nr}(X, \mathbb{Z}) \otimes \mathbb{Q}/\mathbb{Z} \to H^4_{nr}(X, \mathbb{Q}/\mathbb{Z}) \to T^3(X) \to 0.$$

If furthermore $\mathrm{CH}_0(X) = \mathbb{Z}$ (or more generally $\mathrm{CH}_0(X)$ is supported on a subvariety of dimension ≤ 3), one has $H^4_{nr}(X, \mathbb{Z}) = 0$ and one concludes that $H^4_{nr}(X, \mathbb{Q}/\mathbb{Z}) = T^3(X)$.

If we combine this with [23, Remark 6.4], proving that 1-cycles homologous to 0 are algebraically equivalent to 0 on Fano hypersurfaces in projective space, one concludes

Theorem 11.7. *Let $X \subset \mathbb{P}^5$ be a smooth cubic or quartic fourfold. Then* $H^4_{nr}(X, \mathbb{Q}/\mathbb{Z}) = 0$.

An interesting open question is the following:

Question 11.2. *Let $X \subset \mathbb{P}^6$ be a smooth cubic fivefold. Is the group $T^3(X)$ trivial?*

Although we have $H^5(X, \mathbb{Z}) \neq 0$ for X as above, the precise version of Theorem 11.6 (see [27]) shows that the triviality of $T^3(X)$ is equivalent to the vanishing of $H^4_{nr}(X, \mathbb{Q}/\mathbb{Z})$.

11.3 Decomposition of the Diagonal and Universal CH_0-Group

Let us now introduce one tool to study stable rationality: it appears in various forms in [2, 28, 29]. Let X be a smooth projective variety defined over a field K. Recall that $\mathrm{CH}_0(X)$ is the group of 0-cycles of X defined over K, modulo the subgroup of 0-cycles rationally equivalent to 0. Note that unless K is algebraically closed, 0-cycles are not combinations of K-points; some 0-dimensional subschemes defined over K are not union of points defined over K.

Observe now that if K is algebraically closed, say $K = \mathbb{C}$, and X is rationally connected, then we have $\mathrm{CH}_0(X) = \mathbb{Z}$, since all points of X are rationally equivalent. However, starting with a variety X defined over $\mathbb{C}$, we can consider for any field extension $K \supset \mathbb{C}$, (e.g. function fields of varieties Z,) the variety X_K defined over K and its 0-cycles; for example, for each subvariety $Z \subset X$, the generic point of Z gives a point of X_K, where $K = \mathbb{C}(Z)$. The typical case we will be interested in is the case where $Z = X$, so that we are considering the inclusion of X into X, whose graph is the diagonal of X, or rather its restriction to dense Zariski open sets of X, seen as a point of X defined over the field $\mathbb{C}(X)$. The reason to consider these points is the following:

Lemma 11.1. *If X is rational, then* $\mathrm{CH}_0(X)$ *is universally trivial, which means by definition that* $\mathrm{CH}_0(X_K) = \mathbb{Z}$ *for any field extension K of* $\mathbb{C}$.

It turns out that the universal triviality of $\mathrm{CH}_0(X)$ is equivalent to the existence of a Chow theoretic decomposition of the diagonal in $\mathrm{CH}^n(X \times X)$ which is a more geometric formulation of it. Originally, the decomposition of the diagonal appeared in [5, 7], but with rational coefficients. For X defined over $\mathbb{C}$, this is the following statement:

Theorem 11.8. *If* $\mathrm{CH}_0(X) = \mathbb{Z}$, *for some integer $N \neq 0$, there is an equality*

$$N\Delta_X = Z_1 + Z_2 \text{ in } \mathrm{CH}^n(X \times X),\ n = \dim X, \tag{11.2}$$

where $Z_2 = N(X \times x)$ and Z_1 is a n-cycle supported over $D \times X$, for some proper closed algebraic subset $D \subset X$.

Saying that there is a decomposition of the diagonal in $\mathrm{CH}^n(X \times X)$ amounts to saying that we can make $N = 1$ in (11.2). This is equivalent to $\mathrm{CH}_0(X)$ being universally trivial (see Lemma 1.3 in [2]). This follows from the localization formula, which says that a decomposition

$$\Delta_X = Z_1 + Z_2 \text{ in } \mathrm{CH}^n(X \times X),\ n = \dim X \tag{11.3}$$

with $Z_2 = X \times x$ and Z_1 a n-cycle supported over $D \times X$, for some proper closed algebraic subset $D \subset X$, is equivalent to the vanishing of $\Delta_X - X \times x$ in the group

$$\lim_{\overrightarrow{U \subset X}} \mathrm{CH}^n(U \times X) = \mathrm{CH}^n(X_K) = \mathrm{CH}_0(X_K),\ K = \mathbb{C}(X).$$

A decomposition (11.3) implies the vanishing of many birational invariants of X, obtained by letting both sides act on cycle classes, cohomology, intermediate Jacobians, etc. For example, we have (see [10]):

Proposition 11.2. *If X admits a decomposition of the diagonal as in (11.3), then $H^i_{nr}(X, A) = 0$, for any $i > 0$, and any coefficients ring A.*

In [28], an even weaker property was investigated as a potentially interesting birational invariant, namely the existence of an integral cohomological decomposition of the diagonal, that is a decomposition

$$[\Delta_X] = [Z_1] + [Z_2] \text{ in } \mathrm{H}^{2n}(X \times X, \mathbb{Z}),\ n = \dim X \tag{11.4}$$

with $Z_2 = X \times x$ and Z_1 a n-cycle supported over $D \times X$, for some proper closed algebraic subset $D \subset X$. Although this property is a priori much weaker than (11.3), it already has a number of consequences which were essentially stated in [7], but with rational coefficients:

Theorem 11.9. *Let X admit an integral cohomological decomposition of the diagonal (11.4). Then*

1. *$H^i(X, \mathcal{O}_X) = 0$ for $i > 0$.*
2. *$H^{*>0}(X, \mathbb{Z})$ has coniveau ≥ 1 (that is, vanishes away from a proper closed algebraic subset $Y \subset X$).*
3. *Integral degree 4 Hodge classes on X are classes of codimension 2 algebraic cycles.*

When $\dim X = 3$, we have the following improvement (Theorem 11.11) of the theorem above. First of all, let us recall that the intermediate Jacobian $J^3(X)$ of a smooth complex projective variety is the complex torus

$$J^3(X) = H^3(X, \mathbb{C})/(H^3(X, \mathbb{Z}) \oplus F^2H^3(X)). \tag{11.5}$$

When $H^3(X, \mathcal{O}_X) = 0$, this complex torus is an abelian variety (see [8]), and if furthermore $\dim X = 3$, this abelian variety is principally polarized by the unimodular intersection pairing on $H^3(X,\mathbb{Z})/\text{torsion}$. Bloch and Srinivas [7] (see also [19]) prove the following:

Theorem 11.10. *If* $\mathrm{CH}_0(X) = \mathbb{Z}$, *the Griffiths Abel-Jacobi map* $AJ_X : \mathrm{CH}^2(X)_{hom} \to J^3(X)$ *is an isomorphism.*

One question left open by this result is whether there is actually an universal codimension 2 cycle $\mathcal{Z} \in \mathrm{CH}^2(J^3(X) \times X)$ realizing this isomorphism.

Remark 11.1. Note that under the same assumption $H^3(X,\mathcal{O}_X) = 0$, there is an integral Hodge class α of degree 4 on $J^3(X) \times X$ which is determined up to torsion by the property that

$$\alpha_* : H_1(J^3(X),\mathbb{Z}) \to H^3(X,\mathbb{Z})/\text{torsion}$$

is the canonical isomorphism given by the definition (11.5) of $J^3(X)$ as a complex torus and by the fact that α is of type $(1,3)$ in the Künneth decomposition of $H^4(J^3(X) \times X,\mathbb{Z})$. So for a smooth complex projective variety X with $H^3(X,\mathcal{O}_X) = 0$, saying that X admits an universal codimension 2 cycle is equivalent to saying that this degree 4 *integral* Hodge class on $J^3(X) \times X$, is the class of an algebraic cycle (modulo the torsion of $H^4(J^3(X) \times X,\mathbb{Z})$).

We relate in [28] this question to the existence of a cohomological decomposition of the diagonal:

Theorem 11.11. *Assume a smooth projective 3-fold has an integral cohomological decomposition of the diagonal. Then*

1. $H^i(X,\mathcal{O}_X) = 0$ *for* $i > 0$.
2. $H^*(X,\mathbb{Z})$ *has no torsion.*
3. *The even degree integral cohomology of* X *consists of classes of algebraic cycles.*
4. *There exists a universal codimension 2 cycle* $\mathcal{Z} \in \mathrm{CH}^2(J^3(X) \times X)$.
   Conversely, if 1–4 hold and furthermore the following property holds:
5. $J^3(X)$ *has a* 1*-cycle in the minimal class* $\frac{\theta^{g-1}}{(g-1)!}$, *where* $g := \dim J(X)$, *and* $\theta \in H^2(J^3(X),\mathbb{Z})$ *is the class of the natural principal polarization on* $J^3(X)$,

then X *admits an integral cohomological decomposition of the diagonal.*

In the paper [2], the authors also introduce the notion of universal triviality of unramified cohomology. In our Betti cohomology setting, their definition says the following:

Definition 11.7. A smooth complex projective variety X has universally trivial third unramified cohomology with torsion coefficients if for any smooth complex variety B and any class $\alpha \in H^3_{nr}(B \times X,\mathbb{Q}/\mathbb{Z})$, after passing to a dense Zariski open set $B' \subset B$, we have

$$\alpha_{|B' \times X} = pr_1^*\beta,$$

for some class $\beta \in H^3(B',\mathbb{Q}/\mathbb{Z})$.

The following result proved in [29] relates this notion to the existence of an universal codimension 2 cycle.

Theorem 11.12. *Let X be a smooth complex projective variety of dimension n with $\mathrm{CH}_0(X) = \mathbb{Z}$. Assume*

1. *$H^*(X, \mathbb{Z})$ has no torsion and the Künneth components of the diagonal are algebraic.*
2. *The group $H^3_{nr}(X, \mathbb{Q}/\mathbb{Z})$ is trivial (or equivalently by [10], the integral Hodge classes of degree 4 on X are algebraic).*

Then the degree 3 unramified cohomology of X with torsion coefficients is universally trivial if and only if there is an universal codimension 2 cycle $\mathcal{Z} \in \mathrm{CH}^2(J^3(X) \times X)$.

The hypotheses apply to rationally connected threefolds with no torsion in $H^3(X, \mathbb{Z})$ by the results of Sect. 11.2.2. They also apply to cubic hypersurfaces in $\mathbb{P}^5$ by Theorem 11.5.

11.4 Degeneration of Double Solids

In the paper [1], Artin and Mumford constructed a nodal quartic surface $S \subset \mathbb{P}^3$ with the property that the desingularization X of the associated double cover $Y \to \mathbb{P}^3$ ramified along S has some torsion in $H^3(X, \mathbb{Z})$. By Theorem 11.11, X does not admit an integral cohomological decomposition of the diagonal, neither a fortiori a Chow theoretic decomposition of the diagonal. On the other hand, we prove in [29] the following stability result under degeneration:

Theorem 11.13. *Let $\pi : \mathcal{X} \to B$ be a projective morphism of relative dimension $n \geq 2$, where $\mathcal{X}$ is smooth and B is a smooth curve. Assume that the fiber $\mathcal{X}_t$ is smooth for $t \neq 0$, and has at worst ordinary quadratic singularities for $t = 0$. Then*

(i) *If for general $t \in B$, $\mathcal{X}_t$ admits a Chow theoretic decomposition of the diagonal (equivalently, $CH_0(\mathcal{X}_t)$ is universally trivial), the same is true for any smooth projective model $\tilde{\mathcal{X}}_o$ of $\mathcal{X}_o$.*
(ii) *If for general $t \in B$, $\mathcal{X}_t$ admits a cohomological decomposition of the diagonal, and the even degree integral homology of a smooth projective model $\tilde{\mathcal{X}}_o$ of $\mathcal{X}_o$ is algebraic (i.e. generated over $\mathbb{Z}$ by classes of subvarieties), $\tilde{\mathcal{X}}_o$ also admits a cohomological decomposition of the diagonal.*

We now deform the Artin-Mumford surface S (and accordingly Y) which allows us to smooth independently the nodes. We then immediately deduce from Theorems 11.13 and 11.4 which guarantees the second assumption in (ii):

Corollary 11.3. *The very general desingularized nodal double solid X with $k \leq 7$ nodes does not admit a Chow theoretic or cohomological decomposition of the diagonal.*

This implies

Corollary 11.4. *The very general desingularized nodal double solid X with $k \leq 7$ nodes is not stably rational.*

The reason why we state the result for $k \leq 7$ nodes is the fact that only in this case do we know that by deforming the Artin-Mumford surface to a k-nodal quartic surface, the k nodes of the general deformed surface are in general position in $\mathbb{P}^3$. The family of k-nodal surfaces with nodes in general position is easily shown to be irreducible.

On the other hand, it is known by work of Endrass [13] that the desingularized double solids X as above have no torsion in $H^3(X, \mathbb{Z})$. Furthermore, for $k = 7$, the intermediate Jacobian $J^3(X)$ is of dimension 3, hence is (as a ppav) the Jacobian of a curve, hence has a 1-cycle in the class $\theta^2/2!$. By the converse statement in Theorems 11.11 and 11.4, we get:

Theorem 11.14. *The very general desingularized nodal double solid X with 7 nodes does not admit an universal codimension 2 cycle.*

By Theorem 11.12, we finally deduce from Theorem 11.14:

Corollary 11.5. *The very general desingularized nodal double solid X with 7 nodes does not have universally trivial third unramified cohomology.*

Acknowledgements I thank INDAM for inviting me to lecture in Padova and giving me the opportunity to write this note.

References

1. M. Artin, D. Mumford, Some elementary examples of unirational varieties which are not rational. Proc. Lond. Math. Soc. (3) **25**, 75–95 (1972)
2. A. Auel, J.-L. Colliot-Thélène, R. Parimala, Universal unramified cohomology of cubic fourfolds containing a plane (preprint). arXiv:1310.6705
3. L. Barbieri-Viale, On the Deligne–Beilinson cohomology sheaves (prépublication). arXiv:alg-geom/9412006v1
4. A. Beauville, J.-L. Colliot-Thélène, J.-J. Sansuc, P. Swinnerton-Dyer, Variétés stablement rationnelles non rationnelles. Ann. Math. **121**, 283–318 (1985)
5. S. Bloch, *Lectures on Algebraic Cycles*. Duke University Mathematics Series, vol. IV (Mathematics Department, Duke University, Durham, 1980)
6. S. Bloch, A. Ogus, Gersten's conjecture and the homology of schemes. Ann. Sci. Éc. Norm. Supér., Sér. 4, **7**, 181–201 (1974)
7. S. Bloch, V. Srinivas, Remarks on correspondences and algebraic cycles. Am. J. Math. **105**, 1235–1253 (1983)

8. H. Clemens, Ph. Griffiths, The intermediate Jacobian of the cubic threefold. Ann. Math. Second Ser. **95**(2), 281–356 (1972)
9. J.-L. Colliot-Thélène, M. Ojanguren, Variétés unirationnelles non rationnelles: au-delà de l'exemple d'Artin et Mumford. Invent. Math. **97**(1), 141–158 (1989)
10. J.-L. Colliot-Thélène, C. Voisin, Cohomologie non ramifiée et conjecture de Hodge entière. Duke Math. J. **161**(5), 735–801 (2012)
11. T. de Fernex, Birationally rigid hypersurfaces. Invent. Math. **192**, 533–566 (2013)
12. T. de Fernex, D. Fusi, Rationality in families of threefolds. Rend. Circ. Mat. Palermo **62**, 127–135 (2013)
13. S. Endrass, On the divisor class group of double solids. Manuscr. Math. **99**, 341–358 (1999)
14. T. Graber, J. Harris, J. Starr, Families of rationally connected varieties. J. Am. Math. Soc. **16**(1), 57–67 (2003)
15. B. Hassett, Some rational cubic fourfolds. J. Algebr. Geom. **8**(1), 103–114 (1999)
16. B. Hassett, Special cubic fourfolds. Compos. Math. **120**(1), 1–23 (2000)
17. V. Iskovskikh, Yu. Manin, Three-dimensional quartics and counterexamples to the Lüroth problem. Mat. Sb. (N.S.) **86**, 140–166 (1971)
18. J. Kollár, Y. Miyaoka, S. Mori, Rationally connected varieties. J. Algebr. Geom. **1**, 429–448 (1992)
19. J. Murre, Applications of algebraic K-theory to the theory of algebraic cycles, in *Proceedings of the Conference on Algebraic Geometry*, Sitjes, 1983. LNM, vol. 1124 (Springer, Berlin/Heidelberg, 1985), pp. 216–261
20. E. Peyre, Unramified cohomology of degree 3 and Noether's problem. Invent. Math. **171**(1), 191–225 (2008)
21. A.V. Pukhlikov, Birational isomorphisms of four-dimensional quintics. Invent. Math. **87**(2), 303–329 (1987)
22. Ch. Soulé, C. Voisin, Torsion cohomology classes and algebraic cycles on complex projective manifolds. Adv. Math. **198**(1), 107–127 (2005)
23. Z. Tian, H.-R. Zong, One Cycles on Rationally Connected Varieties. Compos. Math. **150**(03), 396–408 (2014)
24. V. Voevodsky, Motivic cohomology with $\mathbb{Z}/l$-coefficients. Ann. Math. (2) **174**(1), 401–438 (2011)
25. C. Voisin, On integral Hodge classes on uniruled and Calabi-Yau threefolds, in *Moduli Spaces and Arithmetic Geometry*. Advanced Studies in Pure Mathematics, vol. 45 (Mathematical Society of Japan, Tokyo, 2006), pp. 43–73
26. C. Voisin, Some aspects of the Hodge conjecture. Jpn. J. Math. **2**, 261–296 (2007)
27. C. Voisin, Degree 4 unramified cohomology with finite coefficients and torsion codimension 3 cycles, in *Geometry and Arithmetic*, ed. by C. Faber, G. Farkas, R. de Jong. Series of Congress Reports (EMS, Zürich, 2012), pp. 347–368
28. C. Voisin, Abel-Jacobi map, integral Hodge classes and decomposition of the diagonal. J. Algebr. Geom. **22**(1), 141–174 (2013)
29. C. Voisin, Unirational threefolds with no universal codimension 2 cycle. arXiv:1312.2122
30. C. Voisin, *Chow Rings, Decomposition of the Diagonal and the Topology of Families*. Annals of Mathematics Studies, vol. 187 (Princeton University Press, Princeton, 2014)
31. S. Zucker, The Hodge conjecture for cubic fourfolds. Compos. Math. **34**(2), 199–209 (1977)

Chapter 12
Unlikely Intersections and Pell's Equations in Polynomials

Umberto Zannier

Abstract This short paper surveys around themes related to a conjecture of R. Pink which, roughly speaking, generalises in particular the well-known Manin-Mumford conjecture (a theorem of M. Raynaud) to abelian varieties varying in families. In a series of works in collaboration with D. Masser we have established this for pencils of abelian surfaces, and in further work with D. Bertrand, Masser and A. Pillay we have considered other commutative algebraic groups of dimension 2.

We shall briefly discuss this progress, and present some applications to solvability of Pell's equations $X^2 - DY^2 = 1$ in polynomials $X = x(t), Y = y(t)$, where $D = D(t)$ is also a polynomial. This is analogue to the classical one for integers, and was studied already by Abel. In this context solvability is no longer ensured by simple conditions on D and may be considered 'exceptional'.

In this paper we shall let $D(t)$ vary in a pencil, and for instance we shall point out how our results on Pink's conjectures imply that for $D(t) = t^6 + t + \lambda$ the Pell's equation is solvable nontrivially only for finitely many complex λ.

12.1 A Special Case of Pink's Conjecture on Unlikely Intersections

Around 2005, R. Pink (see, e.g., [20]) formulated some conjectures on families of semi-abelian varieties which vastly generalised the Manin-Mumford conjecture on torsion points on subvarieties of complex abelian varieties, and touched also the context of Shimura varieties and the André-Oort conjecture; B. Zilber [33] in 2002 had stated similar (somewhat less general) conjectures, with independent motivations from logic. Previous joint work [7] with E. Bombieri and D. Masser

U. Zannier (✉)
Scuola Normale Superiore, Piazza dei Cavalieri 7, 56126 Pisa, Italy
e-mail: u.zannier@sns.it

V. Ancona and E. Strickland (eds.), *Trends in Contemporary Mathematics*, Springer INdAM Series 8, DOI 10.1007/978-3-319-05254-0_12,

had also raised special cases of these issues for the case of multiplicative tori $\mathbb{G}_{\mathrm{m}}^n$ in place of abelian varieties, and proved some statements.[1]

The subject often goes under the name of *Unlikely Intersections*. I shall not touch the general context in any detail in this short survey paper, and refer for instance to the book [30] for some general informations and references.[2] For simplicity I shall limit myself to discuss a very special case of Pink's conjecture, recently proved in joint work with Masser, which hopefully shall be illustrative of some of the main motivations in this topic. Let:

$\mathcal{A}$=an abelian-surface scheme over an (affine) irreducible curve C, defined over $\overline{\mathbb{Q}}$.

So, this is an algebraic family of abelian varieties $\mathcal{A}_\lambda$ of dimension 2, $\lambda \in C$, which are the fibers $\pi^{-1}(\lambda)$ of a map $\pi : \mathcal{A} \to C$ (with a zero section). The total space is thus a threefold.

Note that, for a positive integer n, each fiber $\mathcal{A}_\lambda$ has n^4 points of order (dividing) n. As λ varies, they yield an algebraic curve, a cover of C (reducible for $n > 1$), which is a (torsion) group-subscheme. Letting n vary over $\mathbb{N}$ and taking the union we obtain

$$\mathcal{A}_{tors} = \text{the union of torsion-subschemes.}$$

Hence, this is *a denumerable union of (torsion) curves*. Finally, let

$$X = \text{an algebraic curve in } \mathcal{A}\text{, defined over } \overline{\mathbb{Q}}.$$

In a spirit similar to the Manin-Mumford conjecture, we are interested in the set $X \cap \mathcal{A}_{tors}$, namely in the points on X which are torsion in the appropriate fiber. Since the ambient space has dimension 3, we expect X not to intersect most curves in $\mathcal{A}_{tors}$, unless there are 'special reasons'. This motivates the terminology 'Unlikely Intersections', which in this very example are just the points in the said set. The following result shows that these intersections may be described in 'finite terms':

Theorem 12.1 (Masser, Z.). *The set $X \cap \mathcal{A}_{tors}$ is contained in a finite union of group-subschemes of $\mathcal{A}$ of positive codimension.*

For instance, let us assume that X is irreducible. A 'special reason' occurs when X is contained in such a proper group-subscheme; then the torsion intersections are *not* anymore Unlikely (since now either X is itself torsion or we are intersecting the curve X with other curves *in a surface*). If on the other hand this is not the case, then each intersection of X with a proper group-subscheme is finite and then the result predicts finiteness for $X \cap \mathcal{A}_{tors}$.

---

[1]This was strengthened later by G. Maurin and in further joint work with Bombieri, P. Habegger, Masser.

[2]However one should take into account that some relevant work has been done very recently.

The above conclusion is in turn just a case of a much more general conjecture by Pink. Here we only mention that for instance he considered arbitrary semiabelian schemes $\mathcal{A}$ and subvarieties X of arbitrary dimension, and group-subschemes more general than torsion, but for which the intersections with X remain Unlikely. Again, we refer to [30] for an elementary introduction to the topic and for some references. Here are a few examples.

Example 12.1. Consider the case when $\mathcal{A}_\lambda$ is constant, or else when X is contained in a single fiber $\mathcal{A}_{\lambda_0}$: now we have a single abelian variety to consider (i.e., not depending on $\lambda \in C$) and we are simply seeking the torsion points on a curve X in it. We find just the Manin-Mumford issue in ambient dimension 2.

Example 12.2. A genuine example of the theorem, which is a truly relative version of Manin-Mumford, comes from the following question posed by Masser independently of Pink-Zilber, which originated our work.

To start with, let $\mathcal{L}$ be the Legendre elliptic scheme (over $\mathbb{P}_1 \setminus \{0, 1, \infty\}$). It is defined by $y^2 = x(x-1)(x-\lambda)$ and yields a family of elliptic curves parametrized by λ. Masser, for the sake of example, took the points $P_\lambda = (2, \sqrt{2-\lambda})$ and $Q_\lambda = (3, \sqrt{6(3-\lambda)})$ on $\mathcal{L}_\lambda$ (the signs in the square roots do not matter here), and considered the complex numbers λ for which *both* points become torsion on $\mathcal{L}_\lambda$, namely the set $\{\lambda \in \mathbb{C} : P_\lambda, Q_\lambda \in \mathcal{L}_{tors}\}$.

`Question:` Is this set finite ?

`Answer:` Yes.

We recover this answer from Theorem 12.1, on taking $C = \mathbb{P}_1 \setminus \{0, 1, \infty\}$, $\mathcal{A} = \mathcal{L} \times_C \mathcal{L}$ and X as the curve described by $P_\lambda \times Q_\lambda$, $\lambda \in C$. It is easy to check that no proper group-subscheme contains X (it amounts to prove that the points are generically linearly independent over $\mathbb{Z}$), so the conclusion yields finiteness for the relevant λ. See [15, 16, 30] for details. (We note that there are infinitely many $\lambda \in \overline{\mathbb{Q}}$ for which one alone of the two points becomes torsion, as is proved in the above references; these last intersections are of 'Likely' type.)

In a series of papers [17–19] we considered the possible cases when $\mathcal{A}$ is isogenous to a *square* of an elliptic scheme (as in the Example 12.2), or to a *product* of non-isogenous elliptic schemes, or is *simple* (over any finite cover of the base). These cases present similarities but also different aspects. This also happens with the transition from $\overline{\mathbb{Q}}$ to $\mathbb{C}$, which we found more troublesome than one might perhaps expect and which I shall not touch here.[3]

In place of the curve X one may also speak of (the image of) a section of the scheme (possibly after base change); focusing on the case of *simple* $\mathcal{A}$, the result then may be rephrased in the following fashion:

[3]For the first two cases, the treatment of this transition occupies a substantial part of the corresponding papers, whereas for the 'simple' case this is the object of work still in progress, also with P. Corvaja.

Corollary 12.1. *Let $\mathcal{A}$ be as above, assume it is simple (meaning the generic fiber is geometrically simple) and let $\sigma : C \to \mathcal{A}$ be a non-torsion section, defined over $\overline{\mathbb{Q}}$. Then there are only finitely many $x \in C$ such that $\sigma(x)$ is torsion.*

To deduce this from the above, one takes $X =$ the image of σ. That $\mathcal{A}$ is simple implies that there are no proper group-subschemes other than torsion; hence the conclusion says that $X \cap \mathcal{A}_{tors}$ is contained in a finite union of torsion curves. But since σ is assumed not to be torsion, each intersection of X with a torsion curve is finite, proving the assertion.

This version (in this phrasing) had been actually conjectured by Shou-Wu Zhang already in 1998 [31], prior to Pink or Zilber. (Zhang was mainly interested in certain stronger height statements, but explicitly put forward this corollary as a conjecture.) It may be read as a (sharp) local-global principle.

Remark 12.1. A well-known theorem of J. Silverman predicts (in much more general circumstances) bounded height for the x in question (note that these points are automatically algebraic since σ is defined over $\overline{\mathbb{Q}}$). So the finiteness for the x of bounded degree over $\mathbb{Q}$ becomes Northcott's theorem (see [6]), and hence this type of result may possibly acquire new interest only if there is no a priori restriction on the field of definition of the points. (We also note that this boundedness of the height is an ingredient of our proofs.)

Example 12.3. Define $\mathcal{A}_\lambda$ as the Jacobian of a complete smooth curve H_λ birational to $u^2 = t^6 + t + \lambda$. If C denotes the complement in $\mathbb{A}^1$ of the set of roots of the discriminant of $t^6 + t + \lambda$, H_λ has genus 2 for $\lambda \in C$, and the Jacobians can be assembled to form an abelian-surface family $\mathcal{A}$ over C. One may prove that it is simple (e.g., by specialization, reduction mod p and computation of zeta functions, following a method suggested by M. Stoll, see [8]).

Take now the difference $\infty_+ - \infty_-$ of the two points above $t = \infty$ in H_λ; this is a divisor of degree 0 and yields an element of $\mathcal{A}_\lambda$ and hence a section (over C), which may be checked to be non-torsion. (For this fact, see (a) after Theorem 12.2 below.)

Then we find from the corollary that the difference is torsion only finitely many times. (This sometimes happens: $\lambda = 0$ turns out to be a 'torsion value', with torsion order 5.)

This example shall be relevant for an application to a Pell's equation, presented below.

`About the proof method.` We only say a few words, and again refer to [30] for more: one views the abelian fibers $\mathcal{A}_\lambda$ as complex tori (varying analytically) and then the torsion points become rational points (with respect to coordinates in the lattice bases for the tori).

Now, a torsion point on X yields other ones by conjugation over a number field of definition. Here one can get a lower bound for their number through deep results by Masser, S. David, Masser-G. Wüstholz; note that by conjugation one jumps around the various fibers, so some uniformity is needed. On the other hand, one can estimate efficiently their number from above, since the corresponding rational points lie on a certain *transcendental* real surface, which is obtained as the inverse image of X by the abelian maps. These last estimates stem from work of Bombieri-J. Pila for curves, then of Pila for surfaces, and finally of Pila-A. Wilkie for transcendental varieties of arbitrary dimension. Comparison of estimates yields a contradiction (on the appropriate assumptions) if the torsion order is large enough, concluding the argument. (That the surface in question is 'highly' transcendental follows by a deep result of Y. André, appearing as Thm. 3 in [2].)

The question of looking at rational points on transcendental objects (whereas the original motivations mainly concerned algebraic varieties) was raised especially by P. Sarnak; one should also mention that Sarnak indeed foresaw some principles of this method in the case of tori long ago, in an unpublished paper (P. Sarnak, Torsion points on varieties and homology of abelian covers, manuscript, 1989) related to [21].[4]

Pila and the writer obtained a new proof of the Raynaud's theorem (former Manin-Mumford conjecture) by this method. Pila, J. Tsimerman, E. Ullmo and others extended to further contexts. (See [30] for some references, however not updated to very recent work.)

Effectivity with this method is expected, but not yet proved. (The lower bounds are already effective, but the situation is not yet clear for the upper bounds in the Bombieri-Pila-Wilkie work.) This reflects in an ineffectivity for Theorem 12.1 and its corollaries: for instance, at the moment there is no available procedure to exhibit the finite set in Example 12.2.

Remark 12.2. One of Pink's conjectures concerns general semiabelian schemes; it needs a (subtle!) correction in general: D. Bertrand discovered in 2011 a surprising counterexample for a family of suitable $\mathbb{G}_m$-extensions of a constant CM elliptic curve. (However this fits within another conjecture of Pink in the enlarged context of Shimura varieties.)

Recently, in the paper [4] in collaboration with Bertrand, Masser and A. Pillay, we have studied the general case of semiabelian surfaces; it turns out that Bertrand's counterexample is the only exception to the expected statement, similar to Theorem 12.1.

Masser's student H. Schmidt [23] has carried out the case of extensions of an elliptic curve by $\mathbb{G}_a$. All of this, together with previous well-known results on tori, completes the study of Pink's conjecture for general commutative algebraic group schemes over a curve (defined over $\overline{\mathbb{Q}}$), of relative dimension 2. (There are ten essentially different cases to take into account.)

[4]It is a remarkable content of these papers that estimates for torsion points on subvarieties of tori lead to important information about Betti numbers of abelian covers of a compact variety.

12.2 Pell's Equations in Polynomials

With a view towards an application of the above results, we shall now discuss in brief a function field variant of Pell's equation. As is rather universally known in Number Theory, this is the equation $X^2-DY^2 = 1$, to be solved in *integers* $X, Y \neq 0$,[5] where D is a positive integer, not a perfect square; these are easily seen to be necessary conditions for solvability, which turn out to be also sufficient, a much subtler fact (first formally proved by Lagrange).

In fact, it is well known that the equation was proposed by Fermat (rather than Pell), in 1657, as a challenge to English mathematicians. It was later realized that it actually goes back to antiquity, and an algorithm for producing solutions may be found already in Indian mathematics of the Seventh Century. (See [28] for this and much more.)

Needless to say, the equation turned out to be very important in a number of issues: it yields the borderline example of an affine curve with two points at infinity (a hyperbola) and infinitely many integral points, it represents units in quadratic rings and appears in corresponding class-number formulae; it leads to the structure of integer solutions of arbitrary binary quadratic equations over $\mathbb{Z}$, and to associated algorithms; it also leads to the structure of binary orthogonal groups over $\mathbb{Z}$, and one might continue.

With respect to this celebrated equation, we replace here $\mathbb{Z}$ with $k[t]$, for k a field, to obtain a polynomial analogue. Namely, for $D = D(t) \in k[t]$, we seek nontrivial solutions ($Y \neq 0$) of

$$X^2 - D(t)Y^2 = 1, \qquad X = x(t), Y = y(t) \in k[t], \quad 0 < 2d = \deg D \text{ even.} \tag{12.1}$$

This variant is apparently less known compared to the classical case, but it is old as well, having been studied, e.g., already by Abel [1] in 1826, especially in the context of integration *in finite terms* of algebraic differentials arising from ellipses' and lemniscates' lengths. (This was related to the first examples of Abelian Integrals and of Differential Algebra.)

Replacing the '1' by any nonzero constant is immaterial over $\bar{k}$. In general, a nontrivial solution with any $\mu \in k^*$ in place of 1 leads to a nontrivial solution of (12.1), and the equation represents the units of the quadratic ring $k[t, \sqrt{D(t)}]$.[6] Geometrically, it represents a family of affine hyperbolas over the affine line $\mathbb{A}^1$, and a solution may be viewed as a regular section.

---

[5]The problem of solving the equation in rationals is rather easier: projection to a rational line from the rational point $(1, 0)$ allows to parametrize rationally over $\mathbb{Q}$ the corresponding hyperbola.

[6]All of this easily follows from the cyclic structure of solutions, recalled below. For instance, if $x^2 - Dy^2 = \mu \in k^*$, one may 'square' to get $((x^2 + Dy^2)/\mu)^2 - D(2xy/\mu)^2 = 1$; see also Proposition 12.1 and the observations which follow. Having a non-square $\mu \in k^*$ is somewhat analogue to the so-called 'negative' Pell's equation in the classical case, i.e. with $\mu = -1$.

`Assumptions.` *About the field* k: The issues rather heavily depend on k. We shall disregard the uninteresting case char $k = 2$ and normally we shall (tacitly) work with k *an algebraically closed field of characteristic* 0, like $\overline{\mathbb{Q}}$, $\mathbb{C}$, or $\overline{\mathbb{C}(\lambda)}$. There are however many relevant questions also otherwise. When k is a finite field of char. $\neq 2$ (and only then) one sees that the theory completely parallels the classical one, necessary and sufficient conditions for solvability being that $D(t)$ is non-square in $k[t]$, but its leading term is (analogue of the condition $D > 0$).

About $D(t)$: We shall always assume the necessary conditions that $D(t)$ is non-square of positive even degree $2d$, and normally that it is squarefree (though the square factors lead to interesting issues, contrary to the classical case: see for this Sect. 12.4).

We shall briefly survey on this topic, starting with some basic issues:

`For which` $D(t)$ `do solutions exist?` Let us call 'Pellian' (over k) such polynomials.

Vague as it is, this is for us the most basic question. As remarked, the analogy with the classical case is strict only over finite fields, whereas over fields of characteristic 0 we can have solutions only 'rarely'. Later in Sect. 12.2.2 we shall motivate this claim and give it some more precise meaning; we shall also note how being Pellian is not a 'closed' condition, in the algebraic (or even topological) sense. As to our results, we shall mainly investigate *solvability in 1-dimensional families* (especially the family $D(t) = D_\lambda(t) := t^6 + t + \lambda$, for the sake of example).

`Structure of solutions.` As over $\mathbb{Z}$, a possible nontrivial solution (a, b) generates infinitely many ones by taking powers: $x + y\sqrt{D} = \pm(a \pm b\sqrt{D})^n$, $n \in \mathbb{N}$. It is easy to see (using for instance the proof of Proposition 12.1 below) that all solutions may be obtained from one of minimal degree by these formulae.

`Effectivity.` By this we think of questions like:

How to decide if a given $D(t)$ *is Pellian? How to compute a possible solution?*

All of this is well known to be possible in the classical case.

In our context, this would be easy if we had an a priori bound on the degree of a possible solution $x(t), y(t)$: indeed, viewing the coefficients as unknowns (and working, say, with a fixed $D(t)$), Eq. (12.1) then yields an algebraic system, whose solvability may be checked by known methods. However, it happens that the minimal degree of a solution can be arbitrarily large already for quartic $D(t)$, so surely there is no bound in terms only of d.[7]

This phenomenon makes the issues more subtle. Nevertheless, it turns out that indeed there exist suitable algorithms for the above questions, e.g. in the important case $k = \overline{\mathbb{Q}}$. (On this we shall only say a few words in Remark 12.3(v) below; it is nowadays 'standard', but was seemingly unknown until the late 1960s.)

---

[7]We think here of algebraically closed k; see Remark 12.3(iv) for number fields. This unboundedness is not free of interest in itself and for instance yielded the first counterexample to a certain 'plausible' conjecture in model theory of constructive algebra; this is due to L. van den Dries, Ka. Schmidt and H. Schoutens, see [24].

First examples:

- Quadratic: We have $(T_n(t)/2)^2 - (t^2-4)(U_n(t)/2)^2 = 1$, where T_n, U_n are the Chebyshev polynomials of the first and second kind. (E.g., $T_n(t+t^{-1}) = t^n + t^{-n}$.)

  In the quadratic case there are always solutions over the algebraic closure $\bar{k}$, essentially reducing to this example after an affine substitution $t \to at + b$, to carry $D(t)$ into $c(t^2-4)$.
- Quartic: It may be checked that, e.g., for $D_\lambda(t) = t^4 + t^2 + \lambda t$, $\lambda \in \mathbb{Q}^*$, there are no nontrivial solutions, even over $\mathbb{C}$. On the other hand, $(2t^2+1)^2 - D_0(t) \cdot 2^2 = 1$.

  In the quartic case there are also examples (over $\overline{\mathbb{Q}}$) with any prescribed degree of $y(t)$ in the minimal solution[8] (and also for any $d \geq 2$).
- Sextic: Let us consider the family $D_\lambda(t) = t^6 + t + \lambda$ (relevant later). There are then some cases:

(i) λ a variable: no solution over $k = \overline{\mathbb{C}(\lambda)}$ (easy proof later); indeed,
(ii) $\lambda = 1$, $D_1(t) = t^6 + t + 1$: it may be checked that this is not Pellian over $\mathbb{C}$ [19].
(iii) $\lambda = 0$, $D_0(t) = t^6 + t$: now we have the identity $(2t^5+1)^2 - (t^6+t)(2t^2)^2 = 1$.

The last identity was found (by Masser) with continued fractions. This leads us to discuss briefly this context, together with other ones where this kind of Pell's equation appears.

12.2.1 Some Contexts Involving Pell's Equations in Polynomials

Continued fractions. We can construct a continued fraction for $\sqrt{D(t)}$ similarly to the numerical case, by defining the *integral part* as the polynomial which best approximates the function at $t = \infty$; we obtain $\sqrt{D(t)} = a_0(t) + 1/a_1(t) + 1/\ldots$, where the *partial quotients* $a_i(t)$ are nonconstant polynomials of degree $\leq d$ (over an extension of k at most quadratic).

In the classical case this leads to an eventually periodic continued fraction. In the present context this does not always happen; however, Abel [1] proved that *the c.f. is eventually periodic if and only if $D(t)$ is Pellian, and then the pre-period has length* 1. (See also [27].)

Several attractive number theoretical questions arise in this way (for instance concerning the precise degrees of the partial quotients, and the denominators which appear in their coefficients), on which we cannot pause here.

---

[8]At least for large enough deg y, one may find them even within this family: see [30], p. 92.

Taking truncations of the c.f. we obtain the *Padé approximations* to $\sqrt{D(t)}$, i.e., coprime polynomials P, Q such that $P(t) - \sqrt{D(t)}Q(t) = O(t^{-\deg Q-1})$. In turn, this yields $\deg(P^2 - DQ^2) \leq d - 1$ (observe that $\deg P = \deg Q + d$). So for $d > 1$ we already see from this picture that a solution to Pell's equation seems not automatic, at any rate with this method. However, as in the classical case, all solutions, if there are any, come in this way from the continued fraction (see [27]).

We also recall that all of this has often applications to issues in Diophantine Approximation, on specialising t in the Padé approximations; and here the growth rate of the *height* of the coefficients of P, Q is important. Well, it is a special case of a theorem of Bombieri and P.B. Cohen [5] that *the log-height of the coefficients of P, Q grows linearly in their degree if and only if $D(t)$ is Pellian, otherwise quadratically*, and again the Pell's equation appears.

`Integration.` As mentioned above, Abel was led to the Pell's equation (at least) by the problem of integrating in *finite terms* differentials like $\frac{f(t)dt}{\sqrt{D(t)}}$, f a polynomial. These arise for instance as regular hyperelliptic differentials (when $\deg f \leq d - 2$) or in computing lemniscate lengths (for $f = 1$, $D(t) = t^4 - 1$). By *finite terms* we mean, as in Abel's time, that only a finite number of operations of algebraic, exponential or logarithmic type are allowed.

Abel observed in particular that when the Pell's equation (12.1) has a solution (x, y) then $f(t) = x'(t)/y(t)$ is a polynomial (of degree $d - 1$) and $\int \frac{f(t)dt}{\sqrt{D(t)}} = \log(x + \sqrt{D}y)$ (see also [27]). One may show (using results of Liouville on which we do not pause) that for instance an integral in finite terms exists for an $f(t) \neq 0$ of degree $<2d$ if and only if $D(t)$ is Pellian.

In work in progress with Masser we plan to describe when a pencil of integrals may become solvable in finite terms, as above, under specialisation of the parameter (see Sect. 12.4 for more).

`Extremal polynomials.` We have seen Chebyshev polynomials appearing in one example above; these are *extremal* with respect to the property that their maximum over a suitable interval is minimum, among all polynomials with a given degree and leading coefficient. Similarly, other extremal polynomials, now on the union of d disjoint intervals, appear as X-coordinates of solutions of Pell's equations for a $D(t)$ of degree $2d$. (See, e.g., [3] or [11].)

`Diophantine equations` $f(x) = g(y)$. They appear as describing common values of polynomials f, g, for instance when the variables run through a number field K. In view of Faltings' theorem, the intersection $f(K) \cap g(K)$ of the images may be infinite only if the associated curve has a component of genus 0 or 1. When f, g have coprime degrees a corresponding classification has been carried out in [3] and solutions of a Pell's equation in polynomials appear in one of the infinite families. (Recently, M. Zieve and other ten collaborators [32] have carried out the difficult task of obtaining a complete classification.)

It should be interesting to extend this analysis to components of any given genus.

`Families of continued fractions.` For a non-square polynomial $D(t)$ with integer coefficients, one may consider the integers $|D(n)|$, $n \in \mathbb{N}$, a continued fraction for $\sqrt{|D(n)|}$, and in particular the length of its period, as a function of n. A. Schinzel related this with the solvability of the Pell's equation (12.1), and established in particular that *the length is bounded only if one of* $\pm D(t)$ *is Pellian (over* $\mathbb{Q}$*)*: see for instance [22].[9]

12.2.2 Pell's Equations in Polynomials and Almost-Belyi Maps

Another context relevant for our Pell's equation is that of *Belyi maps*; these are non-constant rational maps $\beta : C \to \mathbb{P}_1$ on a (complete smooth) curve C which are *unbranched* outside $0, 1, \infty$. They are important in various issues, suffices it to recall that *a complex algebraic curve* C *may be defined over* $\overline{\mathbb{Q}}$ *if and only if it admits a Belyi map* (Belyi's theorem, see [6], 12.2).

This context (with $C = \mathbb{P}_1$) shall help us in describing Pellian polynomials (over $\mathbb{C}$), in particular the 'dimension' of their set, once that the degree $2d$ is fixed, as we shall now suppose. To study this issue, it shall first be convenient to take into account the following

Definition: We say that $D_1(t), D_2(t) \in k[t]$ are equivalent (over k) if $D_2(t) = cD_1(at+b)$ for some $a, c \in k^*$, $b \in k$.

Plainly such equivalence does not affect the property of being Pellian. Let now $k = \mathbb{C}$.

Observe that we may normalise $D(t)$ by noting that each equivalence class has, for some $m \leq 2d-2$, at least one and at most $2d-m$ representatives of the shape $t^{2d} + t^m + c_1 t^{m-1} + \ldots + c_m$, counting finitely many families of dimension up to $2d-2$.

For each family, if we fix the degree n of $x(t)$ in a solution, we may see the equation $x(t)^2 - D(t)y(t)^2 = 1$ as defining an (quasi-) affine variety, where the coefficients of x, y, D are the variables. Then a rough counting, which we leave to the interested reader, would suggest that the dimension of the space of normalised $D(t)$ for which there is such a solution should be (at most) $d-1$, i.e. *half* the freedom compared to the whole equivalence classes.

However, to justify rigorously this assertion on these lines seems an intriguing task,[10] and it is in particular to bypass this obstacle that the Belyi-maps viewpoint proves helpful.

For a solution $x(t), y(t)$ as above, let us consider the map $x(t)^2 : \mathbb{P}_1 \to \mathbb{P}_1$, of degree $2n$. The Pell's equation immediately shows that the ramification indices above 1 are all even with the exception of at most $2d$ points (the roots of $D(t)$), whereas above 0 they are automatically even and above ∞ there is total ramification

---

[9] A converse does not hold, as shown e.g. by $D(t) = t^2 + 3$; a proof is possible using the Pell's equation.

[10] One may also expand $\sqrt{D(t)}$ at $t = \infty$ and impose a suitable vanishing at ∞ of $x + \sqrt{D}y$ through a linear system in the coefficients; again, this leads to seemingly complicated determinantal varieties.

(since $x(t)$ is a polynomial). Hence, counting branching as usual, i.e. as the sum of $e-1$ over the ramification indices, we obtain a total branching above $0, 1, \infty$ at least $4n - d - 1$. But the Hurwitz formula for the map x^2 (or a direct argument) yields a total branching $4n - 2$. This means that *the total branching outside* $0, 1, \infty$ *is at most* $d - 1$, and in particular there cannot be more than $d - 1$ further branch points.

Let us fix as above the degree $2d$ of $D(t)$, and let us consider all non-square (or squarefree) $D(t)$ of that degree, and all solutions of the corresponding Pell's equations. In this situation, for growing n, we see that almost all the branching of the map x^2 is concentrated above $0, 1, \infty$, and in this sense we may speak of an *almost-Belyi* map.[11]

Now, to each such map, letting B be the set of its branch points, one may associate naturally a covering space of $\mathbb{P}_1(\mathbb{C}) \setminus B$ and a (monodromy) permutation representation $\pi_1(\mathbb{P}_1(\mathbb{C})\setminus B) \to S_{2n}$. The group on the left is free on $|B|-1 \leq d+1$ generators, which can be chosen in a fairly 'canonical' way. Conversely, it is known that:

(i) Given that the map is a polynomial, knowledge of the branch points and of the representation (up to conjugation in S_{2n}) determines the map, up to automorphisms $t \mapsto at + b$ of the domain $\mathbb{P}_1$.
(ii) Given permutations $\sigma_b \in S_{2n}$, $b \in B$, having product 1, generating a transitive subgroup and such that the sum $\sum(c - 1)$ taken over all of their cycle lengths c, equals $4n - 2$, there exists a rational map $\phi(t) : \mathbb{P}_1 \to \mathbb{P}_1$ of degree $2n$, unbranched outside B and with associated permutation representation sending 'canonical' generators to the σ_i (where the cycle lengths correspond to the ramification indices).

All of this is part of a general theory involving homotopy and Riemann Existence Theorem; see for instance [10, 26] and [29] for general principles, and [9] (which considers some situations similar to ours).

To apply this for our purposes, we have to choose the branch points different from $0, 1, \infty$ and then the permutations. The former choice is arbitrary, and we have $d-1$ free parameters for it. The permutations may be chosen in at most finitely many ways for each given n. We note that a possible choice, with $|B| = d+2$, leading to a Pell's equation (12.1), is $\sigma_\infty = (2n, \ldots, 2, 1)$, $\sigma_0 = (1, 2n)(2, 2n-1)\cdots(n, n+1)$, $\sigma_1 = (1, 2n-1)(2, 2n-2)\cdots(n-d, n+d)$, and $\tau_i = (n-i, n+i)$, $i = 1, 2, \ldots, d-1$, for the remaining $d - 1$ branch points. (We have $\sigma_\infty\sigma_1\tau_1\cdots\tau_{d-1}\sigma_0 = 1$.)

A refinement of this type of combinatorial analysis would give the result that for each $d \geq 2, n \geq d$ there are Pellian $D(t)$ whose minimal solution has degree n (a fact which may be proved also using a subsequent description, as in Remark 12.3(iv)).

[11] All of this is also related to the so-called *abc-inequality* for polynomials: solutions to a Pell's equation yield instances in which the bound is attained up to a summand $\leq d - 1$; see [29]. For $d = 1$ it is attained exactly, the map is fully Belyi and boils down to Chebyshev polynomials, as in one of the above examples.

In any case, we have obtained a (rough) *moduli space* for our problem (a simple example in the context of *Hurwitz Families*, see [10]). Note that the ambiguity coming from the maps $t \mapsto at + b$ of the domain $\mathbb{P}_1$ is in fact taken into account by the above defined equivalence. Then this may be subsumed in the following rough principle:

`branch points + monodromy` $\to$ finitely many possibilities for $D(t)$ up to equivalence.

In turn, since we have only $d - 1$ free parameters for the branch points, we have justified the above expected fact, namely:

For given $\deg D = 2d$*, the equivalence classes of Pellian* $D(t)$ *fall into denumerably many algebraic families of dimension up to* $d - 1$.

In particular, recalling that the equivalence classes form finitely many families of dimension up to $2(d - 1)$, this justifies the assertion that having solutions to (12.1) is somewhat 'exceptional'.

`Density of Pellian polynomials.` All of this, together with arguments stemming from Proposition 12.1 below, also allows to prove that, for a given degree $2d$, Pellian polynomials are dense for the complex topology in the space of polynomials of that degree. On the contrary, independent principles show that for $d > 1$ they are not dense if we work over an ultrametric field $\mathbb{C}_p$.[12]

`Our purposes.` A given $D(t)$ may or may not be Pellian, and we have already mentioned that this is a decidable question, e.g. over $\overline{\mathbb{Q}}$. Then a next question is to look at *pencils* of polynomials, and ask which members of the pencil are Pellian.

Example 12.4. Let us inspect the cases $d = 1, 2, 3$ at the light of the above considerations.

- $d = 1$. This is uninteresting from the above viewpoint: we have no moduli, as already noted in the 'Chebyshev' example above, and we have always solutions over, e.g., $\overline{\mathbb{Q}}$ or $\mathbb{C}$.
- $d = 2$. The equivalence classes now are represented by the families $t^4 + t^2 + \lambda t + \mu$, $t^4 + t + \lambda$ and by $t^4 + 1$. The Pellian polynomials represent denumerably many algebraic curves (or points) within these families. This itself shows that, as remarked above, it may happen (already for quartic D) that (12.1) has no nontrivial solutions; and we shall see (Remark 12.3(iv)) that we may have a minimal solution with $y(t)$ of any prescribed degree.

  The ambient space of normalized $D(t)$ has dimension 2, so we should expect that a pencil could usually intersect the curves corresponding to Pellian elements. Indeed, using Proposition 12.1 below, it may be proved that, e.g., though the pencil $t^4 + t^2 + \lambda t$ is not 'identically Pellian', it contains infinitely many Pellian polynomials, where the degree of the corresponding minimal solutions tends to infinity (see [30], p. 68 and p. 92). It also appears that the set of corresponding λ is dense in $\mathbb{C}$, and consists of algebraic numbers of bounded height in view

[12] A formal proof, found in conversation with Corvaja, has not yet been written down and shall hopefully appear in a future paper.

of a theorem of Silverman. In particular, *there are only finitely many Pellian polynomials in this family which are defined over a number field of bounded degree* (e.g., only $t^4 + t^2$ over $\mathbb{Q}$).

- $d = 3$. Now we have four parameters for the space of equivalence classes, and a denumerable union of algebraic sets, at most 2-dimensional, parametrizing the Pellian ones.

  If as before we take a pencil, we then have Unlikely Intersections: $1 + 2 < 4$. The context now (but not the concept) is different compared to Sect. 12.1; however we shall soon see how to connect the two contexts in this case, providing in particular the explicit example $t^6 + t + \lambda$ of a pencil with only finitely many Pellian polynomials (a simple description of all such pencils is possible, and shall be briefly mentioned after Theorem 12.2 below).[13]

  If we considered higher-dimensional families in place of pencils (again with $d = 3$), then, forgetting certain degenerate cases, it should be sensible to expect infinitely many Pellian polynomials; however this has not yet been verified.

12.3 Connections Between the Pell's Equation and Pink's Conjectures

We start with an observation which has been known since long ago (seemingly, already by Abel, though in different language). We stick to squarefree polynomials $D(t)$, and consider a (hyperelliptic) smooth complete curve H, of genus $g = d-1$, whose function field over k is defined by $u^2 = D(t)$. This equation defines a smooth affine curve in $\mathbb{A}^2$, and H has two more points, above $t = \infty$, denoted $\infty_\pm$. The difference $\delta := \infty_+ - \infty_-$ is a divisor of degree 0 and thus defines a point (again denoted δ) in the Jacobian $J = \mathrm{Pic}^0(H)$ of H. We have:

Proposition 12.1. *The Pell's equation $X^2 - D(t)Y^2 = 1$ has a nonconstant polynomial solution (over $\overline{k}$) if and only if δ is a torsion point of J, and then its order equals the minimal* $\deg x(t)$.

Proof. A proof is very simple. Let $(x(t), y(t))$ be a nontrivial solution of (12.1) and consider the functions $\varphi_\pm := x \pm uy \in k(H)$. They are nonconstant, regular on the affine part of H and satisfy $\varphi_+\varphi_- = 1$. Hence they are also nonzero at finite points, and thus the divisor $\mathrm{div}(\varphi_+)$ of φ_+ is supported at infinity. Since it has degree 0, we must have $\mathrm{div}(\varphi_+) = m(\infty_+ - \infty_-) = m\delta$, for some integer $m \neq 0$. Thus $m\delta$ is a principal divisor and by definition yields 0 in J. The argument may be easily reversed, and also the assertion about the order-degree is clear on looking at poles.

If we do not work over the algebraic closure, things are similar. An easy inspection would add the precision that there exists a solution over k if and only if δ

[13]The above mentioned facts about density suggests that one cannot prove this using merely the complex topology. It seems likely that not even the p-adic ones suffice.

is a divisor defined over k, which in turn holds if and only if the leading coefficient of $D(t)$ is a square in k. Then, the minimal degree of $x(t)$ in a solution over k is either the torsion order of δ or its double.[14]

This Proposition 12.1 (already mentioned in Example 12.4, $d = 2$) creates links with other relevant questions, as in the following

Remark 12.3. (i) The appearance of the Jacobian is no coincidence with Abel's interest in the issue; indeed, we have already alluded about some links with Abelian Integrals.

(ii) A link of the Pell's equation with the Picard group occurs also in the classical case: Dirichlet's celebrated class-number formula for the ring of integers of a real quadratic field $\mathbb{Q}(\sqrt{D})$ in fact 'contains' the minimal solution to the Pell's-type equation $X^2 - DY^2 = \pm 4$. Through the proposition, this analogy becomes strict when k is a finite field.

(iii) As remarked several times, there is a complete analogy between the classical case of Pell's equation and the case of finite k; in particular, there is a 'classical' proof of solvability when the leading coefficient of $D(t)$ is a square in k. The proposition yields another argument; indeed, in this case δ lies in the finite group $J(k)$, and thus has always finite order.

(iv) The proposition, combined with celebrated results of B. Mazur and L. Merel, shows that in the elliptic case ($d = 4$) and if k is a number field, for $D \in k[t]$ a minimal solution of the Pell's equation (if any such solution exists) must have its degree bounded only in terms of $[k : \mathbb{Q}]$ (not otherwise on $D(t)$). This is of course a very deep fact; we do not believe that a similar uniform result is available at the moment for higher genus (not even in the interesting function-field analogues).

On the other hand, if for instance $k = \overline{\mathbb{Q}}$, we see that the degree of the minimal solution can be prescribed arbitrarily, since by varying D among the squarefree quartic polynomials we may prescribe the torsion order of δ in the relevant elliptic curve. (For higher genus this continues to be true, but our proof is more involved, requiring results mentioned in Sect. 12.2.2.)

(v) For fields like, e.g., $k = \overline{\mathbb{Q}}$, the proposition yields an algorithm for checking whether a given polynomial is Pellian, and for computing possible solutions of (12.1); indeed, there are known algorithms for testing whether a given point on an abelian variety over $\overline{\mathbb{Q}}$ is torsion: for instance, one reduces modulo two primes of good reduction, and uses that reduction modulo p preserves the prime to p-part of the torsion order.[15]

[14] If $m\delta = 0$ in J, then some function $\varphi = x + uy \in k[t,u] \setminus k$ has divisor $m\delta$, hence $x^2 - Dy^2$ is a nonzero constant μ; considering $\mu^{-1}\varphi^2$ in place of φ, we may achieve $\mu = 1$; see also footnote 6 above.

[15] In doing this for the Pell's equation, it is computationally convenient to use the proposition twice, i.e. to perform the torsion checking modulo p through the Pell's equation, using continued fractions.

(vi) We may also start with a complete hyperelliptic curve H (say of genus ≥ 2) and reverse the procedure, finding an affine model after removing two points, labelled $\infty_\pm$, related by the canonical involution. Thus we obtain different Pell's equations corresponding to the same H. However, the point δ is not arbitrary in J; inspection shows that it may be chosen arbitrarily in the curve $\{\text{class}(\kappa - 2\xi) : \xi \in H\} \subset J$, where κ is a canonical divisor. All of this may be rephrased in terms of a substitution $t \mapsto (at+b)/(ct+d)$ in the affine equation for H.

Now, Proposition 12.1 combined with Theorem 12.1, specifically applied as in Example 12.3, yields at once the following finiteness result, confirming the claim of Example 12.4, $d = 3$:

Theorem 12.2 ([19]). *There are only finitely many complex λ such that $t^6 + t + \lambda$ is Pellian.*

As in Remark 12.1, the finiteness would be an immediate consequence of a bounded-height result of Silverman, if we restricted to numbers λ of bounded degree over $\mathbb{Q}$ (the relevant ones being automatically algebraic). Also, as pointed out earlier, the proof methods do not lead at the moment to the computability of the finite set in question (which contains 0).

For simplicity we have given here just an instance of what can be proved, but we stress that a similar result follows for all pencils $D_\lambda(t)$ (for λ in a curve C, say defined over $\overline{\mathbb{Q}}$), provided some necessary assumptions are verified, which we are going to illustrate in short, in the shape of obstructions.

`Obstructions to finiteness for Pellian` $D(t) = D_\lambda(t)$ `in a pencil.`

(a) *Existence of identical solutions in* λ, i.e. $D_\lambda(t)$ is Pellian over $\overline{\mathbb{Q}(\lambda)}$, or equivalently $\delta = \delta_\lambda$ is identically torsion. This may be further rephrased by saying that the pencil is contained in one of the Pellian families mentioned in Sect. 12.2.2. In such case of course we obtain solutions for all (but at most finitely many) complex λ_0 by specialization.

In the course of proving Theorem 12.1 one has to verify that this obstruction does not hold for $t^6+t+\lambda$, and let us do this now. Otherwise, since δ_λ is rational over $\overline{\mathbb{Q}}(\lambda)$, we would obtain a nontrivial solution $(x_\lambda(t), y_\lambda(t))$ of the Pell's equation, depending rationally on λ. On clearing denominators we would obtain nonzero polynomials $u, v \in \overline{\mathbb{Q}}[\lambda, t]$ with no common factors and with $u^2 - (t^6 + t + \lambda)v^2 = c$, where $c \in \overline{\mathbb{Q}}[\lambda]$ is nonzero. Now, we have a contradiction: if c is constant, on looking at degrees in λ, otherwise, on specialising λ at a root of c. See also [30], Remark 3.4.2.[16] Note also that this fact amounts to saying that the (finite) set of Theorem 12.2 does not contain any transcendental number.

[16]This kind of argument might not work generally, but one can show that it is always effectively possible to decide about this obstruction.

It is easy to construct examples meeting this obstruction; a simple one is $t^6+\lambda$. A subtler one occurs in a paper of C. McMullen: $t(t^5-10\lambda t^4+35\lambda^2t^3-50\lambda^3t^2+25\lambda^4t-\lambda^{10}-2\lambda^5-1)$. (The minimal solution has $y_\lambda(t)=2(t^2-5\lambda t+5\lambda^2)/(\lambda^5+1)^2$, a quadratic polynomial.)

(b) *The Zariski closure of the set of multiples $n\delta_\lambda$ is a curve Z in J (viewed as a Jacobian over $\overline{\mathbb{Q}}(\lambda)$) and the pair (δ, Z) is not isomorphic to a 'constant' pair (w.r. to λ).*[17] If this happens, the curve Z (being a subgroup of J) must be a union of translates of an elliptic curve and, roughly speaking, we fall in the case of genus 1, when indeed it is known that 'usually' $D_\lambda(t)$ becomes Pellian for infinitely many values of λ. (We omit here a more detailed discussion on the constancy condition.)

Here is an explicit instance, pointed out by Masser:

`Example for this obstruction:` $D_\lambda(t)=t^6+t^2+\lambda$. As above one checks that there are no identical solutions, but if we solve the 'elliptic' Pell's equation $X^2-(t^4+t^2+\lambda_0 t)Y^2=1$ (which may be done for an infinity of $\lambda_0\in\overline{\mathbb{Q}}$), we obtain solutions of our genus 2 Pell by taking $(x_{\lambda_0}(t^2), ty_{\lambda_0}(t^2))$. What happens is that the curve $u^2=D_\lambda(t)$ has maps to the non-isogenous elliptic curves $y^2=t^3+t+\lambda$ $((t,u)\mapsto(t^2,u))$ and $y^2=t^4+t^2+\lambda t$ $((t,u)\mapsto(t^2,tu))$, and δ_λ is sent to the origin on the first curve (hence, in a sense we *lose* half of the torsion condition provided by Proposition 12.1).

Of course this obstruction is automatically excluded if $J=J_\lambda$ is generically simple. We have already remarked that this holds in the case of Theorem 12.2, as may be shown by a criterion of Stoll; this fact is nontrivial. Other examples of simple families (indeed with $\mathrm{End}(J)=\mathbb{Z}$) are given for instance by N. Katz [12] and by Masser [13].[18]

A complete proof of a general form of Theorem 12.2, taking into account the said obstructions, and allowing pencils of polynomials of arbitrary even degree, is completely analogous; it has not yet been written down, but is planned to appear as a part of work in progress. On the contrary, the study of higher-dimensional families of polynomials (of arbitrary even degree) touches more general cases of Pink's conjecture and seems to require new principles.

12.4 Further Issues

In all of the above examples we have considered only squarefree $D(t)$; in the classical case, this is irrelevant for solvability (as long as D is not a square), but in the polynomial case it is not so (except if k is a finite field). As before, there

---

[17] Note that in case (a) the set of multiples is finite, so (a) and (b) could be fused in a single condition.

[18] It is possible to decide effectively about this obstruction as well, using deep tools, like estimates for isogeny degrees of Masser and G. Wüstholz [14]. A direct argument seems not to be available.

is a relation between solvability and a certain point δ being torsion; however in the non-squarefree case the ambient Jacobian must be replaced by other algebraic groups, which appear as *generalized Jacobians* (see [25]).

Let us see some examples, where we indicate in abbreviated form the suitable Jacobian and whether there are or not infinitely many specializations of λ which make the polynomial Pellian. (We leave it to the interested reader to verify the asserted correspondences.)

Example 12.5. We list only a few examples with small degrees.

- $D = (t-\lambda)^2(t^2-1) \to$ roots of unity values of an algebraic function of λ (i.e., $\lambda + \sqrt{\lambda^2-1}$), linked to torsion in $\mathbb{G}_{\mathrm{m}}$, infinitely many values (precisely those of the shape $\cos 2\pi\theta$, $\theta \in \mathbb{Q}$).
- $D = (t-\lambda)^2(t-\lambda-1)^2(t^2-1) \to$ torsion points on the curve $y + y^{-1} = x + x^{-1} + 2$ in $\mathbb{G}_{\mathrm{m}}^2$: finiteness for this example. (All values are $\lambda = -1, 0$, for $x = -1$ or $\pm\sqrt{-1}$.)
- $D = (t-\lambda)^4(t^2-1)$: never solvable; linked to torsion in $\mathbb{G}_{\mathrm{a}} \times \mathbb{G}_{\mathrm{m}}$.
- $D = (t+a)^2(t^4+t^2+\lambda t)$, fixed $a \in \overline{\mathbb{Q}} \to$ torsion in a generalized Jacobian of the elliptic curve $E_\lambda : u^2 = t^4 + t^2 + \lambda t$, a non-split extension of E_λ by $\mathbb{G}_{\mathrm{m}}$ if $a \neq 0$, or by $\mathbb{G}_{\mathrm{a}}$ if $a = 0$; only finitely many values of λ, as may be proved using the results in [4] (see Example 5(i)) or [23]. Recall that, to the contrary, there are infinitely many relevant values for $t^4 + t^2 + \lambda t$.
- $D = (t-\phi(\lambda))^2(t^4+t^2+\lambda t)$: linked either to a non-split extension of E_λ by $\mathbb{G}_{\mathrm{m}}$ or to $E_\lambda \times \mathbb{G}_{\mathrm{m}}$, depending on the algebraic function $\phi \neq 0$. Together with Bertrand, we have recently proved that there is always finiteness. (See also Theorem 4 in [4] for a special case.)

A somewhat different example is given by $D_\lambda(t) = t^4 + t^2 + \lambda t$, where we restrict λ to be a root of unity (a condition which seems not to come from the Pell's equation). We are led to torsion points in a product $E_\lambda \times \mathbb{G}_{\mathrm{m}}$ and one may prove finiteness. This holds even if we impose merely that $|\lambda| = 1$: in this case, by complex conjugation we see that D_λ is Pellian if and only if $D_{\lambda^{-1}}$ is, leading to torsion in $E_\lambda \times E_{\lambda^{-1}}$, which may be dealt with by [18].

There are still further issues connected with these topics. For instance one can consider quadratic equations related to, but more general than, the Pell's. (This already appears in Abel's quoted paper [1]; see also [19] for an 'almost-Pell's equation'.)

Or, again as in [1], one can look at integration of differentials, but this time taken on a pencil of curves. And one may ask about the values of the parameter for which the differential becomes integrable in finite terms, assuming it is not likewise integrable for a generic choice. For example: *The differential* $t^2\mathrm{d}t/\sqrt{t^6+t+\lambda}$ *is integrable in finite terms for only finitely many complex values of* λ, and $\lambda = 0$ is one of them. (This specific fact is directly linked to a Pell's equation and indeed follows from Theorem 12.2 through an old criterion of Liouville.)

Such a general issue, which had been investigated, although in an incomplete way, by J. Davenport, again leads to cases of the Pink's conjecture. A complete study of this is the object of work in progress with David Masser.

Acknowledgements I wish to thank Daniel Bertrand, Pietro Corvaja, David Masser, Andrzej Schinzel and Michael Zieve for helpful conversations and remarks. I am also grateful to Enrico Bombieri, Peter Sarnak, Lucien Szpiro, Gisbert Wüstholz and Shou-Wu Zhang for appreciated invitations to deliver talks which eventually led to the present paper. Further thanks go to INDAM for inviting me to lecture in one of their annual meetings (Giornata INDAM 2008), an occasion which more recently turned into the idea of publishing this paper in their collection. Finally, I thank the ERC for financial support (through the grant 'Diophantine Problems'), concerning the preparation of the cited work with Masser.

References

1. N.H. Abel, Über die Integration der Differential-Formel $\rho dx/\sqrt{R}$, wenn R und ρ ganze Funktionen sind. J. für Math. (Crelle) **1**, 185–221 (1826)
2. Y. André, Mumford-Tate groups of mixed Hodge structures and the theorem of the fixed part. Compositio Math. **82**, 1–24 (1992)
3. R. Avanzi, U. Zannier, Genus one curves defined by separated variables and a polynomial Pell equation. Acta Arith. **XCIX**(3), 227–256 (2001)
4. D. Bertrand, D. Masser, A. Pillay, U. Zannier, Relative Manin-Mumford for semi-abelian surfaces (2013), 40pp. ArXiV : 1307.1008v1
5. E. Bombieri, B.P. Cohen, Siegel's lemma, Padé approximations and Jacobians. Ann. Scuola Normale Sup. Pisa **25**, 155–178 (1997)
6. E. Bombieri, W. Gubler, *Heights in Diophantine Geometry*. New Mathematical Monographs, vol. 4 (Cambidge University Press, Cambridge, 2006)
7. E. Bombieri, D. Masser, U. Zannier, Intersecting a curve with algebraic subgroups of multiplicative groups. Int. Math. Res. Not. **20**, 1119–1140 (1999)
8. J.W.S. Cassels, E.V. Flynn, *Prolegomena to Middlebrow Arithmetic of Curves of Genus 2*. LMS Lecture Notes Series, vol. 230 (Cambridge University Press, Cambridge, 1996)
9. N.D. Elkies, M. Watkins, Polynomial and Fermat-Pell families that attain the Davenport-Mason bound (2013, preprint)
10. M. Fried, Fields of definition of function fields and Hurwitz families – groups as Galois groups. Commun. Algebra **5**, 17–82 (1977)
11. F. Hazama, Twists and generalised Zolotarev polynomials. Pac. J. Math. **203**, 379–393 (2002)
12. N.M. Katz, Monodromy of families of curves: applications of some results of Davenport-Lewis, in *Sem. Th. de Nombres, Paris 1979–80*, ed. by M.-J. Bertin. Progress in Mathematics, vol. 12 (Birkhauser, Boston/Basel/Berlin, 1981), pp. 171–195
13. D. Masser, Specializations of some hyperelliptic Jacobians, in *Number Theory in Progress*. Essays in honor of A. Schinzel, vol. 1819 (Springer, 1999), pp. 324–333
14. D. Masser, G. Wüstholz, Endomorphism estimates for abelian varieties. Math. Z. **215**, 641–653 (1994)
15. D. Masser, U. Zannier, Torsion anomalous points and families of elliptic curves. C. Rendus Acad. Sci. Paris **346**, 491–494 (2008)
16. D. Masser, U. Zannier, Torsion anomalous points and families of elliptic curves. Am. J. Math. **132**, 1677–1691 (2010)
17. D. Masser, U. Zannier, Torsion points on families of squares of elliptic curves. Math. Annalen **352**, 453–484 (2012)

18. D. Masser, U. Zannier, Torsion points on families of products of elliptic curves (2012, preprint, submitted for publication)
19. D. Masser, U. Zannier, Torsion points on families of simple abelian surfaces and Pell's equation over polynomial rings (submitted for publication)
20. R. Pink, A combination of the conjectures of Mordell-Lang and André-Oort, in *Geometric Methods in Algebra and Number Theory*, ed. by F. Bogomolov, Yu. Tschinkel, vol. 253 (Birkhauser, Boston, 2005), pp. 251–282
21. P. Sarnak, S. Adams, Betti numbers of congruence groups (with an appendix by Zeev Rudnick). Isr. J. Math. **88**, 31–72 (1994)
22. A. Schinzel, On some problems of the arithmetical theory of continued fractions II. Acta Arith. **7**, 287–298 (1961/1962); Corrigendum ibid. **47**, 295 (1986)
23. H. Schmidt, PhD thesis in progress, Basel
24. H. Schoutens, Uniform bounds in algebraic geometry and commutative algebra, in *Connections Between Model Theory and Algebraic and Analytic Geometry*, ed. by A. Macintyre. Quaderni di Matematica, vol. 6, (II Universita di Napoli, Napoli, 2000), pp. 43–94
25. J-P. Serre, *Algebraic Groups and Class Fields*. GTM, vol. 117 (Springer, New York, 1988)
26. J-P. Serre, *Topics in Galois Theory* (Jones and Bartlett, Boston, 1992)
27. A.J. van der Poorten, X.C. Tran, Quasi-elliptic integrals and periodic continued fractions. Monatsh. Math. **131**, 155–169 (2000)
28. A. Weil, *Number Theory (from Hammurapi to Legendre)* (Birkhauser, Boston, 1984)
29. U. Zannier, Some remarks on the S-unit equation in function fields. Acta Arith. **54**, 87–97 (1993)
30. U. Zannier, *Some Problems of Unlikely Intersections in Arithmetic and Geometry (with Appendixes by D. Masser)*. Annals of Mathematics Studies, vol. 181 (Princeton University Press, Princeton, 2012)
31. S. Zhang, Small points and Arakelov theory. Doc. Math. Proc. ICM **2**, 217–225 (1998). Extra Vol. II, (electronic)
32. M. Zieve et al., Series of five papers (in preparation)
33. B. Zilber, Exponential sums equations and the Schanuel conjecture. J. Lond. Math. Soc. **65**, 27–44 (2002)

Chapter 13
Birational Geometry of Projective Varieties and Directed Graphs

Paolo Cascini

Dedicated to the memory of Valerio Venturi.

Abstract We give a brief introduction to some of the results in the Minimal Model Programme using the elementary theory of directed graphs.

13.1 Introduction

The main goal of birational geometry is to shed light in the study of projective varieties by studying classes of varieties up to birational isomorphisms, i.e. isomorphisms over a dense Zariski dense of the variety.

In dimension two, i.e. in the case of algebraic surfaces, this study was successfully carried out by the Italian school of Algebraic Geometry, led by Castelnuovo at the beginning of the twentieth century. However since then, it took several decades and work by many mathematicians, that we have a better understanding of the birational geometry of higher dimensional projective varieties today. In particular, Mori's minimal model programme, introduced in the 1980s, predicts the existence of an optimal algorithm to find a good representative of the birational class of each projective variety defined over an algebraically closed field. Although many conjectures in the minimal model programme are still open, the problems were solved in the case of threefolds by the end of the twentieth century and significant progress was made in higher dimension during the last decade.

Recently many books (e.g. see [11, 20, 23]) and survey papers (e.g. see [2, 5, 8, 10, 12, 13, 19]) appeared on this topic to describe the recent progress obtained so far.

P. Cascini (✉)
Department of Mathematics, Imperial College London, 180 Queen's Gate, London SW7 2AZ, UK
e-mail: p.cascini@imperial.ac.uk

V. Ancona and E. Strickland (eds.), *Trends in Contemporary Mathematics*,
Springer INdAM Series 8, DOI 10.1007/978-3-319-05254-0_13,

The goal of these notes is to present the minimal model programme from a slightly different point of view, using directed graphs. Although this approach is not new, our goal is to present the programme in a very simple language, to make it accessible to anyone with only basic knowledge in algebraic geometry.

13.2 Birational Geometry of Projective Varieties

To explain some of the main ideas of the Minimal Model Programme and some of the tools used, we use some basic facts from graph theory. In particular, we describe a directed graph associated to the category of projective varieties. For this reason, we recall some of the basic definitions in graph theory.

Recall that a *directed graph* is a set of vertices connected by oriented edges, i.e. ordered pairs of vertices. Two edges are said to be *consecutive* if the ending vertex of one coincides with the starting vertex of the other. A *chain* in a directed graph is a sequence of distinct vertices connected by consecutive edges. A *cycle* in a directed graph is a sequence of consecutive ordered edges, starting and ending at the same vertex. A *tree* is a directed graph which does not contain any cycle. Note that the topological space underlying the tree is not necessarily simply connected, e.g. it could contain two distinct vertices and two edges connecting the two vertices with the same orientation.

Given two vertices X and Y of a directed graph, we say that Y is *below* X if we can find a chain starting from X and ending in Y. Clearly, If Y is below X, then we say that X is *above* Y. An *end-point* for a directed graph, is a vertex which does not admit any other vertex below it.

13.2.1 Projective Surfaces

The easiest example in the study of directed graphs associated to the birational geometry of projective varieties is given by the category of smooth projective surfaces. To this end, we consider the *directed graph* whose vertices are smooth projective surfaces defined over an algebraically closed field k and whose edges are proper birational morphisms. The connected component containing a projective surface X corresponds to the birational class of X. We now look at some easy properties of this component. First, it is easy to check that this graph is a tree. Indeed, if X, Y are non-isomorphic projective surfaces connected by an edge, i.e. if there exists a non-trivial projective morphism $f: X \to Y$, then the second Betti number of X is greater than the one of Y. Thus, the claim follows easily.

Note that there are always infinitely many vertices above a vertex associated to a projective surface X, as it is always possible to blow-up an infinite sequence of points to obtain an infinite chain above X. On the other hand, using the inequality on the second Betti number described above, it is easy to check that starting from a vertex X, it is always possible to find an end-point below X. More specifically, there exists no infinite chain starting from X. Thus, we can think of the end-point Y to be a good representative of the connected class of X. We will see that also in higher dimension, one the main goals of the minimal model programme is to find the end-point of a connected component associated to a projective variety X.

We now show that projective surfaces can be divided into two large classes. The same dichotomy is expected to hold also in higher dimension.

First, we assume that X is a smooth projective surface such that $h^0(X, mK_X) > 0$ for some positive integer m. Then the subgraph obtained by considering the vertices below X and the corresponding edges is finite. In addition, there exists a unique vertex which is an end-point for the connected component containing X. Such a vertex Y is called the *minimal model* of X and, by Castelnuovo theorem, it is characterised by the fact that it does not admit any smooth rational curve E of self-intersection -1. Alternatively, Y is the only surface in the connected component of X such that K_Y is nef, i.e. $K_Y \cdot C \geq 0$ for any curve C in Y.

We now assume that X is a smooth projective surface such that $h^0(X, mK_X) = 0$ for all positive integer m. In this case, X is *uniruled*, i.e. it is covered by rational curves. It is possible to show that although the graph below X might be finite, there are always infinitely many end-points for the connected component of X. For example, if $X = \mathbb{P}^2$ is the two-dimensional projective space over the field k, then the connected component containing the vertex associated to X, corresponds to the set of all the smooth *rational surfaces*. Clearly X is an end-point of such a graph, but also each Hirzebruch surface $\mathbb{F}_n = \mathbb{P}(\mathcal{O}_{\mathbb{P}^1} \oplus \mathcal{O}_{\mathbb{P}^1}(-n))$, with $n \in \mathbb{N}$, is such. Even worse, there are surfaces which admit infinitely many vertices below it; the classic example is $\mathbb{P}^2$ blown-up at nine general points. Finally, note that not all the projective surfaces which admit a Mori fiber space is an end-point for the directed graph we have constructed (e.g. the blow-up of $\mathbb{P}^2$ at one point admits a Mori fiber space, but it corresponds to a vertex which is not an end-point).

13.2.2 Higher Dimensional Projective Varieties

The goal of the minimal model program is to generalise the study of the directed graph we have seen in the previous section, to projective varieties of any dimension. Although we expect a similar behaviour, we will see that there are many problems arising as we go from dimension two to dimension three or higher.

In order to define the directed graph associated to projective varieties over an algebraically closed field k, we first try to understand what the edges are. A simple analysis in birational geometry shows that it is not suitable to consider only proper birational morphism within projective varieties, as otherwise we might get end-points of the graph which do not admit any special property and it would be hard to find a suitable characterisation of a good representative of the birational class of a projective variety. Thus, to solve this issue, we consider birational map with some extra properties. First, a birational map within normal projective varieties

$\varphi: X \dashrightarrow Y$ is called a *contraction* if φ^{-1} does not contract any divisor. In other word, we do not want to consider maps like the blow-up of a proper subvariety, as this would not improve the understanding of our projective variety. Unfortunately, it is not enough to consider these maps as edges of our directed graph. Indeed, even in dimension three, it is possible to find pairs of non-isomorphic projective manifolds, which are isomorphic in codimension one, i.e. there exists X and Y which are isomorphic only after removing finitely many curves from both of them. In this case, X and Y would be part of a cycle, as the birational morphism connecting X and Y and its inverse are both contraction. Therefore, we need to be more restrictive in the choice of those maps that define the edges of our directed graph.

Let $\varphi: X \dashrightarrow Y$ be a birational contraction. Then φ is said to be *K-negative* if there exist proper maps $p: W \to X$ and $q: W \to Y$ such that

$$p^* K_X = q^* K_Y + E$$

where E is an effective $\mathbb{Q}$-divisor and the support of E is equal to the union of all the exceptional divisors contracted by q.[1] It is then easy to show that if $\varphi: X \dashrightarrow Y$ is a K-negative birational contraction then its inverse is not K-negative. More in general, it is possible to check that if we define an edge of our graph to be a K-negative birational contraction, then the graph is a tree as it does not admit any cycle.

We now define the vertices of our new directed graph. Although, it would be tempting to consider only smooth projective varieties, as above it is possible to show that the end-points of our graph will not satisfy any useful property. On the other hand, the Negativity Lemma (e.g. see [20]) implies that if the canonical divisor of a projective variety is nef then the vertices associated to X is an end-point of the graph. Thus, it is natural to ask under what condition, a projective variety corresponding to the end-point of the directed graph admits a nef canonical divisor. To this end, we need to consider projective varieties with some mild singularities. First, for simplicity, we assume that all the varieties we consider are *$\mathbb{Q}$-factorial*, i.e. we assume that X is normal and any Weil divisor S is such that mS is Cartier for some positive integer m. Moreover, we assume that X is *terminal*, which means that X admits a smooth variety Y above X. Note that this is equivalent in assuming that there exists a smooth projective variety Y and a proper birational morphism $f: Y \to X$ such that

$$K_Y = f^* K_X + E$$

for some f-exceptional $\mathbb{Q}$-divisor $E \geq 0$. Thus, terminal projective varieties appear naturally in the graph that we are considering. We can finally construct our directed graph: the vertices are terminal projective varieties defined over an algebraically closed field k and the edges are K-negative birational contractions.

---

[1] Note that this definition is slightly different than the one given in [3].

At this point, it is natural to ask if the same properties described in the case of surfaces, would hold for this directed graph. In particular, if X is a terminal projective variety, we can ask if there might exist an infinite chain starting from X. It is possible to show that this coincides with the following famous open problem:

Conjecture 13.1 (Termination of flips). Let X be a terminal projective variety. Then there exists no infinite sequence of flips

$$X = X_0 \dashrightarrow X_1 \dashrightarrow X_2 \dashrightarrow \dots$$

starting from X.

A *flip* is a special K-negative birational contraction which is an isomorphism in codimension one (see [20] for more details). Existence of flips over the complex numbers was proven in [3, 15] and later on in [6, 9]. Termination of flips was proven by Shokurov in dimension three and under some assumptions in higher dimension. In particular, one of the main achievements in this direction is obtained by combining the results in [1] and [16].

We now go back to the study of our directed graph and we can ask if the dichotomy within uniruled and non-uniruled surfaces extends to the case of higher dimensional projective varieties. More specifically, we expect that if $h^0(X, mK_X) > 0$ for some positive integer m, then there exists always an end-point below X, represented by a terminal projective variety Y with the property that K_Y nef. As in the case of surfaces, Y will be called a *minimal model* of X.

Thus, we have the following:

Conjecture 13.2 (Existence of Minimal Models). Let X be a terminal projective variety with non-trivial canonical ring

$$R(X, K_X) = \bigoplus_{m \geq 0} H^0(X, mK_X).$$

Then X admits a minimal model, i.e. there exists a K-negative birational map

$$\varphi: X \dashrightarrow Y$$

into a terminal projective variety Y, such that K_Y is nef.

If X is a variety of general type, i.e. if K_X is big, then the conjecture holds in any dimension [3]. Note that even in this case, Conjecture 13.1 is still open. Thus, the sub-graph of all the varieties below a given terminal projective variety of general type could contain an infinite chain, such that each vertex of this chain is connected to an end-point after finitely many consecutive edges.

If X is uniruled, then a very similar picture as in the case of surfaces holds. Although the graph given by the terminal projective varieties below X might be infinite, we do have a good description of the end-points of the connected component containing X. Indeed, these vertices correspond to varieties Y which admit a *Mori fibre space* [3], i.e. a non-trivial map $\eta: Y \to Z$ onto a lower dimensional projective variety, such that the general fiber F is a (possibly singular) Fano variety, i.e. the anti-canonical divisor $-K_F$ is ample. Note that Y might admit more than one structure as a Mori fiber space (e.g. the simplest example is $\mathbb{P}^1 \times \mathbb{P}^1$).

Although the picture for uniruled varieties is quite well-understood, we still need to understand whether there is no other connected component of our directed graph, corresponding to projective varieties which do not belong to the two classes of varieties described above (i.e. non uniruled varieties with trivial canonical ring). Currently, this is one of the most important open problem in the Minimal Model Programme. For simplicity, we will only discuss it in the case of complex projective varieties:

Conjecture 13.3 (Weak Abundance Conjecture). Let X be a terminal complex projective variety with trivial canonical ring, i.e.

$$R(X, K_X) = \mathbb{C}.$$

Then X is uniruled.

Note that in dimension three, all the conjectures described above (i.e. Conjectures 13.1–13.3) hold in full generality for complex projective varieties (e.g. see [20, 21]).

13.3 Birational Geometry of Log Pairs

It is immediate to see from the arguments in the previous section, that the canonical divisor plays a very important role in the study of the birational geometry of a projective variety over any algebraically closed field k. In addition, if $k = \mathbb{C}$, several generalisations of Kodaira's vanishing theorem, such as Kawamata-Viehweg vanishing, had a huge impact in birational geometry (e.g. in [6], it was shown that finite generation of the canonical ring follows almost directly from these results). This is another evidence of the fact that the canonical divisor is an essential tool in birational geometry. On the other hand, there are at least two main problems if we work in this generality. First, if S is a normal hypersurface in a projective variety X, then the adjunction formula [21] implies that

$$(K_X + S)|_S = K_S + \mathrm{Diff}_S$$

where Diff_S is an effective $\mathbb{Q}$-divisor on S, i.e. a linear combination of prime divisors of S with positive rational coefficients. Secondly, even in the simple case of a minimal elliptic surface $\pi: X \to C$ over a smooth curve C, we have that

$$K_X = \pi^*(K_S + D)$$

for some effective $\mathbb{Q}$-divisor D on C. Thus, it is natural to consider a larger category, which includes not only varieties but pairs (X, Δ) where X is a projective variety and Δ is an effective $\mathbb{Q}$-divisor. It is thus convenient to replace varieties by log pairs (X, Δ), and the canonical divisor by $K_X + \Delta$.

Therefore, our new goal is to define the right generalisation of the directed graph that we want to consider. To this end, we consider log pairs with Kawamata log terminal singularities: this is the most suitable condition in terms of the singularities of a log pair in the minimal model programme: a pair (X, Δ) is said to be *Kawamata log terminal* if for any proper birational morphism $\varphi: Y \to X$, we may write

$$K_Y + \Delta_Y = \varphi^*(K_X + \Delta)$$

where Δ_Y is a (non-necessarily effective) $\mathbb{Q}$-divisor whose coefficients are strictly less than 1. Note that this definition does not assume resolution of singularities.

We can now define the edges of our new directed graph. Let (X, Δ) and (Y, Δ) be two Kawamata log terminal pairs. An edge from (X, Δ) to (Y, Δ') is a birational contraction $f: X \dashrightarrow Y$ such that $f_*\Delta = \Delta'$ and f is $(K + \Delta)$-negative, i.e. there exist proper maps $p: W \to X$ and $q: W \to Y$ from a projective variety W which resolve the indeterminacy of f and such that

$$p^*(K_X + \Delta) = q^*(K_Y + \Delta') + E$$

where $E \geq 0$ is an effective $\mathbb{Q}$-divisor whose support coincides with the union of all the exceptional divisors of q. In particular, under these assumptions, if the log pair (Y, Δ) is below (X, Δ), then $H^0(X, m(K_X + \Delta)) \neq 0$ for some positive integer m if and only if the same property holds for (Y, Δ').

Thus, as in the absolute case (i.e. when $\Delta = 0$), given a Kawamata log terminal pair (X, Δ), we can investigate the graph given by pairs below (X, Δ). In particular, it is expected that there are no infinite chains below (X, Δ). As in Conjecture 13.1, this corresponds to *termination of log-flips*. Furthermore, it is expected that there exists always an end-point below (X, Δ). If $H^0(X, m(K_X + \Delta)) \neq 0$ for some positive integer m, then an end-point (Y, Γ) is characterised by the property that $K_Y + \Gamma$ is nef. The pair (Y, Γ) is called a *minimal model* of (X, Δ). In [3], it was proven that if Δ is big, then (X, Δ) always admits a minimal model.

We now consider the case of Kawamata log terminal pairs (X, Δ) such that $H^0(X, m(K_X + \Delta)) = 0$ for all positive integers m. Similarly as in the absolute case, it is expected that an end-point (Y, Δ') below (X, Δ) admits a *Mori fiber space*, which is a non-trivial morphism $\eta: Y \to Z$ such that $-(K_X + \Delta)|_F$ is ample for the general fiber F of η. In [3], it was proven that if $(K_X + \Delta)$ is not pseudo-effective (i.e. if there exists an ample $\mathbb{Q}$-divisor such that $K_X + \Delta + A$ is not big) then there exists an end-point below (X, Δ), which admits a Mori fiber space.

We now consider a special case, which illustrates the fact that the point of view of directed graphs is useful to understand the birational geometry of a Kawamata log terminal pair (X, Δ). Assume that Δ is big and that $K_X + \Delta$ is not pseudo-effective. Since Δ is big, the non-vanishing theorem proven in [3] implies that the assumption that $K_X + \Delta$ is not pseudo-effective coincides with the a-priori stronger assumption that $H^0(X, m(K_X + \Delta)) = 0$ for any positive integer m. Thus, it is natural to ask if (X, Δ) admits infinitely many end-points below (X, Δ). Note that the picture in the absolute case (i.e. when $\Delta = 0$) is very different, as we have already showed that there might be infinitely many edges starting from the vertex associated to $(X, 0)$. On the other hand, assuming that Δ is big, it is possible to show, by using boundedness of the length of extremal rays [17], that there are only finitely many edges starting from (X, Δ). Thus, König's Lemma implies that the sub-graph given by all the vertices below (X, Δ) is finite if and only if there are no infinite chains starting from (X, Δ) (see [22, Lemma 6.7] for more details). Clearly, if the sub-graph below (X, Δ) is finite, there are only finitely many end-point below (X, Δ).

13.4 Flops and Sarkisov Programme

We have seen that in general projective varieties X (and Kawamata log terminal pairs (X, Δ)) are divided into two large families, depending on the existence of a global section of mK_X (respectively $m(K_X + \Delta)$) for some positive integer m. If there is no such a section, than the picture is more complicated, as the vertices might have infinitely many edges starting from it.

Now it is natural to investigate the relations within the end-points of the directed graphs we constructed. First, assume that X is an end-point in the absolute case (i.e. with $\Delta = 0$), represented by a terminal projective variety such that K_X is nef. Then by a result of Kawamata [18], any other end-point Y in the connected component of X is connected to X by a composition of flops (see [20] for a definition of flop). Note that a flop $\varphi: X \dashrightarrow Y$ is a special K-trivial isomorphism in codimension one, i.e. if $p: W \to X$ and $q: W \to Y$ are proper morphisms which resolve the indeterminacy locus of φ, then

$$p^* K_X = q^* K_Y.$$

A similar picture holds for log pairs (see [18] for more details).

It is conjectured that any two terminal projective varieties which are isomorphic in codimension one and K-equivalent (i.e. they admit a K-trivial birational map which is an isomorphism in codimension one) are connected by a finite sequence of flops (e.g. see [24]). At the moment, the conjecture is open even in the case of smooth projective varieties of dimension three.

Finally, if X is a terminal projective variety of general type, then the number of end-points in the connected component of the graph containing X is always finite [3].

We now consider the case of uniruled projective varieties. We have seen that even in dimension two, the number of end-points in each connected component of the graph which contains a uniruled projective variety is infinite. It is therefore natural to ask about the relation within birational pairs of projective varieties which admit a Mori fiber space. More specifically let $\eta: X \to Z$ and $\eta': X \to Z'$ be two Mori fiber spaces, with X and X' terminal projective varieties which admit a birational map $\psi: X \dashrightarrow Y$. Then the goal of the Sarkisov programme is to show that φ can be decomposed into a sequence of *Sarkisov links*, which are elementary transformations obtained as compositions of flops and divisorial contractions (e.g. see [14] for more details). The programme was successfully carried out in [4, 7] in dimension three and in [14] in full generality.

Acknowledgements These notes were inspired by many discussions with colleagues and friends. In particular, the author would like to thank J. McKernan for uncountable many conversations on this topic. The author is partially supported by an EPSRC grant.

References

1. C. Birkar, Ascending chain condition for log canonical thresholds and termination of log flips. Duke Math. J. **136**(1), 173–180 (2007)
2. C. Birkar, Lectures on birational geometry. arXiv:1210.2670 (2012)
3. C. Birkar, P. Cascini, C. Hacon, J. McKernan, Existence of minimal models for varieties of log general type. J. Am. Math. Soc. **23**(2), 405–468 (2010)
4. A. Bruno, K. Matsuki, Log Sarkisov program. Int. J. Math. **8**(4), 451–494 (1997)
5. P. Cascini, V. Lazić, The minimal model program revisited, in *Contributions to Algebraic Geometry*, ed. by P. Pragacz, O. Zariski (European Mathematical Society, Zurich, 2012), pp. 169–187
6. P. Cascini, V. Lazić, New outlook on the minimal model program, I. Duke Math. J. **161**(12), 2415–2467 (2012)
7. A. Corti, Factoring birational maps of threefolds after Sarkisov. J. Algebr. Geom. **4**(2), 223–254 (1995)
8. A. Corti, Finite generation of adjoint rings after Lazić: an introduction, in *Classification of Algebraic Varieties*, ed. by A. Corti. EMS Series of Congress Reports (EMS Publishing House, Zurich, 2011), pp. 197–220
9. A. Corti, V. Lazić, New outlook on the minimal model program, II. Math. Ann. **356**(2), 617–633 (2013)
10. A. Corti, P. Hacking, J. Kollár, R. Lazarsfeld, M. Mustaţă, Lectures on flips and minimal models, in *Analytic and Algebraic Geometry*. Volume 17 of IAS/Park City Mathematics Series (American Mathematical Society, Providence, 2010), pp. 557–583
11. O. Debarre, *Higher-Dimensional Algebraic Geometry*. Universitext (Springer, New York, 2001)
12. S. Druel, Existence de modèles minimaux pour les variétés de type général (d'après Birkar, Cascini, Hacon et McKernan). Astérisque, (326):Exp. No. 982, vii, 1–38 (2010), 2009. Séminaire Bourbaki, vol. 2007/2008
13. A. Grassi, Birational geometry old and new. Bull. Am. Math. Soc. (N.S.) **46**(1), 99–123 (2009)
14. C. Hacon, J. McKernan, The Sarkisov program. J Algebr. Geom. **22**(2), 389–405 (2009)
15. C. Hacon, J. McKernan, Existence of minimal models for varieties of log general type II. J. Am. Math. Soc. **23**(2), 469–490 (2010)
16. C. Hacon, J. McKernan, C. Xu, ACC for log canonical thresholds Ann. Math. **180**(2), 523–571 (2014)
17. Y. Kawamata, On the length of an extremal rational curve. Invent. Math. **105**, 609–611 (1991)

18. Y. Kawamata, Flops connect minimal models. Publ. Res. Inst. Math. Sci. **44**(2), 419–423 (2008)
19. Y. Kawamata, Finite generation of a canonical ring, in *Current Developments in Mathematics, 2007* (International Press, Somerville, 2009), pp. 43–76
20. J. Kollár, S. Mori, *Birational Geometry of Algebraic Varieties*. Volume 134 of Cambridge Tracts in Mathematics (Cambridge University Press, Cambridge/New York, 1998)
21. J. Kollár et al., *Flips and Abundance for Algebraic Threefolds*. (Société Mathématique de France, Paris, 1992). Papers from the second summer seminar on algebraic geometry held at the University of Utah, Salt Lake City, Aug 1991, Astérisque No. 211 (1992)
22. B. Lehmann, A cone theorem for nef curves. J. Algebr. Geom. **21**(3), 473–493 (2012)
23. K. Matsuki, *Introduction to the Mori program*. Universitext (Springer, New York, 2002)
24. C.-L. Wang, K-equivalence in birational geometry and characterizations of complex elliptic genera. J. Algebr. Geom. **12**(2), 285–306 (2003)

Chapter 14
Dynkin and Extended Dynkin Diagrams

Idun Reiten

Abstract In this paper we give a survey of some of the occurrences of Dynkin and extended Dynkin diagrams in algebra. It is based on my lecture at the INdAM day in 2009, with some later developments included.

14.1 Introduction

Most of this paper is based on a lecture which I gave at the INdAM day in 2009. The aim is to discuss various occurrences of Dynkin and extended Dynkin diagrams in algebra, with no claim of being complete. Often the occurrence can be traced back to elementary considerations of (sub)additive functions and quadratic forms, so we start in Sect. 14.2 with a discussion of these concepts. We illustrate with easy examples.

In Sect. 14.3 we deal with path algebras kQ of a finite connected acyclic quiver Q, that is, a finite connected oriented graph with no oriented cycles, where k is an algebraically closed field. The quiver Q being Dynkin corresponds to a finiteness condition for kQ. In Sect. 14.4 we discuss almost split sequences and AR-quivers, with focus on special classes of finite dimensional algebras and invariant rings of Krull dimension two. Here we apply the elementary considerations from Sect. 14.2.

In Sect. 14.5 we deal with the theory of cluster algebras, initiated by Fomin and Zelevinsky. Also here the Dynkin diagrams appear in connection with some finiteness conditions. In Sect. 14.6 we discuss quiver mutation, introduced by Fomin and Zelevinsky in connection with their definition of cluster algebras. When Q is acyclic, and has at least three vertices, the associated mutation class is finite if and only if the acyclic quiver we start with is Dynkin or extended Dynkin.

I. Reiten (✉)
Institutt for matematiske fag, NTNU, N-7491 Trondheim, Norway
e-mail: idunr@math.ntnu.no

V. Ancona and E. Strickland (eds.), *Trends in Contemporary Mathematics*, Springer INdAM Series 8, DOI 10.1007/978-3-319-05254-0_14,

In the last section we consider some algebraic objects associated with finite connected acyclic quivers. We have already mentioned path algebras associated with such a quiver, and in addition there are associated Coxeter groups and preprojective algebras. Also in the last two cases Q being Dynkin is equivalent to a finiteness condition. For the algebraic objects associated with the same Q, there are some interesting connections which we discuss.

There is some overlap with [32] in Sects. 14.2–14.4. For completeness, it was convenient to include some of these preliminary results. We refer to [32] for a more detailed discussion.

14.2 Elementary Considerations

In this section we discuss some elementary properties of graphs, which lead to the considerations of Dynkin and extended Dynkin diagrams (or graphs).

We start with listing the (simply laced) Dynkin and extended Dynkin diagrams.

Dynkin Diagrams: A_n, D_n, E_6, E_7, E_8 Extended Dynkin Diagrams: $\widetilde{A}_n, \widetilde{D}_n, \widetilde{E}_6, \widetilde{E}_7, \widetilde{E}_8$

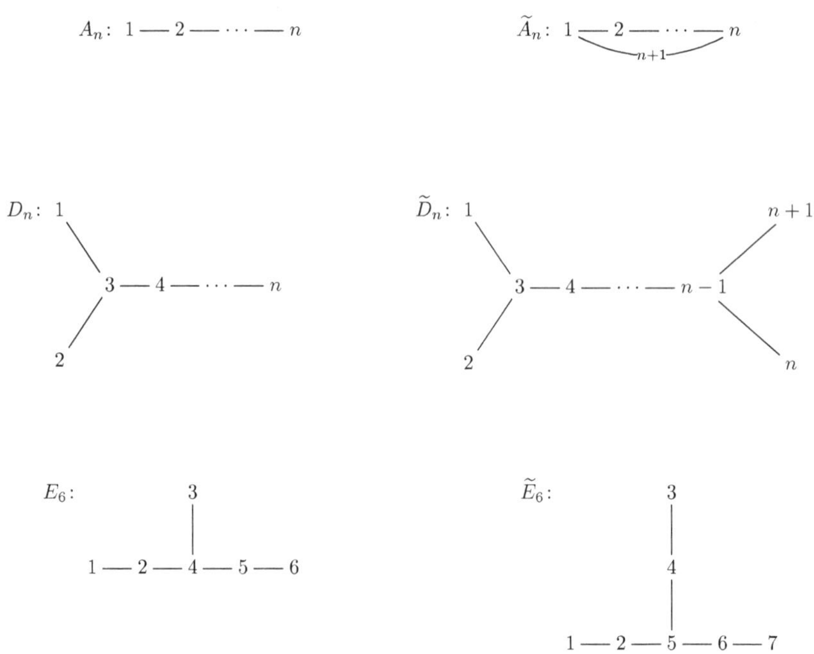

E_7:

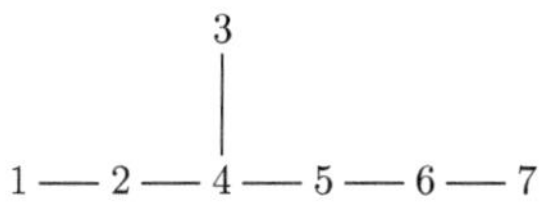

$\widetilde{E}_7$:

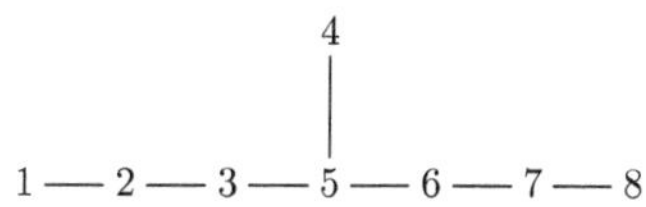

E_8:

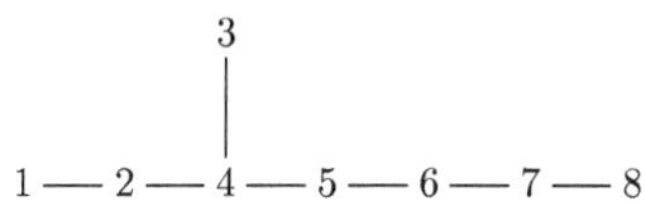

$\widetilde{E}_8$:

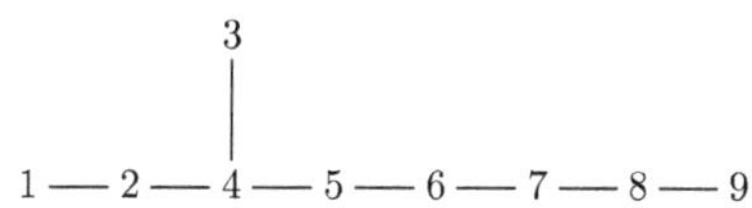

14.2.1 Additive and Subadditive Functions

Let Γ be a connected graph, with no loops •⊃. But multiple edges are allowed, so for example • —— • ═ •. We denote the vertices by $1, \dots, n$. A function $a\colon \Gamma_0 \longrightarrow \mathbb{N}$, where Γ_0 denotes the vertices of Γ and $\mathbb{N}$ the positive integers, is **additive** if for each $i = 1, \dots, n$ we have

$$2a(i) = \sum_{i-j} a(j)$$

The function is **subadditive** if $2a(i) \geq \sum_{i-j} a(j)$ for all $i = 1, \dots, n$.

Interesting questions are then the following:

- Which graphs admit an additive function ?
- Which graphs admit a function which is subadditive, but not additive ?

We illustrate with some examples.

Example 14.1. (i) For the graph • —— •, 3 —— 4 gives a subadditive function which is not additive: $2 \cdot 4 > 3$, $2 \cdot 3 > 4$. There is no additive function since $2a_2 = a_1$ and $2a_1 = a_2$ is not possible for positive integers a_1 and a_2.

(ii) • ═ • has the additive function 1 ═ 1, and has no subadditive function which is not additive.

(iii) The graph

has no (sub)additive function. For $2a_1 \geq 3a_2$, $2a_2 \geq 3a_1$ is not possible for positive integers a_1, a_2, a_3.

(iv) 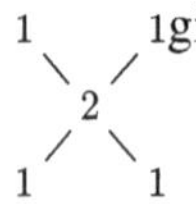

gives an additive function for 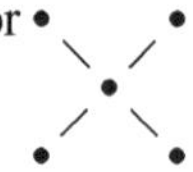.

The general answer is the following (see [26]).

Proposition 14.1. *Let Γ be a finite connected graph with no loops.*

(*a*) *There is a subadditive not additive function for Γ if and only if Γ is a Dynkin diagram.*

(*b*) *There is an additive function for Γ if and only if Γ is an extended Dynkin diagram.*

14.2.2 Quadratic Forms

Let again Γ be a connected graph without loops, and with vertices $1, \ldots, n$. We denote by $\mathbb{R}$ the real numbers. Then we have an associated quadratic form $q_\Gamma \colon \mathbb{R}^n \longrightarrow \mathbb{R}$, given by

$$q_\Gamma(x_1, \ldots, x_n) = \sum_{i=1}^{n} x_i^2 - \sum_{i-j} x_i x_j.$$

We have the following natural questions:

- For which graphs is q_Γ positive definite, that is, $q_\Gamma(x_1, \ldots, x_n) > 0$ for $(x_1, \ldots, x_n) \neq (0, \ldots, 0)$?
- For which graphs is q_Γ positive semidefinite, that is, $q_\Gamma(x_1, \ldots, x_n) \geq 0$ for all $(x_1, \ldots, x_n)$, but not positive definite ?

We illustrate with some examples.

Example 14.2. (i) For $1 — 2$ we have $q_\Gamma(x_1, x_2) = x_1^2 + x_2^2 - x_1 x_2 = (x_1 - \frac{1}{2}x_2)^2 + \frac{3}{4}x_2^2 > 0$ when $(x_1, x_2) \neq (0, 0)$. Hence q_Γ is positive definite.

(ii) For $1 = 2$ we have $q_\Gamma(x_1, x_2) = x_1^2 + x_2^2 - 2x_1 x_2 = (x_1 - x_2)^2 \geq 0$, so q_Γ is positive semidefinite, but not positive definite since $q_\Gamma(1, 1) = 0$.

(iii) For

we have $q_\Gamma(x_1, x_2) = x_1^2 + x_2^2 - 3x_1 x_2 = (x_1 - x_2)^2 - x_1 x_2$. Then $q_\Gamma(1, 1) = -1$, so Γ is neither positive definite nor positive semidefinite.

In general the following is known.

Proposition 14.2. (*a*) *For a finite connected graph Γ without loops, then q_Γ is positive definite if and only if Γ is Dynkin.*

(*b*) *For a finite connected graph* Γ *without loops, then* q_Γ *is positive semidefinite, but not positive definite, if and only if* Γ *is extended Dynkin.*

Note that it follows that the two sets of conditions discussed in this section are actually equivalent (see [32]). Many of the occurrences of Dynkin or extended Dynkin diagrams in mathematics can be traced back to these elementary conditions, and we shall see examples of this.

14.3 Path Algebras of Quivers

In this section we start with a finite connected acyclic quiver Q, and associate with it the path algebra kQ, where k is an algebraically closed field. Then it turns out that Q being Dynkin is equivalent to some finiteness condition for kQ.

As a vector space over k, the paths in Q, including the trivial paths e_i associated with the vertices i, form a k-basis for kQ. Multiplication of two paths is given by composition when possible, and is defined to be 0 otherwise. This induces a multiplication on kQ. Then kQ is a finite dimensional algebra since Q is acyclic.

For example, let Q be the quiver $1 \xrightarrow{\alpha} 2 \xrightarrow{\beta} 3$. Then $\{e_1, e_2, e_3, \alpha, \beta, \beta\alpha\}$ are the paths in Q. For the multiplication of paths (where we start on the right) we have for example $\alpha \cdot e_1 = \alpha$, $\beta \cdot \alpha = \beta\alpha$, $\alpha \cdot \beta = 0$. In this case there is only a finite number of (finite dimensional) indecomposable kQ-modules, namely 6, as we shall explain below.

There is another more concrete way of describing the (finite dimensional) kQ-modules, namely as (finite dimensional) representations of quivers rep Q. We illustrate with the same example. A representation of $1 \xrightarrow{\alpha} 2 \xrightarrow{\beta} 3$ can be written as

$$V_1 \xrightarrow{f_\alpha} V_2 \xrightarrow{f_\beta} V_3$$

where V_1, V_2, V_3 are finite dimensional vector spaces over k, and f_α, f_β are linear transformations. A map from

$$V_1 \xrightarrow{f_\alpha} V_2 \xrightarrow{f_\beta} V_3$$

to

$$V_1' \xrightarrow{f_\alpha'} V_2' \xrightarrow{f_\beta'} V_3'$$

is a triple (u, v, w) of linear transformations $u \colon V_1 \longrightarrow V_1'$, $v \colon V_2 \longrightarrow V_2'$, $w \colon V_3 \longrightarrow V_3'$ such that the following diagram commutes

$$\begin{array}{ccccc} V_1 & \xrightarrow{f_\alpha} & V_2 & \xrightarrow{f_\beta} & V_3 \\ \downarrow{\scriptstyle u} & & \downarrow{\scriptstyle v} & & \downarrow{\scriptstyle w} \\ V_1' & \xrightarrow{f_\alpha'} & V_2' & \xrightarrow{f_\beta'} & V_3' \end{array}$$

A direct sum of two representations is defined in the natural way. Then we obtain the category rep Q of representations of Q. We have the following relationship to mod kQ.

Proposition 14.3. *For a finite connected acyclic quiver Q we have an equivalence of categories*

$$\operatorname{mod} kQ \xrightarrow{\simeq} \operatorname{rep} Q$$

A famous result of Gabriel from the early seventies is the following [23].

Theorem 14.1. *Let Q be a finite connected acyclic quiver and k an algebraically closed field. Then there is only a finite number of indecomposable (finite dimensional) kQ-modules (or equivalently of indecomposable (finite dimensional) representations of Q) if and only if the underlying graph $|Q|$ of Q is Dynkin.*

So we see that the underlying graph of a quiver Q being Dynkin is strongly connected with a finiteness condition for kQ.

Gabriel also proved that when Q is a quiver such that $|Q|$ is Dynkin, there is a 1–1 correspondence between the positive roots for $|Q|$ from Lie theory and the dimension vectors for indecomposable representations of Q. Inspired by this, Tits gave an argument using geometry for the fact that if kQ is of finite representation type (that is, there is only a finite number of indecomposable kQ-modules), then $q_{|Q|}$ is positive definite (see [9]). Hence he obtained a new proof for the fact that if kQ is of finite representation type, then Q is Dynkin. And Bernstein-Gelfand-Ponomarev were inspired to give a new proof of Gabriel's theorem [9]. This work has had a major influence on later work on the representation theory of finite dimensional algebras. Also the extended Dynkin diagrams appear here. They correspond to the class of tame path algebras (see [32] and the references cited there for more details).

14.4 Almost Split Sequences and AR-Quivers

Dynkin diagrams also appear in a different way for some classes of finite dimensional algebras, and for some classes of commutative rings of Krull dimension two. This is in both cases through considering almost split sequences introduced in [4] and the associated AR-quiver, which we define below (see [1, 6, 33]).

Let A be a finite dimensional k-algebra. For any indecomposable (finite dimensional) A-module X which is not projective, there is a special kind of exact sequence

$$0 \longrightarrow Z \xrightarrow{f} Y \xrightarrow{g} X \longrightarrow 0$$

called an **almost split sequence**, which is uniquely determined by X up to isomorphism. It is not a split sequence, Z is indecomposable, and for any non-isomorphism $s: U \longrightarrow X$, where U is indecomposable, there is some A-homomorphism $t: U \longrightarrow Y$ such that $gt = s$. Such sequences exist for any indecomposable non-projective A-module. An indecomposable projective module is a direct summand of the A-module A and an indecomposable injective A-module is a direct summand of the A-module $\mathrm{Hom}_k(A, k)$. The k-algebra A is said to be **selfinjective** when the projective and the injective modules coincide.

There is a quiver, called the **AR-quiver**, associated with any finite dimensional k-algebra via the almost split sequences. There is a vertex corresponding to each indecomposable A-module, up to isomorphism, and if the middle term Y of an almost split sequence is isomorphic to $Y_1^{s_1} \oplus \cdots \oplus Y_t^{s_t}$, where all Y_i are indecomposable nonisomorphic, then there are s_i arrows from Y_i to X and from Z to Y_i, for each $i = 1, \ldots, t$. There is an operation τ on the AR-quiver, sending the right term of an almost split sequence to the left term.

We illustrate with the following.

Example 14.3. When $A = k[x]/(x^4)$, we have the indecomposable modules A, $k[x]/(x^3)$, $k[x]/(x^2)$ and $k[x]/(x)$. The almost split sequences are of the form

$$0 \longrightarrow k[x]/(x) \longrightarrow k[x]/(x^2) \longrightarrow k[x]/(x) \longrightarrow 0$$

$$0 \longrightarrow k[x]/(x^2) \longrightarrow k[x]/(x^3) \oplus k[x]/(x) \longrightarrow k[x]/(x^2) \longrightarrow 0$$

$$0 \longrightarrow k[x]/(x^3) \longrightarrow k[x]/(x^2) \oplus A \longrightarrow k[x]/(x^3) \longrightarrow 0$$

The AR-quiver is:

$$k[x]/(x) \rightleftarrows k[x]/(x^2) \rightleftarrows k[x]/(x^3) \rightleftarrows A$$

If we remove A, which is projective, and replace $\rightleftarrows$ by ——, and replace the other indecomposable modules by their length (or k-dimension), then we get $1 \text{——} 2 \text{——} 3$, which gives a subadditive not additive function. This explains why the graph $\bullet \text{——} \bullet \text{——} \bullet$ is Dynkin.

For an arbitrary selfinjective algebra of finite representation type, a Dynkin diagram occurs in a similar way. Now we get a Dynkin diagram where the vertices correspond to τ-orbits of indecomposable modules [34] (see [26, 35] for the idea of using dimension vectors).

We explain that also for a class of two-dimensional invariant rings there is an AR-quiver which is closely related to Dynkin and extended Dynkin diagrams (see [2, 5]). Let $R = k[[X, Y]]$ be the power series ring in two variables, where we now assume that the algebraically closed field k has characteristic 0. Let $G \subset \mathsf{SL}(2, k)$

where G is a finite group, and $R = k[[X,Y]]^G$ the corresponding invariant ring. Let $\mathsf{CM}(R)$ be the finitely generated (maximal) Cohen-Macaulay modules over R. This is a certain subcategory of the category of finitely generated R-modules. Also for $\mathsf{CM}(R)$ we have existence of almost split sequences [3], and there are only a finite number of indecomposable modules in $\mathsf{CM}(R)$ in this case. The definition of almost split sequences is similar to the definition for finite dimensional algebras. As for the finite dimensional algebra $k[x]/(x^4)$, we have that the left and right term of an almost split sequence are isomorphic, and the projective R-module R occurs neither on the left nor on the right of an almost split sequence. But in this case, contrary to what holds for $k[x]/(x^4)$, there is a special exact sequence

$$0 \longrightarrow R \xrightarrow{f} E \xrightarrow{g} R$$

called the **fundamental exact sequence**, with properties similar to properties of almost split sequences, except for $g\colon E \longrightarrow R$ being surjective. Also in this setting there is a function r (here called rank), such that $r(U)+r(W)$ for an exact sequence $0 \longrightarrow U \longrightarrow V \longrightarrow W \longrightarrow 0$ in $\mathsf{CM}(R)$. In addition we have $r(E) = 2r(R)$. Then by using the criterion for additive and subadditive functions, we get an extended Dynkin diagram when replacing $\rightleftarrows$ by —— and we get a Dynkin diagram when the vertex corresponding to R is removed. Here all the Dynkin and extended Dynkin diagrams appear for some choice of the finite group G.

The occurrence of Dynkin and extended Dynkin diagrams here is related to the occurrence for McKay quivers and for resolutions of singularities (see [30]).

14.5 Cluster Algebras of Finite Type

Cluster algebras were introduced by Fomin and Zelevinsky in [21], and they have since then had an enormous influence on many different areas of mathematics, including quiver representations. The definition is somewhat complicated, so we do not give it in full generality, but instead we indicate the main ideas by treating a special case. Cluster algebras are included here since one of the main theorems of Fomin and Zelevinsky on the subject was that cluster algebras are of finite type if and only if there is an associated Dynkin quiver [22].

It is convenient to first introduce the concept of quiver mutation, which is an essential ingredient in the definition of cluster algebras. This concept also appears in the physics literature.

We start with a finite quiver Q which has no loops $\bullet\circlearrowright$ or two-cycles $\rightleftarrows$, and vertices labeled $1,\ldots,n$. For each vertex $i = 1,\ldots,n$ we define $\mu_i(Q)$ as follows. We reverse all arrows entering or leaving i. For each pair of vertices (r,s) in Q where we have $a > 0$ arrows from r to i and $b > 0$ arrows from i to s in Q, we draw ab arrows from r to s. Then remove all 2-cycles to get $\mu_i(Q)$.

We illustrate with some examples.

Example 14.4. (i) Let Q be the quiver $1 \longrightarrow 2 \longrightarrow 3$.
Then $\mu_1(Q)\colon 1 \longleftarrow 2 \longrightarrow 3$ and $\mu_2(Q)$: 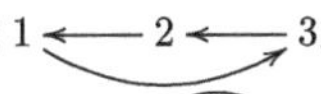.

(ii) If Q is the quiver 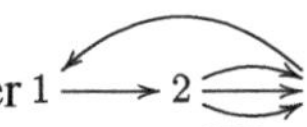 then $\mu_2(Q)$: 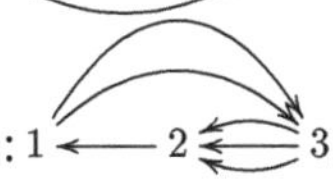.

(iii) Let Q be the quiver 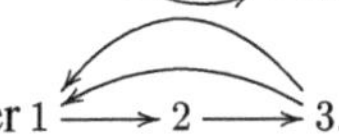.
Then $\mu_2(Q)$ is 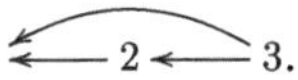.

Now we illustrate how to define a cluster algebra through an example.

Example 14.5. We start with a finite acyclic quiver $Q\colon 1 \longrightarrow 2$. Let $F = \mathbb{Q}(x_1, x_2)$ be the function field in two variables over the field of rational numbers $\mathbb{Q}$. Then we have the **initial seed** $(\{x_1, x_2\},\ 1 \longrightarrow 2\ )$, where $\{x_1, x_2\}$ is a free generating set of F over $\mathbb{Q}$, and Q is the quiver with two vertices we started with. This is an **acyclic cluster algebra** since one of the quivers is acyclic (and it has no "coefficients").

We apply a mutation μ_1 to this initial seed to get a new seed. The new quiver is obtained by quiver mutation of the old one at vertex 1. The new 2-tuple in F is obtained from the previous seed as follows. We replace x_1 with x_1', where $x_1 x_1' = m_1 + m_2$. Here m_1 and m_2 are two monomials, where m_1 is determined by the end points of the arrows leaving 1, and m_2 by the end points of the arrows entering 1. Since the only arrow leaving 1 has end point 2, and there is exactly one arrow, we get $m_1 = x_2^1 = x_2$. We get $m_2 = 1$ since there is no arrow entering 1. Hence it follows that $x_1 x_1' = 1 + x_2$, so that $x_1' = \frac{1+x_2}{x_1}$. Then we obtain the new seed $(\{\frac{1+x_2}{x_1}, x_2\},\ 1 \longleftarrow 2\ )$. Applying μ_1 again, we get back to the original seed. Applying μ_2, we replace x_2 by a new cluster variable x_2'', obtained by $x_2 x_2'' = \frac{1+x_2}{x_1} + 1 = \frac{1+x_1+x_2}{x_1}$, so that $x_2'' = \frac{1+x_1+x_2}{x_1 x_2}$. Continuing we get the picture

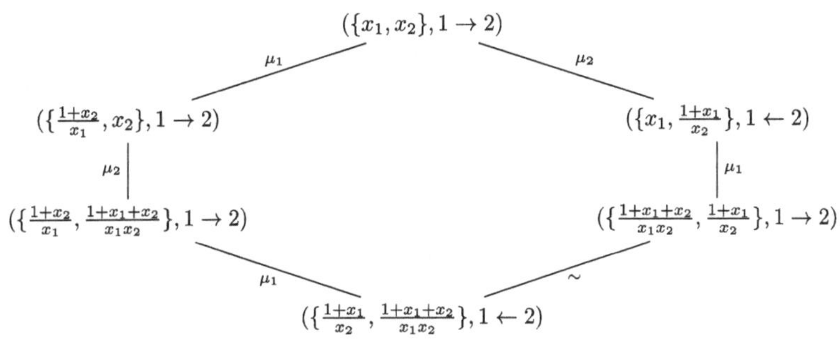

The 2-element subsets are the **clusters**, and the elements $x_1, x_2, \frac{1+x_2}{x_1}, \frac{1+x_1}{x_2}, \frac{1+x_1+x_2}{x_1 x_2}$ appearing in the clusters are the **clusters variables**. There are five seeds, five clusters and five cluster variables in this case.

By definition the cluster algebra is of finite type if there is only a finite number of cluster variables. The following was proved in [22], here stated for the socalled skew symmetric case.

Theorem 14.2. *A cluster algebra determined by a quiver is of finite type if and only if one of the quivers appearing in the seeds is a Dynkin quiver.*

When the quiver Q in the initial seed for the cluster algebra is acyclic (and there are no socalled coefficients) this result can be explained as follows. There is a bijection between the cluster variables not in the initial seed and the indecomposable rigid kQ-modules, that is, the indecomposable X with $\mathrm{Ext}^1_{kQ}(X,X) = 0$ (see [14, 15, 17, 18]). It is then enough to observe that kQ has only a finite number of indecomposable rigid modules if and only if the path algebra kQ is of finite representation type, which is well known. Hence the cluster algebra is of finite type if and only if kQ is of finite representation type.

14.6 Finite Mutation Classes

We have mentioned that there is only a finite number of cluster variables if and only if one of the quivers in the mutation class is Dynkin. However, having finite mutation classes is true more generally. But we still have a nice connection with the Dynkin and extended Dynkin diagrams in the acyclic case [11].

Theorem 14.3. *Let Q be a finite connected acyclic quiver with $n \geq 3$ vertices. Then the mutation class of Q is finite if and only if the underlying graph $|Q|$ of Q is a Dynkin or extended Dynkin diagram.*

This has been generalized in [20] to give a complete characterization of when we have finite mutation type.

The proof of Theorem 14.3 in [11] used the cluster category $\mathscr{C}_Q$ associated with a finite acyclic quiver Q with n vertices, as introduced in [12]. This is a triangulated category [28], and it is defined as an orbit category $\mathsf{D}^b(\mathrm{mod}\, kQ)/\tau^{-1}[1]$ of the bounded derived category of $\mathrm{mod}\, kQ$, where τ denotes the AR-translation in $\mathsf{D}^b(\mathrm{mod}\, kQ)$ coming from what is called almost split triangles. Roughly speaking, the indecomposable objects in $\mathscr{C}_Q$ are given by the indecomposable objects in $\mathrm{mod}\, kQ$, together with n additional ones. There are special objects in $\mathscr{C}_Q$ called **cluster-tilting objects** T, defined by $\mathrm{Ext}^1_{kQ}(T,T) = 0$ and $|T| = n$, where $|T|$ denotes the number of nonisomorphic indecomposable summands of T. An alternative description is that they are induced by tilting modules over kQ or over a path algebra derived equivalent to kQ. A kQ-module T is a tilting kQ-module if $\mathrm{Ext}^1_{kQ}(T,T) = 0$ and $|T| = n$. The endomorphism algebras $\mathrm{End}_{\mathscr{C}_Q}(T)$ are by definition the **cluster tilted algebras** [13]. The following description of the mutation class of Q could then be used to prove Theorem 14.3 [15].

Theorem 14.4. *The quivers in the mutation class of an acyclic quiver Q are exactly the quivers of the cluster-tilted algebras associated with the cluster category $\mathscr{C}_Q$.*

The cluster tilted algebras are closely related to the tilted algebras, which are by definition the algebras of the form $\mathrm{End}_{kQ}(T)$, where T is a tilting kQ-module.

14.7 Coxeter Groups, Preprojective Algebras and Path Algebras

When we have a finite connected acyclic quiver Q, we have seen that we can associate with it a finite dimensional path algebra kQ, and that kQ is of finite (representation) type if and only if Q is Dynkin. We can also associate with Q a group, called the **Coxeter group** (see [10]). In addition we have the preprojective algebra Π_Q associated with Q.

We start with giving the relevant definitions. Let $1, \ldots, n$ be the vertices of Q. Then the associated Coxeter group W_Q has generators $s_1, \ldots, s_n$. The relations are given as follows: $s_i^2 = 1$ for all $i = 1, \ldots, n$; $s_i s_j = s_j s_i$ if there is no arrow between i and j; for $i \neq j, s_i s_j s_i = s_j s_i s_j$ if there is exactly one arrow between i and j. We have the following (see [10]).

Theorem 14.5. *Let Q be as above. The Coxeter group W_Q is finite if and only if Q is a Dynkin quiver.*

The preprojective algebra Π_Q is defined as follows. Denote by $\overline{Q}$ the quiver where for each arrow in Q we have added a new arrow a^* in the opposite direction, to get from

$$i \xrightarrow{a} j \quad \text{to} \quad i \underset{a^*}{\overset{a}{\rightleftarrows}} j$$

Then $\Pi_Q = k\overline{Q}/\sum_{a \in Q}(aa^* - a^*a)$. We have the following (see [19, 25, 34] and [7, 8, 29]).

Theorem 14.6. *Let Q be as above.*

(*a*) *The preprojective algebra Π_Q is finite dimensional if and only if Q is Dynkin.*
(*b*) *The preprojective algebra Π_Q is noetherian if and only if Q is Dynkin or extended Dynkin.*

For a fixed quiver Q as above we then have both a Coxeter group, a preprojective algebra and a path algebra. It is interesting to see if there is some connection between the group W_Q and the algebras Π_Q and kQ, as we now discuss.

We first recall some results from [16, 24]. Let $w \in W_Q$. Let $\underline{w} = s_{i_1} \ldots s_{i_t}$ be a reduced expression for the element w in W_Q, that is, t is smallest possible. For each $i = 1, \ldots, n$, there is a maximal ideal $I_i = \Pi_Q(1 - e_i)\Pi_Q$ in Π_Q. Associated with the reduced expression we have the ideal $I_{\underline{w}} = I_{i_1} \ldots I_{i_t}$. This turns out to be independent of the reduced expression, so we can write I_w (see [16, 27]). Associated with $w \in W_Q$ we then also have the algebra Π_Q/I_w, which is finite dimensional. This algebra is Gorenstein of dimension at most one, and $\mathsf{Sub}(\Pi_Q/I_w)$, the submodules of finite direct sums of copies of Π_Q/I_w, has the associated stable category $\underline{\mathsf{Sub}}(\Pi_Q/I_w)$, which is 2-Calabi-Yau [16]. Such categories are interesting in connection with categorification of cluster algebras.

Similarly it is natural to look for a connection between kQ and W_Q, suggested by the fact that they are constructed from the same Dynkin quiver. This is illustrated by some recent work in [31].

Theorem 14.7. (*a*) *There is a natural bijection between elements in* W_Q *and cofinite quotient closed subcategories* $\mathscr{C}_w$ *of* mod kQ.

(*b*) *For* $w \in W_Q$, *then* $\mathscr{C}_w$ *is the additive category generated by the indecomposable summands of the ideal* I_w *in* Π_Q *as a* kQ*-module, together with the preinjective and regular* kQ*-modules if* Q *is not Dynkin.*

So in the above result both the path algebra, the Coxeter group and the preprojective algebra associated to Q are involved.

References

1. I. Assem, D. Simson, A. Skowroński, *Elements of the Representation Theory of Associative Algebras. Vol. 1. Techniques of Representation Theory*. London Mathematical Society Student Texts, vol. 65 (Cambridge University Press, Cambridge, 2006). pp. x+458
2. M. Auslander, Rational singularities and almost split sequences. Trans. Am. Math. Soc. **293**, 511–531 (1986)
3. M. Auslander, Isolated singularities and existence of almost split sequences, in *Representation Theory, II*, Ottawa, 1984. Lecture Notes in Mathematics, vol. 1178 (Springer, Heidelberg, 1986), pp. 194–242
4. M. Auslander, I. Reiten, Representation theory of Artin algebras. III. Almost split sequences. Commun. Algebra **3**, 239–294 (1975)
5. M. Auslander, I. Reiten, Almost split sequences for rational double points. Trans. Am. Math. Soc. **302**, 87–97 (1987)
6. M. Auslander, I. Reiten, S.O. Smalø, *Representation Theory of Artin Algebras*. Cambridge Studies in Advanced Mathematics, vol. 36 (Cambridge University Press, Cambridge, 1995), pp. xiv+423
7. D. Baer, Homological properties of wild hereditary Artin algebras, in *Representation Theory, I*, Ottawa, 1984. Lecture Notes in Mathematics, vol. 1177 (Springer, New York, 1986), pp. 1–12
8. D. Baer, W. Geigle, H. Lenzing, The preprojective algebra of a tame hereditary Artin algebra. Commun. Algebra **15**, 425–457 (1987)
9. I.N. Bernstein, I.M. Gelfand, V.A. Ponomarev, Coxeter functors, and Gabriel's theorem (Russian). Uspehi Mat. Nauk **28**, 19–33 (1973)
10. A. Björner, F. Brenti, *Combinatorics of Coxeter Groups*. Graduate Texts in Mathematics, vol. 231 (Springer, New York, 2005), pp. xiv+363
11. A.B. Buan, I. Reiten, Acyclic quivers of finite mutation type. Int. Math. Res. Not. **12804**, 10 (2006)
12. A.B. Buan, R.J. Marsh, M. Reineke, I. Reiten, G. Todorov, Tilting theory and cluster combinatorics. Adv. Math. **204**, 572–618 (2006)
13. A.B. Buan, R.J. Marsh, I. Reiten, Cluster-tilted algebras. Trans. Am. Math. Soc. **359**, 323–332 (2007)
14. A.B. Buan, R.J. Marsh, I. Reiten, G. Todorov, Clusters and seeds in acyclic cluster algebras, with an appendix coauthored in addition by P. Caldero and B. Keller. Proc. Am. Math. Soc. **135**, 3049–3060 (2007)
15. A.B. Buan, R.J. Marsh, I. Reiten, Cluster mutation via quiver representations. Comment. Math. Helv. **83**, 143–177 (2008)

16. A.B. Buan, O. Iyama, I. Reiten, J. Scott, Cluster structures for 2-Calabi-Yau categories and unipotent groups. Compos. Math. **145**, 1035–1079 (2009)
17. P. Caldero, B. Keller, From triangulated categories to cluster algebras. II. Ann. Sci. École Norm. Sup. (4) **39**, 983–1009 (2006)
18. P. Caldero, B. Keller, From triangulated categories to cluster algebras. Invent. Math. **172**, 169–211 (2008)
19. V. Dlab, C.M. Ringel, The preprojective algebra of a modulated graph, in *Representation Theory, II. Proceedings of Second International Conference*, Carleton University, Ottawa, 1979. Lecture Notes in Mathematics, vol. 832 (Springer, Berlin/Heidelberg, 1980), pp. 216–231
20. A. Felikson, M. Shapiro, P. Tumarkin, Skew-symmetric cluster algebras of finite mutation type. J. Eur. Math. Soc. (JEMS) **14**, 1135–1180 (2012)
21. S. Fomin, A. Zelevinsky, Cluster algebras. I. Foundations. J. Am. Math. Soc. **15**, 497–529 (2002)
22. S. Fomin, A. Zelevinsky, Cluster algebras. II. Finite type classification. Invent. Math. **154**, 63–121 (2003)
23. P. Gabriel, Unzerlegbare Darstellungen. I. Manuscr. Math. **6**, 71–103 (1972)
24. C. Geiss, B. Leclerc, J. Schröer, Kac-Moody groups and cluster algebras. Adv. Math. **228**, 329–433 (2011)
25. I.M. Gelfand, V.A. Ponomarev, Model algebras and representations of graphs. Funct. Anal. Appl. **13**, 157–166 (1980)
26. D. Happel, U. Preiser, C.M. Ringel, Vinberg's characterization of Dynkin diagrams using subadditive functions with application to DTr-periodic modules, in *Representation Theory, II. Proceedings of Second International Conference*, Carleton University, Ottawa, 1979. Lecture Notes in Mathematics, vol. 832 (Springer, Berlin/Heidelberg, 1980), pp. 280–294
27. O. Iyama, I. Reiten, Fomin-Zelevinsky mutation and tilting modules over Calabi-Yau algebras. Am. J. Math. **130**, 1087–1149 (2008)
28. B. Keller, On triangulated orbit categories. Doc. Math. **10**, 551–581 (2005)
29. H. Lenzing, Homological transfer from finitely presented to infinite modules, in *Abelian Group Theory*. Lecture Notes in Mathematics, vol. 1006 (Springer, Berlin/Heidelberg, 1983), pp. 734–761
30. J. McKay, Graphs, singularities, and finite groups, in *The Santa Cruz Conference on Finite Groups*, University of California, Santa Cruz, 1979. Proceedings of Symposia in Pure Mathematics, vol. 37 (American Mathematical Society, Providence, 1980), pp. 183–186
31. S. Oppermann, I. Reiten, H. Thomas, Quotient closed subcategories of quiver representations (2012). arXiv:1205.3268, Compos. Math. (to appear)
32. I. Reiten, Dynkin diagrams and the representation theory of algebras. Not. Am. Math. Soc. **44**, 546–556 (1997)
33. C.M. Ringel, *Tame Algebras and Integral Quadratic Forms*. Lecture Notes in Mathematics, vol. 1099 (Springer, Berlin, 1984), pp. xiii+376
34. C. Riedtmann, Algebren, Darstellungsköcher, Überlagerungen und zurück. Comment. Math. Helv. **55**, 199–224 (1980)
35. G. Todorov, Almost split sequences for TrD-periodic modules, in *Representation Theory, II. proceedings of the Second International Conference*, Carleton University, Ottawa, 1979. Lecture Notes in Mathematics, vol. 832 (Springer, Berlin/Heidelberg, 1980), pp. 600–631

Chapter 15
Tracking Control of 1D Scalar Conservation Laws in the Presence of Shocks

Rodrigo Lecaros and Enrique Zuazua

Abstract We analyze a model tracking problem for a 1D scalar conservation law. It consists in optimizing the initial datum so to minimize a weighted distance to a given target during a given finite time horizon.

Even if the optimal control problem under consideration is of classical nature, the presence of shocks is an impediment for classical methods, based on linearization, to be directly applied.

We adapt the so-called alternating descent method that exploits the generalized linearization that takes into account both the sensitivity of the shock location and of the smooth components of solutions. This method was previously applied successfully on the inverse design problem and that of identifying the nonlinearity in the equation.

Partially supported by Grants MTM2008-03541 and MTM2011-29306 of MICINN Spain, Project PI2010-04 of the Basque Government, ERC Advanced Grant FP7-246775 NUMERIWAVES and ESF Research Networking Programme OPTPDE. The first author was partially supported by Basal-CMM project.
This work was finished while the second author was visiting the Laboratoire Jacques Louis Lions with the support of the Paris City Hall "Research in Paris" program.

R. Lecaros (✉)
BCAM – Basque Center for Applied Mathematics, Mazarredo 14, E-48009 Bilbao, Basque Country, Spain

CMM – Centro de Modelamiento Matemático, Universidad de Chile (UMI CNRS 2807), Avenida Blanco Encalada 2120, Casilla 170-3, Correo 3, Santiago, Chile
e-mail: rlecaros@dim.uchile.cl

E. Zuazua
BCAM – Basque Center for Applied Mathematics, Mazarredo 14, E-48009 Bilbao, Basque Country, Spain

Ikerbasque – Basque Foundation for Science, Alameda Urquijo 36-5, Plaza Bizkaia, 48011 Bilbao, Basque Country, Spain
e-mail: zuazua@bcamath.org; www.bcamath.org/zuazua/

V. Ancona and E. Strickland (eds.), *Trends in Contemporary Mathematics*, Springer INdAM Series 8, DOI 10.1007/978-3-319-05254-0_15,

The efficiency of the method in comparison with more classical discrete methods is illustrated by several numerical experiments.

15.1 Introduction

There is an extensive literature on the control, optimization and inverse design of partial differential equations. But, when dealing with nonlinear models, most often, the analysis requires to linearize the system under consideration. This is why most of the existing results do not apply to hyperbolic conservation laws since the shock discontinuities of solutions are an impediment to linearize the system under consideration in a classical manner.

This paper is devoted to analyze this issue for 1D scalar conservation laws. To complement previous related works by our group we consider the tracking problem in which the goal is to optimally choose the initial datum so that the solution is as close as possible to a prescribed trajectory. Our previous works are devoted to the problem of inverse design in which the goal is to determine the initial datum so that the solution achieves a given target at the final time (see, for instance, [9, 10] or the nonlinearity entering in the conservation law [8]).

To be more precise, given a finite time $T > 0$, a target function $u^d \in L^2(\mathbb{R} \times (0,T))$, and a positive weight function $\rho \in L^\infty(\mathbb{R} \times (0,T))$ with compact support in $\mathbb{R} \times (0,T)$, we consider the functional cost to be minimized J, over a suitable class of initial data $\mathcal{U}_{ad}$, defined by

$$J(u^0) = \frac{1}{2}\int_0^T \int_{\mathbb{R}} \rho(x,t)|u(x,t) - u^d(x,t)|^2 dx\, dt, \tag{15.1}$$

where $u : \mathbb{R}_x \times \mathbb{R}_t \to \mathbb{R}$ is the unique entropy solution of the scalar conservation law

$$\partial_t u + \partial_x(f(u)) = 0, \text{ in } \mathbb{R} \times (0,T); \quad u(x,0) = u^0(x), \ x \in \mathbb{R}. \tag{15.2}$$

Thus, the problem under consideration reads: To find $u^{0,min} \in \mathcal{U}_{ad}$ such that

$$J(u^{0,min}) = \min_{u^0 \in \mathcal{U}_{ab}} J(u^0). \tag{15.3}$$

Here the flux $f : \mathbb{R} \to \mathbb{R}$ is assumed to be smooth: $f \in C^1(\mathbb{R},\mathbb{R})$. The initial datum u^0 will be assumed to belong to a suitable admissible class $\mathcal{U}_{ad}$ to ensure the existence of a minimizer. As a preliminary fact we remind that for $u^0 \in L^1(\mathbb{R}) \cap L^\infty(\mathbb{R}) \cap BV(\mathbb{R})$, there exists an unique entropy solution in the sense of Kružkov (see [15]) in the class $C^0([0,T]; L^1(\mathbb{R})) \cap L^\infty(\mathbb{R} \times [0,T]) \cap L^\infty([0,T]; BV(\mathbb{R}))$.

As we will see, the existence of minimizers can be established under some natural assumptions on the class of admissible data $\mathcal{U}_{ad}$, using the well-posedness and compactness properties of solutions of the conservation law (15.2). The uniqueness of the minimizers is false, in general, due, in particular, to the possible presence of discontinuities in the solutions of (15.2).

One of our main goals in this paper is to compare how the discrete approach and the alternating descent methods perform in this new prototypical optimization problem. This alternating descent method was introduced and developed in the previous works by our group to deal with and exploit the sensitivity of shocks. In other words, the solutions of the conservation law are seen as a multi-physics object integrated by both the solution itself but also by the geometric location of the shock. Perturbing the initial datum of the conservation law produces perturbations on both components. Accordingly it is natural to analyze and employ both sensitivities to develop descent algorithms leading to an efficient computation of the minimizer. This is the basis of the alternating descent method [9].

As we shall see, in agreement with the results achieved in previously considered examples, the alternating descent method performs better than the classical discrete one which consists simply in applying a classical gradient descent algorithm to a discrete version of the functional and the conservation law.

This paper is limited to the one dimensional case but the alternating descent method can also be extended to the multi-dimensional frame, although this requires a much more careful geometric treatment of shocks since, for instance in $2-d$, these are curves evolving in time whose location has to be carefully determined and their motion and perturbation handled carefully to avoid numerical instabilities (see [16]).

There are other possible methods that could be employed to deal with these problems. In the present $1-d$ setting, for instance, whenever the flux is convex, one could use the techniques developed in [1] based on the Lax-Oleinik representation formula of solutions. One could also employ the nudging method developed by J. Blum and coworkers [2, 3]. It would be interesting to compare in a systematic manner the results obtained in this article with those that could be achieved by these other methods.

The rest of this paper is organized as follows. In Sect. 15.2 we formulate the problem under consideration more precisely and prove the existence of minimizers.

In Sect. 15.3 we introduce the discrete approximation of the continuous optimal control problem. We prove the existence of minimizers for the discrete problem and their Γ-convergence towards the continuous ones as the mesh-size parameters tend to zero.

As we shall see, purely discrete approaches based on the minimization of the resulting discrete functionals by descent algorithms lead to very slow iterative processes. We thus need to introduce an alternated descent algorithm that takes into account the possible presence of shock discontinuities in solutions. For doing this the first step is to develop a careful sensitivity analysis. This is done in Sect. 15.4.

In Sect. 15.5 we present the alternating descent method which combines the advantages of both the discrete approach and the sensitivity analysis in the presence of shocks.

In Sect. 15.6 we explain how to implement both descent algorithms: The discrete approach consists mainly in applying a descent algorithm to the discrete version J_Δ of the functional J, while the alternating descent method, by the contrary, is a continuous method based on the analysis of the previous section on the sensitivity of shocks.

In Sect. 15.7 we present some numerical experiments illustrating the overall efficiency of the alternating descent method. We conclude discussing some possible extensions of the results and methods presented in the paper.

15.2 Existence of Minimizers

In this section we prove that, under certain conditions on the set of admissible initial data $\mathcal{U}_{ad}$, there exists at least one minimizer of the functional J, given in (15.1). To do this, we consider the class of admissible initial data $\mathcal{U}_{ad}$ as:

$$\mathcal{U}_{ad} = \{u^0 \in L^\infty(\mathbb{R}) \cap BV(\mathbb{R}),\ \text{supp}(u^0) \subset K,\ \|u^0\|_{L^\infty} + TV(u^0) \leq C\}, \tag{15.4}$$

where $K \subset \mathbb{R}$ is a given compact set and $C > 0$ a given constant. Note, however, that the same theoretical results and descent strategies we shall develop here can be applied to a much wider class of admissible sets.

Theorem 15.1. *Assume that* $u^d \in L^2(\mathbb{R} \times (0,T))$, *let* $\mathcal{U}_{ad}$ *be defined in* (15.4) *and* f *be a* C^1 *function. Then the minimization problem,*

$$\min_{u^0 \in \mathcal{U}_{ab}} J(u^0), \tag{15.5}$$

has at least one minimizer $u^{0,min} \in \mathcal{U}_{ad}$. *Moreover, uniqueness is false in general.*

The proof follows the classical strategy of the Direct Method of the Calculus of Variations. We refer the reader to [9] for a similar proof in the case where the functional J is replaced by the one in which the distance to a given target at the final time $t = T$ is minimized. Note that, in particular, for the class $\mathcal{U}_{ad}$ of admissible initial data considered, solutions enjoy uniform BV bounds allowing to prove the needed compactness properties to pass to the limit. In the case of convex fluxes, using the well-known one-sided Lipschitz condition, the class of admissible initial data can be further extended.

We also observe that, as indicated in [9], the uniqueness of the minimizer is in general false for this type of optimization problems.

15.3 The Discrete Minimization Problem

The purpose of this section is to show that discrete minimizers obtained by a numerical approximation of (15.1) and (15.2), converge to a minimizer of the continuous problem as the mesh-size tends to zero. This justifies the usual engineering practice of replacing the continuous functional and model by discrete ones to compute an approximation of the continuous minimizer.

Let us introduce a mesh in $\mathbb{R} \times [0, T]$ given by $(x_i, t^n) = (i\Delta x, n\Delta t)$ ($i = -\infty, \dots, \infty$; $n = 0, \dots, N+1$, so that $(N+1)\Delta t = T$), and let u_i^n be a numerical approximation of $u(x_i, t^n)$ obtained as solution of a suitable discretization of Eq. (15.2).

Let us consider the following approximation of the functional J in (15.1):

$$J_\Delta(u_\Delta^0) = \frac{\Delta x \, \Delta t}{2} \sum_{n=0}^{N} \sum_{i=-\infty}^{\infty} \rho_i^n \left(u_i^n - (u^d)_i^n\right)^2, \tag{15.6}$$

where $u_\Delta^0 = \{u_i^0\}$ is the discrete initial datum and $u_\Delta^d = \{(u^d)_i^n\} = \Pi_\Delta u^d$ is the discretization of the target u^d at (x_i, t^n), Π_Δ being a discretization operator. A common choice consists in taking

$$\Pi_\Delta u^d = (u^d)_i^n = \frac{1}{\Delta x \Delta t} \int_{x_{i-1/2}}^{x_{i+1/2}} \int_{t^{n-1/2}}^{t^{n+1/2}} u^d(x,t)\, dx\, dt, \tag{15.7}$$

where $x_{i\pm 1/2} = x_i \pm \Delta x/2$ and $t^{n\pm 1/2} = t^n \pm \Delta t/2$.

Moreover, we introduce an approximation of the class of admissible initial data $\mathcal{U}_{ad}$ denoted by $\mathcal{U}_{ad,\Delta}$ and constituted by sequences $\varphi_\Delta = \{\varphi_i\}_{i\in\mathbb{Z}}$ for which the associated piecewise constant interpolation function, that we still denote by φ_Δ, defined by

$$\varphi_\Delta(x) = \varphi_i, \quad x \in (x_{i-1/2}, x_{i+1/2}),$$

satisfies $\varphi_\Delta \in \mathcal{U}_{ad}$. Obviously, $\mathcal{U}_{ad,\Delta}$ coincides with the class of discrete vectors with support on those indices i such that $x_i \in K$ and for which the discrete L^∞ and TV-norms are bounded above by the same constant C.

Let us consider $S_\Delta : l^1(\mathbb{Z}) \to l^1(\mathbb{Z})$, an explicit numerical scheme for (15.2), where

$$u_\Delta^n = S_\Delta^n u_\Delta^0, \tag{15.8}$$

is the approximation of the entropy solution $u(\cdot, t) = S(t)u^0$ of (15.2), i.e. $u_\Delta^n \simeq S(t)u^0$, with $t = n\Delta t$. Here $S : L^1(\mathbb{R}) \cap L^\infty(\mathbb{R}) \cap BV(\mathbb{R}) \to L^1(\mathbb{R}) \cap L^\infty(\mathbb{R}) \cap BV(\mathbb{R})$, is the semigroup (solution operator)

$$S : u_0 \to u = Su_0,$$

which associates to the initial condition $u_0 \in L^1(\mathbb{R}) \cap L^\infty(\mathbb{R}) \cap BV(\mathbb{R})$ the entropy solution u of (15.2).

For each $\Delta = \Delta t$ (with $\lambda = \Delta t/\Delta x$ fixed, typically given by the corresponding CFL-condition for explicit schemes), it is easy to see that the discrete analogue of Theorem 15.1 holds. In fact this is automatic in the present setting since $\mathcal{U}_{ad,\Delta}$ only involves a finite number of mesh-points. But passing to the limit as $\Delta \to 0$ requires a more careful treatment. In fact, for that to be done, one needs to assume that the scheme under consideration, (15.8), is a contracting map in $l^1(\mathbb{Z})$.

Thus, we consider the following discrete minimization problem: Find $u_\Delta^{0,min}$ such that

$$J_\Delta(u_\Delta^{0,min}) = \min_{u_\Delta^0 \in \mathcal{U}_{ad,\Delta}} J_\Delta(u_\Delta^0). \tag{15.9}$$

The following holds

Theorem 15.2. *Assume that u_Δ^n is obtained by a numerical scheme* (15.8), *which satisfies the following:*

- *For a given $u^0 \in \mathcal{U}_{ad}$, $u_\Delta^n = S_\Delta^n \Pi_\Delta u^0$ converges to $u(x,t)$, the entropy solution of* (15.2). *More precisely, if $n\Delta t = t$ and $T_0 > 0$ is any given number,*

$$\max_{0 \le t \le T_0} \|u_\Delta^n - u(\cdot,t)\|_{L^1(\mathbb{R})} \to 0, \quad \text{as} \quad \Delta \to 0. \tag{15.10}$$

- *The map S_Δ is L^∞-stable, i.e.*

$$\|S_\Delta u_\Delta^0\|_{L^\infty} \le \|u_\Delta^0\|_{L^\infty}. \tag{15.11}$$

- *The map S_Δ is a contracting map in $l^1(\mathbb{Z})$ i.e.*

$$\|S_\Delta u_\Delta^0 - S_\Delta v_\Delta^0\|_{L^1} \le \|u_\Delta^0 - v_\Delta^0\|_{L^1}. \tag{15.12}$$

Then:

- *For all Δ, the discrete minimization problem* (15.9) *has at least one solution $u_\Delta^{0,min} \in \mathcal{U}_{ad,\Delta}$.*
- *Any accumulation point of $u_\Delta^{0,min}$ with respect to the weak-$*$ topology in L^∞, as $\Delta \to 0$, is a minimizer of the continuous problem* (15.5).

The proof of this result can be developed as in [9]. We also refer to [16] for an extension to the multi-dimensional case.

This convergence result applies for 3-point conservative numerical approximation schemes, where S_Δ is given by

$$(S_\Delta v_\Delta)_i = v_i - \frac{\Delta t}{\Delta x}\left(g(v_{i+1}, v_i) - g(v_i, v_{i-1})\right), \tag{15.13}$$

and g is the numerical flux. This scheme is consistent with the corresponding Eq. (15.2) when $g(u,u) = f(u)$.

When the discrete semigroup $S_\Delta(u,v,w) = v - \lambda(g(u,v) - g(v,w))$, with $\lambda = \Delta t/\Delta x$ is monotonic increasing with respect to each argument, the scheme is also monotone. This ensures the convergence to the weak entropy solutions of the continuous conservation law, as the discretization parameters tend to zero, under a suitable CFL condition (see Ref. [12], Chap. 3, Th. 4.2).

All this analysis and results apply to the classical Godunov, Lax-Friedfrichs and Engquist-Osher schemes, the corresponding numerical flux being:

$$g^G(u,v) = \begin{cases} \min_{w\in[u,v]} f(w), \text{ if } u \le v, \\ \max_{w\in[u,v]} f(w), \text{ if } u \ge v, \end{cases} \tag{15.14}$$

$$g^{LF}(u,v) = \frac{(f(u)+f(v))}{2} - \frac{(v-u)}{2\lambda^x}, \tag{15.15}$$

$$g^{EO}(u,v) = \frac{f(u)+f(v)-\int_u^v |f'(\tau)|d\tau}{2}. \tag{15.16}$$

See Chapter 3 in [12] for more details.

These 1D methods satisfy the conditions of Theorem 15.2.

15.4 Sensitivity Analysis: The Continuous Approach

We divide this section in three subsections. Specifically, in the first one we consider the case where the solution u of (15.2) has no shocks. In the second and third subsections we analyze the sensitivity of the solution and the functional in the presence of a single shock.

15.4.1 Sensitivity Without Shocks

In this subsection we give an expression for the sensitivity of the functional J with respect to the initial datum based on a classical adjoint calculus for smooth solutions. First we present a formal calculus and then we show how to justify it when dealing with a classical smooth solution for (15.2).

Let $C_0^1(\mathbb{R})$ be the set of C^1 functions with compact support and let $u^0 \in C_0^1(\mathbb{R})$ be a given initial datum for which there exists a classical solution $u(x,t)$ of (15.2) that can be extended to a classical solution in $t \in [0, T+\tau]$ for some $\tau > 0$. Note that this imposes some restrictions on u^0 other than being smooth.

Let $\delta u^0 \in C_0^1(\mathbb{R})$ be any possible variation of the initial datum u^0. Due to the finite speed of propagation, this perturbation will only affect the solution in a bounded set of $\mathbb{R}\times[0,T]$. This simplifies the argument below that applies in a much more general setting provided solutions are smooth enough.

Then, for $\varepsilon > 0$ sufficiently small, the solution $u^\varepsilon(x,t)$ corresponding to the initial datum

$$u^{\varepsilon,0}(x) = u^0(x) + \varepsilon\delta u^0(x), \tag{15.17}$$

is also a classical solution in $(x,t) \in \mathbb{R} \times (0,T)$ and $u^\varepsilon \in C^1(\mathbb{R} \times [0,T])$ can be written as

$$u^\varepsilon = u + \varepsilon \delta u + o(\varepsilon), \text{ with respect to the } C^1 \text{ topology}, \tag{15.18}$$

where δu is the solution of the linearized equation,

$$\partial_t \delta u + \partial_x (f'(u)\delta u) = 0, \text{ in } \mathbb{R} \times (0,T); \quad \delta u(x,0) = \delta u^0(x), \ x \in \mathbb{R}. \tag{15.19}$$

Let δJ be the Gateaux derivative of J at u^0 in the direction δu^0. We have,

$$\delta J(u^0)[\delta u^0] = \int_0^T \int_{\mathbb{R}} \rho(x,t)(u(x,t) - u^d(x,t))\delta u(x,t)\, dx\, dt \tag{15.20}$$

where δu solves the linearized system above (15.19). Now, we introduce the adjoint system,

$$-\partial_t p - f'(u)\partial_x p = \rho\,(u - u^d), \text{ in } \mathbb{R} \times (0,T); \quad p(x,T) = 0, \ x \in \mathbb{R}. \tag{15.21}$$

Multiplying the equations satisfied by δu by p, integrating by parts, and taking into account that p satisfies (15.21), we easily obtain

$$\int_0^T \int_{\mathbb{R}} \rho(x,t)(u(x,t) - u^d(x,t))\delta u(x,t)\, dx\, dt = \int_{\mathbb{R}} p(x,0)\delta u^0(x)\, dx. \tag{15.22}$$

Thus, δJ in (15.20) can be written as,

$$\delta J(u^0)[\delta u^0] = \int_{\mathbb{R}} p(x,0)\delta u^0(x) dx. \tag{15.23}$$

This expression provides an easy way to compute a descent direction for the continuous functional J, once we have computed the adjoint state. We just take:

$$\delta u^0(x) = -p(x,0). \tag{15.24}$$

Under the assumptions above on $u^0, u, \delta u$ and p can be obtained from their data $u^0(x)$, $\delta u^0(x)$ and $\rho\,(u - u^d)$ by using the characteristic curves associated to (15.2). For the sake of completeness we briefly explain this below.

The characteristic curves associated to (15.2) are defined by

$$x'(t) = f'(u(x(t),t)), \quad t \in (0,T), \quad x(0) = x^0. \tag{15.25}$$

They are straight lines whose slopes depend on the initial data:

$$x(t) = x^0 + tf'(u^0(x^0)), \quad t \in (0,T).$$

As we are dealing with classical solutions, u is constant along such curves and, by assumption, two different characteristic curves do not meet each other in $\mathbb{R} \times [0, T + \tau]$. This allows to define u in $\mathbb{R} \times [0, T + \tau]$ in a unique way from the initial data.

For $\varepsilon > 0$ sufficiently small, the solution $u^\varepsilon(x,t)$ corresponding to the initial datum (15.17) has similar characteristics to those of u. This allows guaranteeing that two different characteristic lines do not intersect for $0 \leq t \leq T$ if $\varepsilon > 0$ is small enough. Note that u^ε may possibly be discontinuous for $t \in (T, T + \tau]$ if u^0 generates a discontinuity at $t = T + \tau$ but this is irrelevant for the analysis in $[0, T]$ we are carrying out. Therefore $u^\varepsilon(x,t)$ is also a classical solution in $(x,t) \in \mathbb{R} \times [0, T]$ and it is easy to see that the solution u^ε can be written as (15.18) where δu satisfies (15.19).

System (15.19) can be solved again by the method of characteristics. In fact, as u is a regular function, the first equation in (15.19) can be written as

$$\partial_t \delta u + f'(u)\partial_x \delta u = -\partial_x(f'(u))\,\delta u, \tag{15.26}$$

i.e.

$$\frac{d}{dt}\delta u(x(t),t) = -\partial_x(f'(u))\,\delta u, \tag{15.27}$$

where $x(t)$ are the characteristic curves defined by (15.25). Thus, the solution δu along a characteristic line can be obtained from δu^0 by solving this differential equation, i.e.

$$\delta u(x(t),t) = \delta u^0(x^0)\exp\left(-\int_0^t \partial_x(f'(u))(x(s),s)ds\right).$$

Finally, the adjoint system (15.21) is also solved by characteristics, i.e.

$$-\frac{d}{dt}p(x(t),t) = \rho(x(t),t)(u(x(t),t) - u^d(x(t),t)).$$

This yields the steepest descent direction in (15.24) for the continuous functional:

$$p(x^0,0) = u^0(x^0)\int_0^T \rho(x(s),s)ds - \int_0^T \rho(x(s),s)u^d(x(s),s)ds.$$

Remark 15.1. Note that for classical solutions the Gateaux derivative of J at u^0 is given by (15.23) and this provides an obvious descent direction for J at u^0, given by (15.24). However this fact is not very useful in practice since, even when we initialize the iterative descent algorithm with a smooth u^0, we cannot guarantee that the solution remains classical along the iterative process.

15.4.2 Sensitivity of the State in the Presence of Shocks

Inspired in several results on the sensitivity of solutions of conservation laws in the presence of shocks in one-dimension (see [4–7, 13, 17]), we focus on the particular case of solutions having a single shock. But the analysis can be extended to consider more general one-dimensional systems of conservation laws with a finite number of noninteracting shocks. We introduce the following hypothesis:

Hypothesis 15.1. *Assume that $u(x,t)$ is a weak entropy solution of* (15.2) *with a discontinuity along a regular curve $\Sigma = \{(\varphi(t),t),\ t \in (0,T)\}$, which is Lipschitz continuous outside Σ. In particular, it satisfies the Rankine-Hugoniot condition on Σ*

$$\varphi'(t)[u]_{\Sigma^t} = [f(u)]_{\Sigma^t}.$$

Here we have used the notation: $[v]_{\Sigma^t} = \lim_{\varepsilon \searrow 0} v(\varphi(t)+\varepsilon,t) - v(\varphi(t)-\varepsilon,t)$, for the jump at $\Sigma^t = (\varphi(t),t)$ of any piecewise continuous function v with a discontinuity at Σ^t.

Note that Σ divides $\mathbb{R} \times (0,T)$ into two parts: Q_- and Q_+, the sub-domains of $\mathbb{R} \times (0,T)$ to the left and to the right of Σ respectively.

As we will see, in the presence of shocks, to deal correctly with optimal control and design problems, the state of the system needs to be viewed as constituted by the pair (u,φ) combining the solution of (15.2) and the shock location φ. This is relevant in the analysis of sensitivity of functions below and when applying descent algorithms.

We adopt the functional framework based on the generalized tangent vectors (see [7] and Definition 4.1 in [9]).

Let u^0 be the initial datum, that we assume to be Lipschitz continuous to both sides of a single discontinuity located at $x = \varphi^0$, and consider a generalized tangent vector $(\delta u^0, \delta\varphi^0) \in L^1(\mathbb{R}) \times \mathbb{R}$ for all $0 \le T$. Let $u^{0,\varepsilon}$ be a path which generates $(\delta u^0, \delta\varphi^0)$. For ε sufficiently small, the solution $u^\varepsilon(\cdot,t)$ of (15.2) is Lipschitz continuous with a single discontinuity at $x = \varphi^\varepsilon(t)$, for all $t \in [0,T]$. Therefore, $u^\varepsilon(\cdot,t)$ generates a generalized tangent vector $(\delta u(\cdot,t), \delta\varphi(t)) \in L^1(\mathbb{R}) \times \mathbb{R}$. Moreover, in [9] it is proved that it satisfies the following linearized system:

$$\partial_t \delta u + \partial_x (f'(u)\delta u) = 0, \qquad \text{in } Q_- \cup Q_+ \tag{15.28}$$

$$\frac{d}{dt}\left([u]_{\Sigma^t}\delta\varphi\right) = [f'(u)\delta u]_{\Sigma^t} - [\delta u]_{\Sigma^t}\frac{d}{dt}\varphi, \quad t \in (0,T) \tag{15.29}$$

$$\delta u(x,0) = \delta u^0(x), \qquad \{x < \varphi^0\} \cup \{x > \varphi^0\} \tag{15.30}$$

$$\delta\varphi(0) = \delta\varphi^0. \tag{15.31}$$

15.4.3 Sensitivity of the Cost in the Presence of Shocks

In this section we study the sensitivity of the functional J with respect to variations associated with the generalized tangent vectors defined in the previous section. We first define an appropriate generalization of the Gateaux derivative of J.

Definition 15.1. Let $J : L^1(\mathbb{R}) \to \mathbb{R}$ be a functional and $u^0 \in L^1(\mathbb{R})$ be Lipschitz continuous with a discontinuity in Σ^0, an initial datum for which the solution of (15.2) satisfies hypothesis (15.1). J is Gateaux differentiable at u^0 in a generalized sense if for any generalized tangent vector $(\delta u^0, \delta\varphi^0)$ and any family $u^{0,\varepsilon}$ associated to $(\delta u^0, \delta\varphi^0)$ the following limit exists,

$$\delta J = \lim_{\varepsilon \to 0} \frac{J(u^{0,\varepsilon}) - J(u^0)}{\varepsilon},$$

and it depends only on (u^0, φ^0) and $(\delta u^0, \delta\varphi^0)$, i.e. it does not depend on the particular family $u^{0,\varepsilon}$ which generates $(\delta u^0, \delta\varphi^0)$. The limit is the generalized Gateaux derivative of J in the direction $(\delta u^0, \delta\varphi^0)$.

The following result easily provides a characterization of the generalized Gateaux derivative of J in terms of the solution of the associated adjoint system (15.33)–(15.38).

Proposition 15.1. *The Gateaux derivative of J can be written as follows*

$$\delta J(u^0)[\delta u^0, \delta\varphi^0] = \int_{\mathbb{R}} p(x,0)\delta u^0(x)dx - q(0)[u]_{\Sigma^0}\delta\varphi^0, \tag{15.32}$$

where the adjoint state pair (p, q) satisfies the system

$$-\partial_t p - f'(u)\partial_x p = \rho(u - u^d), \quad \text{in } Q_- \cup Q_+ \tag{15.33}$$

$$[p]_{\Sigma^t} = 0, \quad t \in (0,T) \tag{15.34}$$

$$q(t) = p(\varphi(t), t), \quad t \in (0,T) \tag{15.35}$$

$$-\frac{d}{dt}q = \frac{(1 + (\dot{\varphi})^2)^{1/2}[\rho(u - u^d)^2]_{\Sigma^t}}{2[u]_{\Sigma^t}}, \quad t \in (0,T) \tag{15.36}$$

$$p(x,T) = 0, \quad \{x < \varphi(T)\} \cup \{x > \varphi(T)\} \tag{15.37}$$

$$q(T) = 0. \tag{15.38}$$

Let us briefly comment the result of Proposition 15.1 before giving its proof.

System (15.33)–(15.38) has a unique solution. In fact, to solve the backward system (15.33)–(15.38) we first define the solution q on the shock Σ from the conditions for q (15.36) and (15.38). This determines the value of p along the shock.

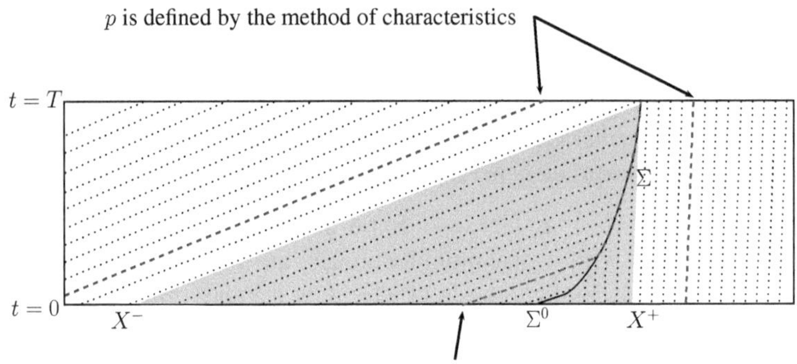

Fig. 15.1 Characteristic lines entering on a shock and how they may be used to build the solution of the adjoint system both away from the shock and on its region of influence

We then propagate this information, together with (15.33) and (15.37), to both sides of Σ, by characteristics (see Fig. 15.1 where we illustrate this construction).

Formula (15.32) provides an obvious way to compute a first descent direction of J at u^0. We just take

$$(\delta u^0, \delta\varphi^0) = (-p(\cdot, 0), q(0)[u]_{\Sigma^0}). \tag{15.39}$$

Here, the value of $\delta\varphi^0$ must be interpreted as the optimal infinitesimal displacement of the discontinuity of u^0.

In [9], when considering the inverse design problem, it was observed that the solution p of the corresponding adjoint system at $t = 0$ was discontinuous, with two discontinuities, one in each side of the original location of the discontinuity at Σ^0. This was a reason not to use this descent direction and for introducing the alternating descent method. In the present setting, however, the adjoint state p obtained is typically continuous. This is due to the fact that p at both side of the discontinuity is defined by the method of characteristics and that, on the region of influence of the characteristics emanating from the shock, the continuity is preserved by the fact that, on one hand, $q = q(t)$ itself is continuous as the primitive of an integrable function and that the data for p and q at $t = T$ are continuous too. Despite of this, as we shall see, the implementation of the alternating descent direction method is worth since it significantly improves the results obtained by the purely discrete approach.

We finish this section with the proof of Proposition 15.1.

Proof (of Proposition 15.1). A straightforward computation shows that J is Gateaux differentiable in the generalized sense of Definition 15.1 and that the generalized Gateaux derivative of J in the direction of the generalized tangent vector $(\delta u^0, \delta\varphi^0)$ is given by

$$\delta J(u^0)[\delta u^0, \delta\varphi^0] = \int_0^T \int_{\{x>\varphi(t)\}\cup\{x<\varphi(t)\}} \rho(x,t)(u(x,t) - u^d(x,t))\delta u(x,t)\,dx\,dt$$
$$- \int_\Sigma \left[\rho\frac{(u-u^d)^2}{2}\right]_{\Sigma^t} \delta\varphi(t)\,d\Sigma(t), \quad (15.40)$$

where the pair $(\delta u, \delta\varphi)$ solves the linearized problem (15.28)–(15.30) with initial data $(\delta u^0, \delta\varphi^0)$.

Let us now introduce the adjoint system (15.33)–(15.38). Multiplying the equations of δu by p, and integrating we get

$$0 = \int_0^T \int_{\{x>\varphi(t)\}\cup\{x<\varphi(t)\}} \left(\partial_t\delta u + \partial_x(f'(u)\delta u)\right) p\,dx\,dt$$
$$= -\int_0^T \int_{\{x>\varphi(t)\}\cup\{x<\varphi(t)\}} \delta u\left(\partial_t p + f'(u)\partial_x p\right)\,dx\,dt$$
$$+ \int_{\{x>\varphi(T)\}\cup\{x<\varphi(T)\}} \delta u(x,T)p(x,T)dx - \int_{\{x>\varphi^0\}\cup\{x<\varphi^0\}} \delta u^0(x)p(x,0)dx$$
$$- \int_\Sigma \left([\delta u\,p]_{\Sigma^t} n^t_\Sigma + [f'(u)\delta u\,p]_{\Sigma^t}\cdot n^x_\Sigma\right) d\Sigma(t), \quad (15.41)$$

where (n^x, n^t) are the Cartesian components of the normal vector to the curve Σ.

Therefore, since p satisfies the adjoint equation (15.33), (15.37), from (15.40) we obtain

$$\delta J(u^0)[\delta u^0, \delta\varphi^0] = \int_{\{x>\varphi^0\}\cup\{x<\varphi^0\}} \delta u^0(x)p(x,0)dx$$
$$+ \int_\Sigma \left([\delta u\,p]_{\Sigma^t} n^t_\Sigma + [f'(u)\delta u\,p]_{\Sigma^t}\cdot n^x_\Sigma\right) d\Sigma(t) - \int_\Sigma \left[\rho\frac{(u-u^d)^2}{2}\right]_{\Sigma^t} \delta\varphi(t)\,d\Sigma(t). \quad (15.42)$$

The last two terms in the right hand side of (15.42) will determine the conditions that p must satisfy on the shock.

Observe that for any functions f, g we have

$$[fg]_{\Sigma^t} = \overline{f}[g]_{\Sigma^t} + \overline{g}[f]_{\Sigma^t},$$

where $\overline{g}$ represents the average of g to both sides of the shock Σ^t, i.e.

$$\overline{g}(t) = \frac{1}{2}\lim_{\varepsilon\searrow 0}\left(g(\varphi(t)+\varepsilon,t) + g(\varphi(t)-\varepsilon,t)\right), \quad \forall t \in (0,T).$$

Thus we have

$$\int_\Sigma \left([\delta u\, p]_{\Sigma^t} n^t_\Sigma + [f'(u)\delta u\, p]_{\Sigma^t}\cdot n^x_\Sigma\right) d\Sigma = \int_\Sigma [p]_{\Sigma^t}\left(\overline{\delta u} n^t_\Sigma + \overline{f'(u)\delta u}\cdot n^x_\Sigma\right) d\Sigma$$
$$+ \int_\Sigma \overline{p}\left([\delta u]_{\Sigma^t} n^t_\Sigma + [f'(u)\delta u]_{\Sigma^t}\cdot n^x_\Sigma\right) d\Sigma. \tag{15.43}$$

Now, we note that the Cartesian components of the normal vector to Σ are given by

$$n^t = \frac{-\varphi'(t)}{\sqrt{1+(\varphi'(t))^2}}, \quad n^x = \frac{1}{\sqrt{1+(\varphi'(t))^2}},$$

and $d\Sigma(t) = \sqrt{1+(\varphi'(t))^2}dt$. Using (15.28), (15.34), we have

$$\int_\Sigma \left([\delta u\, p]_{\Sigma^t} n^t_\Sigma + [f'(u)\delta u\, p]_{\Sigma^t} n^x_\Sigma\right) d\Sigma = \int_0^T \overline{p}\left(-[\delta u]_{\Sigma^t}\varphi'(t) + [f'(u)\delta u]_{\Sigma^t}\right) dt$$
$$= \int_0^T \overline{p}\frac{d}{dt}\left([u]_{\Sigma^t}\delta\varphi\right) dt. \tag{15.44}$$

Therefore, replacing (15.35), (15.36), (15.38) and (15.44) in (15.42), we obtain (15.32).

This concludes the proof. □

15.5 The Alternating Descent Method

Here we explain how the *alternating descent method* introduced in [9] can be adapted to the present optimal control problem (15.3).

First let us introduce some notation. We consider two points

$$X^- = \varphi(T) - Tf'(u^-(\varphi(T),T)), \quad X^+ = \varphi(T) - Tf'(u^+(x,T)), \tag{15.45}$$

(p,q) the solutions of (15.33)–(15.38) and

$$p^{0,-} = \lim_{x\nearrow X^-} p(x,0), \quad p^{0,+} = \lim_{x\searrow X^+} p(x,0).$$

Now, we introduce the two classes of descent directions we shall use in our descent algorithm.

First directions: With this set of directions, we mainly modify the profile of u^0. We set the first directions $d_1 = (\delta u^0, \delta\varphi^0)$, given by:

$$\delta u^0 = \begin{cases} -p(x,0), \; x < X^-, \\ -p^{0,-}, \quad x \in (X^-, \varphi^0), \\ -p^{0,+}, \quad x \in (\varphi^0, X^+), \\ -p(x,0), \; x > X^+. \end{cases}$$

$$\delta\varphi^0 = \begin{cases} 0, & \text{if } \int_{X^-}^{X^+} p(x,0)\delta u^0(x)dx \le 0, \\ \dfrac{\int_{X^-}^{X^+} p(x,0)\delta u^0(x)dx}{[u^0]_{\Sigma^0} q(0)}, & \text{if } \int_{X^-}^{X^+} p(x,0)\delta u^0(x)dx > 0, \text{ and } q(0) \neq 0. \end{cases} \tag{15.46}$$

We note that, if $q(0) = 0$ and $\int_{X^-}^{X^+} p(x,0)\delta u^0(x)dx > 0$, these directions are not defined.

Second directions: They are aimed to move the shock without changing the profile of the solution to both sides. Then $d_2 = (\delta u^0, \delta\varphi^0)$, with:

$$\delta u^0 \equiv 0, \quad \delta\varphi^0(x) = [u^0]_{\Sigma^0} q(0). \tag{15.47}$$

We observe that d_1 given by (15.46) satisfies

$$\delta J(u^0)[d_1] = -\int_{x \notin [X^-, X^+]} |p(x,0)|^2 dx$$
$$-p^{0,-}\int_{X^-}^{\varphi^0} p(x,0)dx - p^{0,+}\int_{\varphi^0}^{X^+} p(x,0)dx - [u^0]_{\Sigma^0} q(0)\delta\varphi^0 \le 0. \tag{15.48}$$

Note that in those cases where d_1 is not defined, as indicated above, this descent direction is simply not employed in the descent algorithm.

And for d_2 given by (15.47) we have

$$\delta J(u^0)[d_2] = -([u^0]_{\Sigma^0} q(0))^2 \le 0. \tag{15.49}$$

Thus, the two classes of descent directions under consideration have three important properties:

1. They are both descent directions.
2. They allow to split the design of the profile and the shock location.
3. They are true generalized gradients and therefore keep the structure of the data without increasing its complexity.

15.6 Numerical Approximation of the Descent Direction

We have computed the gradient of the continuous functional J in several cases (u smooth or having shock discontinuities) but, in practice, one has to work with discrete versions of the functional J. In this section we discuss two possible ways of searching discrete descent directions based either on the discrete or the continuous approaches and in the later on the alternating descent method.

The discrete approach consists mainly in applying directly a descent algorithm to the discrete version J_Δ of the functional J by using its gradient. The alternating descent method, by the contrary, is a method inspired on the continuous analysis of the previous section in which the two main classes of descent directions that are, first, identified and later discretized.

Let us first discuss the discrete approach.

15.6.1 The Discrete Approach

Let us consider the approximation of the functional J by J_Δ defined (15.1) and (15.6) respectively. We shall use the Engquist-Osher scheme which is a 3-point conservative numerical approximation scheme for (15.2). More explicitly we consider:

$$u_i^{n+1} = u_i^n - \frac{\Delta t}{\Delta x}\left(g(u_{i+1}^n, u_i^n) - g(u_i^n, u_{i-1}^n)\right), \quad i \in \mathbb{Z},\ n = 0, \ldots, N, \qquad (15.50)$$

where g is the numerical flux defined in (15.16).

The gradient of the discrete functional J_Δ requires computing one derivative of J_Δ with respect to each node of the mesh. This can be done in a cheaper way using the discrete adjoint state. We illustrate it for the Engquist-Osher numerical scheme. However, as the discrete functionals J_Δ are not necessarily convex the gradient methods could possibly provide sequences that do not converge to a global minimizer of J_Δ. But this drawback and difficulty appears in most applications of descent methods in optimal design and control problems.

As we will see, in the present context, the approximations obtained by discrete gradient methods are satisfactory, although convergence is slow due to unnecessary oscillations that the descent method introduces.

The gradient of J_Δ, rigorously speaking, requires the linearization of the numerical scheme (15.50) used to approximate Eq. (15.2). Then the linearization corresponds to

$$\begin{aligned}\delta u_i^{n+1} &= \delta u_i^n - \frac{\Delta t}{\Delta x}\left(\partial_a g(u_{i+1}^n, u_i^n)\delta u_{i+1}^n - \partial_b g(u_i^n, u_{i-1}^n)\delta u_{i-1}^n\right)\\ &\quad -\frac{\Delta t}{\Delta x}\left(\partial_b g(u_{i+1}^n, u_i^n) - \partial_a g(u_i^n, u_{i-1}^n)\right)\delta u_i^n,\\ &\quad i \in \mathbb{Z},\ \ n = 0, \ldots, N. \end{aligned} \tag{15.51}$$

In view of this, the discrete adjoint system of (15.51) can also be written for the differentiable flux functions:

$$\begin{aligned} p_i^{N+1} &= 0,\ \ i \in \mathbb{Z},\\ p_i^n &= p_i^{n+1} + \frac{\Delta t}{\Delta x}\partial_b g(u_{i+1}^n, u_i^n)\left(p_{i+1}^{n+1} - p_i^{n+1}\right)\\ &\quad + \frac{\Delta t}{\Delta x}\partial_a g(u_i^n, u_{i-1}^n)\left(p_i^{n+1} - p_{i-1}^{n+1}\right) + F_i^n,\\ &\quad i \in \mathbb{Z},\ \ n = 0, \ldots, N, \end{aligned} \tag{15.52}$$

where

$$F_i^n = \Delta t \rho_i^n \left(u_i^n - (u^d)_i^n\right),\ \ i \in \mathbb{Z},\ \ n = 0, \ldots, N. \tag{15.53}$$

In fact, when multiplying the equations in (15.51) by p_i^{n+1} and adding in $i \in \mathbb{Z}$ and $n = 0, \ldots, N$, the following identity is easily obtained,

$$\Delta x \sum_{i \in \mathbb{Z}} p_i^0 \delta u_i^0 = \Delta x \sum_{n=0}^{N} \sum_{i \in \mathbb{Z}} F_i^n \delta u_i^n. \tag{15.54}$$

This is the discrete version of formula (15.22) which allows us to simplify the derivative of the discrete cost functional.

Thus, for any variation δu_Δ^0, the Gateaux derivative of the cost functional defined in (15.6) is given by

$$\delta J_\Delta = \Delta x \Delta t \sum_{n=0}^{N} \sum_{i \in \mathbb{Z}} \rho_i^n \left(u_i^n - (u^d)_i^n\right) \delta u_i^n, \tag{15.55}$$

where $\delta u_{i,j}^n$ solves the linearized system (15.51). If we consider p_i^n the solution of (15.52) with (15.53), we obtain that δJ_Δ in (15.55) can be written as,

$$\delta J_\Delta = \Delta x \sum_{i \in \mathbb{Z}} p_i^0 \delta u_i^0,$$

and this allows to obtain easily the steepest descent direction for J_Δ by considering

$$\delta u_\Delta^0 = -p_\Delta^0. \tag{15.56}$$

Remark 15.2. We observe that for the Enquis-Osher's flux (15.16), the system (15.52) is the upwind method for the continuous adjoint system. We do not address here the problem of the convergence of this adjoint scheme towards the solution of the continuous adjoint system. Of course, this is an easy matter when u is smooth but it is far from being trivial when u has shock discontinuities. Whether or not this discrete adjoint system, as $\Delta \to 0$, allows reconstructing the complete adjoint system, with the inner Dirichlet condition along the shock (15.33)–(15.38), constitutes an interesting problem for future research. We refer to [14] and [18] for preliminary work on this direction in one-dimension.

15.6.2 The Alternating Descent Method

Now we describe the implementation of the alternating descent method.

The main idea is to approximate a minimizer of J alternating between two directions of descent: First we perturb the initial datum u^0 using the direction (15.46), which principally changes the profile of u^0. Second we move the shock curve without altering the profile of u^0 at both sides of Σ, using the direction (15.47).

More precisely, for a given initialization u_Δ^0 and target function u_Δ^d, we compute Σ_Δ^0, the jump-point of u_Δ^0,u_Δ and p_Δ^0 as the solutions of (15.51), (15.52) respectively.

As numerical approximation of the adjoint state $q(0)$ we take the value of p_Δ^0 over the point Σ_Δ^0, i.e.

$$q_\Delta^0 = p_\Delta^0(\Sigma_\Delta^0) = p_i^0, \quad \Sigma_\Delta^0 \in (x_{i-1/2}, x_{i+1/2}).$$

The main advantage of this method introduced in [9] is that for an initial datum u^0 with a single discontinuity, the descent directions are generalized tangent vectors, i.e. they introduce Lipschitz continuous variations of u^0 at both sides of the discontinuity and a displacement of the shock position. In this way, the new datum obtained modifying the old one, in the direction of this generalized tangent vector, has again a single discontinuity. The method can be applied in a much more general context in which, for instance, the solution has various shocks since the method is able both to generate shocks and to destroy them, if any of these facts contributes to the decrease of the functional. This method is in some sense close to those employed in shape design in elasticity in which topological derivatives (that would correspond to controlling the location of the shock in our method) are combined with classical shape deformations (that would correspond to simply shaping the solution away from the shock in the present setting) [11].

As mentioned above, in the context of the tracking problem we are considering, the solutions of the adjoint system do not increase the number of shocks. Thus, a direct application of the continuous method, without employing this alternating variant, could make sense. Note that the purely discrete method is a limited version of the continuous one since one simply employs the descent direction indicated by p^0 without paying attention to the value q^0 corresponding to the shock sensitivity. However, the numerical experiments we have done with the full continuous method do not improve the results obtained by the discrete one since the definition of the location of the shocks is lost along the iterative process.

15.7 Numerical Experiments

In this section we present some numerical experiments which illustrate the results obtained in an optimization model problem with each one of the numerical methods described in the previous section.

We have chosen as computational domain the interval $(-4, 4)$ and we have taken as boundary conditions in (15.50), at each time step $t = t^n$, the value of the initial data at the boundary. This can be justified if we assume that the initial datum u^0 is constant in a sufficiently large inner neighborhood of the boundary $x = \pm 4$ (which depends on the size of the L^∞-norm of the data under consideration and the time horizon T), due to the finite speed of propagation. A similar procedure is employed for the adjoint equation.

We underline once more that the solutions obtained with each method may correspond to global minima or local ones since the gradient algorithm does not distinguish them.

In the experiments we consider the Burgers' equation, i.e. $f(z) = z^2/2$,

$$\partial_t u + \partial_x \left(\frac{u^2}{2}\right) = 0, \quad u(x,0) = u^0(x). \tag{15.57}$$

The weight function ρ, under consideration in the experiments is given by

$$\rho(x,t) = \begin{cases} 1 & t \in (T/2, T) \\ 0 & \text{otherwise.} \end{cases}$$

And the time horizon $T = 1$.

To compare the efficiency of the different methods we consider a fixed $\Delta x = 1/20, \lambda = \Delta t/\Delta x = 2/3$ (which satisfies the CFL condition). We then analyze the number of iterations that each method needs to attain a prescribed value of the functional.

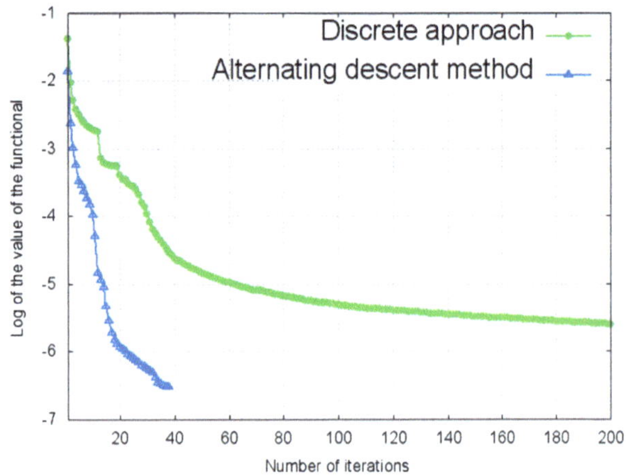

Fig. 15.2 Experiment 1. $\log(J)$ versus the number of iterations in the descent algorithm for the discrete and the alternating descent methods

15.7.1 Experiment 1

We first consider a piecewise constant target profile u^d given by the solution of (15.57) with the initial condition $(u^d)^0$ given by

$$(u^d)^0(x) = \begin{cases} 0.7 & x \in [-2, 1] \\ 0 & \text{otherwise.} \end{cases} \tag{15.58}$$

Note that, in this case, (15.58) yields a particular solution of the optimization problem and the minimum value of J vanishes.

We solve the optimization problem (15.3) with the above described different methods starting from the following initialization for u^0:

$$u^0(x) = \begin{cases} 0.5 & x \leq 0 \\ -0.1 & x > 0, \end{cases} \tag{15.59}$$

which also has a discontinuity but located on a different point.

In Fig. 15.2 we plot $\log(J)$ with respect to the number of iterations, for both, the purely discrete method and the alternating descent one. We see that the latter stabilizes in fewer iterations.

In Figs. 15.3 and 15.4, we present the minimizers obtained by the methods above, and the associated solutions, Figs. 15.5 and 15.6.

The initial datum u^0 obtained by the alternating descent method (Fig. 15.4) is a good approximation of (15.58). The solution given by the discrete approach (Fig. 15.3) presents added spurious oscillations. Furthermore, the discrete method is much slower and does not achieve the same level of accuracy since the functional J_Δ does not decrease so much.

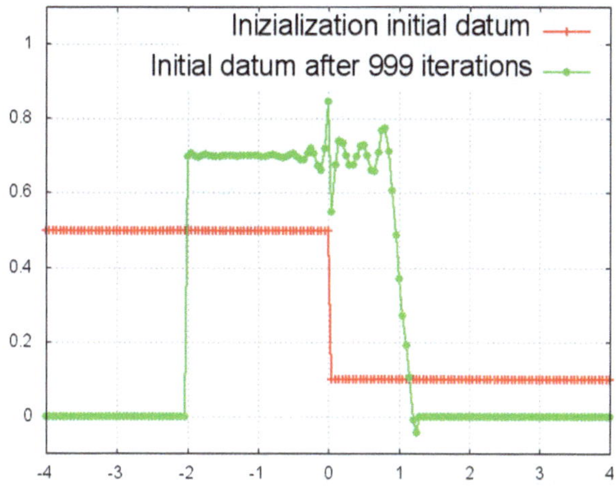

Fig. 15.3 Experiment 1: u^0, discrete method, iteration $k = 999$

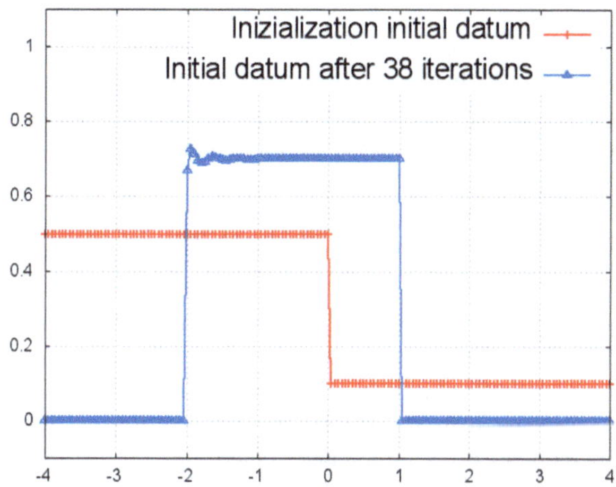

Fig. 15.4 Experiment 1: u^0, alternating descent method, iteration $k = 38$

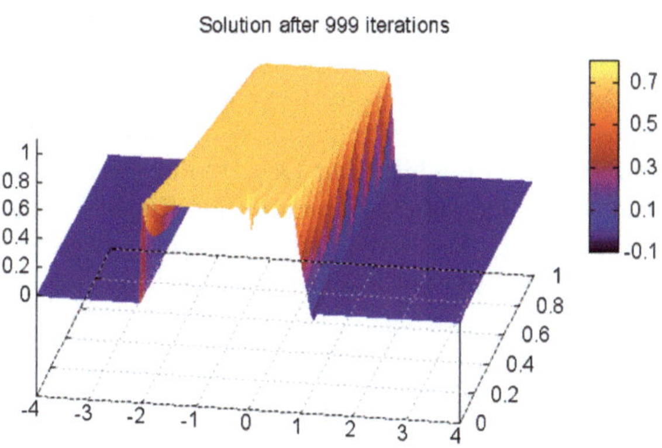

Fig. 15.5 Experiment 1: Solution $u(x,t)$, discrete method, iteration $k = 999$

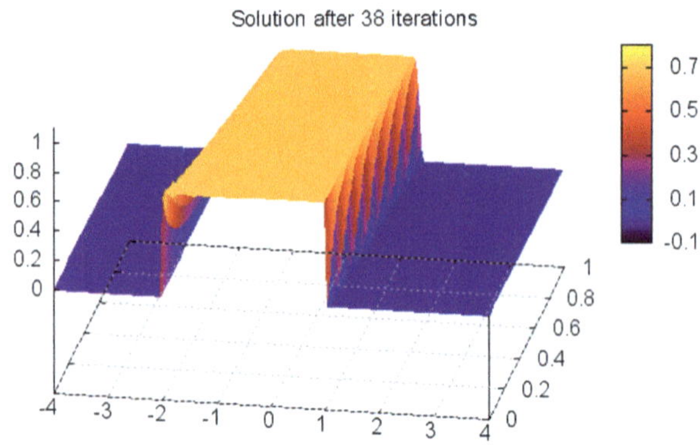

Fig. 15.6 Experiment 1: Solution $u(x,t)$, alternating descent method, iteration $k = 38$

15.7.2 Experiment 2

The previous experiment indicates that the alternating descent method performs significantly better. In order to show that this is a systematic fact, which arises independently of the initialization of the method, we consider the target u^d given by the solution of (15.57) with the initial condition $(u^d)^0$ given by

$$(u^d)^0(x) = \begin{cases} 0.5 & x \le 0.5 \\ 0 & \text{otherwise}, \end{cases} \tag{15.60}$$

but this time we compare the performance of both methods starting from different initializations.

The obtained numerical results are presented in Fig. 15.7.

We see that, regardless the initialization considered, the alternating descent method performs significantly better.

We observe that in the five experiments the alternating descent method perfumes better ensuring the descent of the functional in much fewer iterations and yielding smoother, less oscillatory approximation of the minimizer.

Note also that the discrete method, rather than yielding discontinuous approximations of the minimizer as the alternating descent method does, it produces an initial datum with a Lipschitz front. Observe that these are two different configurations that can lead to the same evolution for the Burgers equation after some time, once the front develops the discontinuity. This is in agreement with the fact that the functional to be minimized is only active in the time-interval $T/2 \le t \le T$.

15.8 Conclusions and Perspectives

In this paper we have adapted and presented the alternating descent method for a tracking problem for a $1-d$ scalar conservation law, the goal being to identify an optimal initial datum so that the solution gets as close as possible to a given

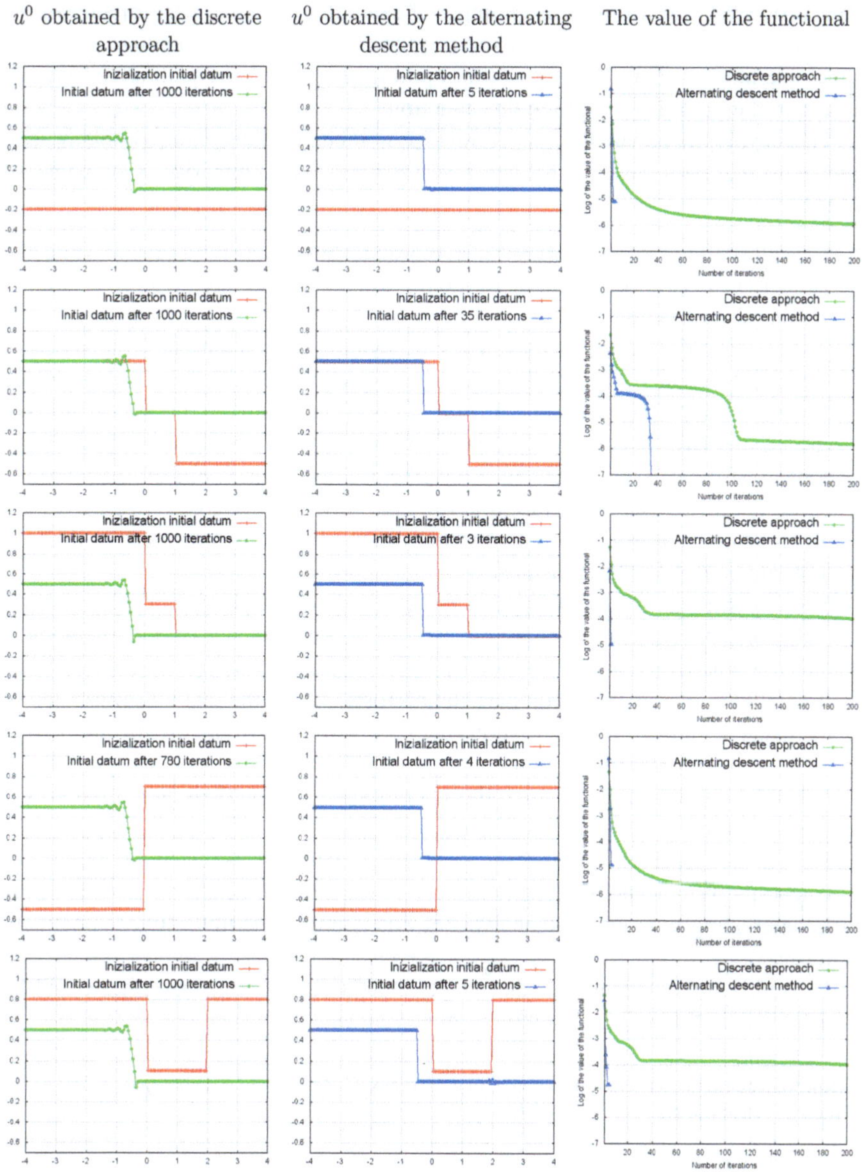

Fig. 15.7 Experiment 2. We present a comparison of the results obtained with both methods starting out of five different initialization configurations. In the left column we exhibit the results obtained with the discrete method. In the second one those achieved by the alternating descent method. In the last one we plot the evolution of the functional with both methods

trajectory. We have exhibited in a number of examples the better performance of the alternating descent method with respect to the classical discrete method. Thus, the paper extends previous works on other optimization problems such as the inverse design or the optimization of the flux function.

The developments in this paper raise a number of interesting problems and questions that will deserve further investigation. We summarize here some of them:

- The alternating descent method and the discrete one do not seem to yield the same minimizer. This should be further investigated in a more systematic manner in other experiments.

  The minimizer that the discrete method yields seems to replace the shock of the initial datum by a Lipschitz function which eventually develops the same dynamics within the time interval $T/2 \leq t \leq T$ in which the functional we have chosen is active. This is an agreement with the behavior of these methods in the context of the classical problem of inverse design in which one aims to find the initial datum so that the solution at the final time takes a given value. It would be interesting to prove analytically that the two methods may lead to different minimizers in some circumstances.
- The experiments in this paper concern the case where the target u^d is exactly reachable. It would be interesting to explore the performance of both methods in the case where the target u^d is not a solution of the underlying dynamics.
- It would be interesting to compare the performance of the methods presented in the paper with the direct continuous method, without introducing the alternating strategy. Our preliminary numerical experiments indicate that the continuous method, without implementing the alternating strategy, does not improve the results that the purely discrete strategy yields.
- In [16] the problem of inverse design has been investigated adapting and extending the alternating descent method to the multi-dimensional case. It would be interesting to extend the analysis of this paper to the multidimensional case too.
- It would worth to compare the results in this paper with those that could be achieved by the nudging method as in [2, 3].

Acknowledgements The authors would like to thank C. Castro, F. Palacios and A. Pozo for stimulating discussions.

References

1. A. Adimurthi, S.S. Ghoshal, V. Gowda, Optimal controllability for scalar conservation laws with convex flux. 12 Jan 2012
2. D. Auroux, J. Blum, Back and forth nudging algorithm for data assimilation problems. Comptes Rendus Math. **340**(12), 873–878 (2005)

3. D. Auroux, M. Nodet, The back and forth nudging algorithm for data assimilation problems: theoretical results on transport equations. ESAIM: Control Optim. Calc. Var. **18**(2), 318–342, 7 (2012)
4. C. Bardos, O. Pironneau, A formalism for the differentiation of conservation laws. C. R. Math. Acad. Sci. Paris **335**(10), 839–845 (2002)
5. F. Bouchut, F. James, One-dimensional transport equations with discontinuous coefficients. Nonlinear Anal. **32**(7), 891–933 (1998)
6. F. Bouchut, F. James, Differentiability with respect to initial data for a scalar conservation law, in *Hyperbolic Problems: Theory, Numerics, Applications, Vol. I (Zürich, 1998)*. Volume 129 of International Series of Numerical Mathematics (Birkhäuser, Basel, 1999), pp. 113–118
7. A. Bressan, A. Marson, A variational calculus for discontinuous solutions of systems of conservation laws. Commun. Partial Differ. Equ. **20**(9–10), 1491–1552 (1995)
8. C. Castro, E. Zuazua, Flux identification for 1-d scalar conservation laws in the presence of shocks. Math. Comput. **80**(276), 2025–2070 (2011)
9. C. Castro, F. Palacios, E. Zuazua, An alternating descent method for the optimal control of the inviscid Burgers equation in the presence of shocks. Math. Models Methods Appl. Sci. **18**(03), 369–416 (2008)
10. C. Castro, F. Palacios, E. Zuazua, Optimal control and vanishing viscosity for the Burgers equation, in *Integral Methods in Science and Engineering*, ed. by C. Constanda, M. Pérez, vol. 2 (Birkhäuser, Boston, 2010), pp. 65–90
11. S. Garreau, P. Guillaume, M. Masmoudi, The topological asymptotic for PDE systems: the elasticity case. SIAM J. Control Optim. **39**(6), 1756–1778 (2000)
12. E. Godlewski, P. Raviart, *Hyperbolic Systems of Conservation Laws*. Mathématiques and Applications (Ellipses, Paris, 1991)
13. E. Godlewski, P.A. Raviart, The linearized stability of solutions of nonlinear hyperbolic systems of conservation laws. A general numerical approach. Math. Comput. Simul. **50**(1–4), 77–95 (1999). Modelling'98, Prague
14. L. Gosse, F. James, Numerical approximations of one-dimensional linear conservation equations with discontinuous coefficients. Math. Comput. **69**(231), 987–1015 (2000)
15. S.N. Kružkov, First order quasilinear equations with several independent variables. Mat. Sb. (N.S.) **81**(123), 228–255 (1970)
16. R. Lecaros, E. Zuazua, Control of $2D$ scalar conservation laws in the presence of shocks. (2014, preprint)
17. S. Ulbrich, Adjoint based derivative computations for the optimal control of discontinuous solutions of hyperbolic conservation laws. Syst. Control Lett. **48**(3–4), 313–328 (2003). Optimization and control of distributed systems
18. X. Wen, S. Jin, Convergence of an immersed interface upwind scheme for linear advection equations with piecewise constant coefficients. I. L^1-error estimates. J. Comput. Math. **26**(1), 1–22 (2008)

Chapter 16
Finite Simple Groups of Small Essential Dimension

Arnaud Beauville

Abstract We discuss the notion of essential dimension of a finite group (over $\mathbb{C}$) and explain its relation with birational algebraic geometry. We show how this leads to a (partial) classification of simple finite groups of essential dimension ≤ 3.

16.1 Introduction

The *essential dimension* of a finite group G has been introduced in the seminal paper [5]. We have to refer to that paper for motivation (see also the survey paper [2]); in a very informal way, the essential dimension of G (over a field k) is the minimum number of parameters needed to define a general Galois extension L/K with Galois group G and $K \supset k$.

Since then the notion of essential dimension has been put in a much larger context (see [14] for a recent survey). In this note we go back to the original problem, in the simplest case where k is the field of complex numbers. In that case the definition is quite concrete, and the computation of the essential dimension becomes an interesting problem of classical algebraic geometry. We will explain how one can use results on the birational geometry of unirational varieties to classify the simple finite groups of essential dimension ≤ 2, and obtain partial results in dimension 3.

A. Beauville (✉)
Laboratoire J.-A. Dieudonné, Université de Nice Sophia Antipolis, UMR 7351 du CNRS, Parc Valrose, F-06108 Nice cedex 2, France
e-mail: arnaud.beauville@unice.fr

V. Ancona and E. Strickland (eds.), *Trends in Contemporary Mathematics*, Springer INdAM Series 8, DOI 10.1007/978-3-319-05254-0_16,

16.2 Definitions and Basic Properties

Let G be a finite group, and X a complex algebraic variety with a faithful action of G. We will say that X is *G-linearizable* if there exists a complex representation V of G and a rational dominant G-equivariant map $V \dashrightarrow X$. The *essential dimension* $\mathrm{ed}(G)$ of G (over $\mathbb{C}$) is the minimal dimension of all linearizable G-varieties.

Let us mention immediately three obvious consequences of the definition, which we will use frequently in the sequel:

- One has $\mathrm{ed}(G) \geq 0$, and $\mathrm{ed}(G) = 0$ if and only if $G = \{1\}$.
- If H is a subgroup of G, $\mathrm{ed}(H) \leq \mathrm{ed}(G)$.
- Let $u : X \dashrightarrow Y$ be a G-equivariant birational map; then X is G-linearizable if and only if Y is. In particular, to compute $\mathrm{ed}(G)$ it suffices to consider smooth projective varieties.

Examples 16.1. (a) A vector space V with a faithful linear action of G is a G-linearizable variety, hence $\mathrm{ed}(G) \leq \dim(V)$. Therefore, if we denote by $\mathrm{rd}(G)$ the minimal dimension of a faithful representation of G ("*representation dimension*" of G), we have $\mathrm{ed}(G) \leq \mathrm{rd}(G)$.

(b) In the situation of (a), the group G acts on the projective space $\mathbb{P}(V)$, and the rational map $V \dashrightarrow \mathbb{P}(V)$ is G-equivariant. The action of G on $\mathbb{P}(V)$ is faithful if and only if $G \subset \mathrm{GL}(V)$ contains no nontrivial homothety; this is the case in particular if the center of G is trivial. In this situation we have $\mathrm{ed}(G) \leq \mathrm{rd}(G) - 1$.

(c) Here is a more elaborate example. The permutation action of the symmetric group $\mathfrak{S}_n$ on $\mathbb{C}^n$ extends to $(\mathbb{P}^1)^n$. On the other hand the group $H = \mathrm{PGL}_2(\mathbb{C})$ acts also on $(\mathbb{P}^1)^n$; the Zariski open subset $(\mathbb{P}^1)^n_{\neq}$ of points with distinct coordinates is stable under both actions. For $n \geq 3$ H acts freely on $(\mathbb{P}^1)^n_{\neq}$; the quotient $X_n = (\mathbb{P}^1)^n_{\neq}/H$ is an algebraic variety of dimension $n-3$. The action of $\mathfrak{S}_n$ on $(\mathbb{P}^1)^n_{\neq}$ commutes with that of H, hence induces an action on X_n, which is faithful for $n \geq 5$; the projection $\mathbb{C}^n \dashrightarrow X_n$ is $\mathfrak{S}_n$-equivariant. We conclude that *the essential dimension of $\mathfrak{S}_n$ is $\leq n-3$* for $n \geq 5$.

The following result will provide our main tool to prove that a variety is *not* G-linearizable:

Proposition 16.1 ([13]). *Let A be a finite abelian group, X an A-linearizable projective variety. There is a point of X fixed by A.*

The proof is so simple that we cannot resist to copy it from [13]. Since a linear action on a vector space fixes the origin, the Proposition follows from a more general result:

Lemma 16.1. *Let X, Y be two A-varieties, with X smooth and Y proper, $f : X \dashrightarrow Y$ a rational A-equivariant map. If X has a point fixed by A, so does Y.*

Proof. The proof is by induction on $\dim(X)$, the case $\dim(X) = 0$ being clear.

Let $x \in X$ be a point fixed by A; let $B_x(X)$ be the blow up of x in X, and E the exceptional divisor. The action of A extends to $B_x(X)$ and E. Since Y is proper, the rational map $B_x(X) \dashrightarrow Y$ is defined outside a subset of codimension ≥ 2, so it induces a A-equivariant rational map $E \dashrightarrow Y$. Since an abelian group acting on a projective space has always a fixed point, A fixes a point of E, hence of Y by the induction hypothesis. ■

Corollary 16.1. *We have* $\mathrm{ed}(A) = \mathrm{rd}(A)$. *In particular, the essential dimension of* $(\mathbb{Z}/n)^r$ *is* r.

Proof. Let X be a smooth A-linearizable projective variety, x a point of X fixed by A. The group A acts on the tangent space $T_x(X)$, and this action is isomorphic in a neighborhood of 0 to the action of A on X in a neighborhood of x. Therefore the representation of A on $T_x(X)$ is faithful; thus $\mathrm{rd}(A) \leq \dim(X)$, hence $\mathrm{rd}(A) \leq \mathrm{ed}(A)$. The opposite inequality is obvious (Example 16.1.(a)). Finally the equality $\mathrm{rd}((\mathbb{Z}/n)^r) = r$ is an easy exercise. ■

The equality $\mathrm{ed}(G) = \mathrm{rd}(G)$ holds more generally for a p-group [12]; the proof uses much more sophisticated techniques.

Let us conclude this section with the list of groups of essential dimension 1:

Proposition 16.2 ([5]). *The finite groups of essential dimension* 1 *are the cyclic groups and the dihedral groups* D_n *for* n *odd.*

Proof. We have $\mathrm{rd}(\mathbb{Z}/m) = 1$, and $\mathrm{rd}(D_n) = 2$; when n is odd the center of D_n is trivial. Thus $\mathrm{ed}(\mathbb{Z}/m) = \mathrm{ed}(D_n) = 1$ by Example 16.1.(a) and (b).

Let G be a finite group with $\mathrm{ed}(G) = 1$, and X a G-linearizable smooth projective curve. Then $X \cong \mathbb{P}^1$, so G is a subgroup of $\mathrm{PGL}_2(\mathbb{C})$, hence isomorphic to $\mathbb{Z}/n, D_n, \mathfrak{A}_4, \mathfrak{S}_4$ or $\mathfrak{A}_5$. Now except $\mathbb{Z}/m$ and D_n for n odd, all these groups contain a copy of $(\mathbb{Z}/2)^2$. We conclude with Corollary 16.1.

16.3 Groups of Essential Dimension 2

The groups of essential dimension 2 have been classified in [10]; the list is already quite large. We will restrict ourselves to the class of simple groups.

Proposition 16.3. *The simple finite groups of essential dimension* 2 *are* $\mathfrak{A}_5$ *and* $\mathrm{PSL}_2(\mathbb{F}_7)$.

Of course this follows immediately from [10]; but the proof is significantly easier for simple groups.

Proof. We have $\mathrm{ed}(\mathfrak{A}_5) = \mathrm{ed}(\mathfrak{S}_5) = 2$ by Example 16.1.(c) and Proposition 16.2; the group $\mathrm{PSL}_2(\mathbb{F}_7)$ has a faithful representation of dimension 3 (which can be realized as $H^0(C, K_C)$, where C is the Klein quartic curve), hence it has essential dimension 2 by Example 16.1.(*b*) and Proposition 16.2.

Let G be a finite group with $\mathrm{ed}(G) = 2$, and X a G-linearizable smooth projective surface. By Castelnuovo's theorem X is rational, so G is a finite subgroup of the Cremona group $\mathrm{Cr}_2 = \mathrm{Bir}(\mathbb{P}^2)$. The classification of these subgroups has been worked out in the nineteenth century (Kantor, Wiman), and completed in [8]. Let us state the result in the case of interest for us, namely that of simple groups; here again, the proof is much easier in that case than for general groups.

Theorem ([8]). *The simple finite subgroups of* Cr_2 *are cyclic or isomorphic to* $\mathfrak{A}_5$, $\mathfrak{A}_6$ *or* $\mathrm{PSL}_2(\mathbb{F}_7)$. ∎

Thus our task now is to eliminate the group $\mathfrak{A}_6$. This group appears only once in the list, as a group of automorphisms of $\mathbb{P}^2$ (the so-called Valentiner group). The inverse image $\tilde{\mathfrak{A}}_6$ of $\mathfrak{A}_6$ in SL(3) is a central extension of $\mathfrak{A}_6$ by $\mathbb{Z}/3$.

One way of describing this extension is to view $\mathfrak{A}_6$ as the subgroup of $\mathrm{PGL}_3(\mathbb{F}_4)$ preserving the set formed by the six points

$$(1,0,0)\ ,\ (0,1,0)\ ,\ (0,0,1)\ ,\ (1,1,1)\ ,\ (1,\alpha,\beta)\ ,\ (1,\beta,\alpha)\ ,$$

where $\mathbb{F}_4 = \{0, 1, \alpha, \beta\}$; then $\tilde{\mathfrak{A}}_6$ is the pull back of $\mathfrak{A}_6$ in $\mathrm{GL}_3(\mathbb{F}_4)$ (see [7], 4.2). The elements

$$u = \begin{pmatrix} 1 & 0 & 0 \\ 0 & \alpha & 0 \\ 0 & 0 & \beta \end{pmatrix} \qquad v = \begin{pmatrix} 0 & 1 & 0 \\ 0 & 0 & 1 \\ 1 & 0 & 0 \end{pmatrix}$$

of $\mathrm{GL}_3(\mathbb{F}_4)$ belong to $\tilde{\mathfrak{A}}_6$, and their commutator is a nontrivial element of its center. We conclude with the following lemma:

Lemma 16.2. *Let V be a complex vector space, G a finite subgroup of* $\mathrm{PGL}(V)$, *$\tilde{G}$ its inverse image in* $\mathrm{GL}(V)$. *Assume that there exist elements u, v of $\tilde{G}$ such that their commutator (u, v) is a nontrivial homothety. Then $\mathbb{P}(V)$ is not G-linearizable.*

Proof. There is no line in V stable under u and v, since otherwise u and v would commute on that line. Therefore the images of u and v in $\mathrm{PGL}(V)$ do not fix a common point of $\mathbb{P}(V)$. Since they commute, Proposition 16.1 shows that $\mathbb{P}(V)$ is not G-linearizable. ∎

Remark. Assume moreover that G is simple, and that the action of G on $\mathbb{P}(V)$ does not come from a linear action on V. Then it follows from [3] that the hypothesis of the lemma is always satisfied. Thus $\mathbb{P}(V)$ is never G-linearizable in that case.

16.4 Groups of Essential Dimension 3

16.4.1 Prokhorov's List

The classification of all finite groups of essential dimension 3 is definitely out of reach at this moment. On one hand the classification of finite subgroups of the Cremona group Cr_3 seems untractable; moreover, a G-linearizable threefold is unirational, but there is no reason to expect it to be rational.

However when we restrict our attention to simple groups, using a remarkable theorem of Prokhorov we get a partial result:

Proposition 16.4. *The simple groups of essential dimension* 3 *are* $\mathfrak{A}_6$ *and possibly* $\mathrm{PSL}_2(\mathbb{F}_{11})$.

Proof. We have $\mathrm{ed}(\mathfrak{A}_6) = 3$ by Example 16.1.(c) and the previous classification. The group $\mathrm{PSL}_2(\mathbb{F}_{11})$ admits a faithful representation of dimension 5, so its essential dimension is 3 or 4 by Example 16.1.(b); the exact value is not known (see Sect. 16.4.5).

To prove the converse, let G be a finite simple group with $\mathrm{ed}(G) = 3$. By definition there exists a G-linearizable projective threefold X. This implies in particular that X is rationally connected. Such pairs (G, X) have been classified in [16]. We have the following possibilities for G:

1. $G = \mathfrak{A}_5, \mathfrak{A}_6, \mathrm{PSL}_2(\mathbb{F}_7), \mathrm{PSL}_2(\mathbb{F}_{11})$.
2. $G = \mathfrak{A}_7, \mathrm{PSp}_4(\mathbb{F}_3), \mathrm{SL}_2(\mathbb{F}_8)$.

The groups of the first row have already been dealt with. We will deal separately with the three remaining cases.

16.4.2 The Group $\mathfrak{A}_7$

Proposition 16.5 ([9]). *The essential dimension of* $\mathfrak{A}_7$ *is* 4.

Proof. The group $\mathfrak{A}_7$ appears twice in Prokhorov's list:

- $\mathfrak{A}_7$ acts by permutation of coordinates on the variety X given by $\sum X_i = \sum X_i^2 = \sum X_i^3 = 0$ in $\mathbb{P}^6$.
- $\mathfrak{A}_7$ embeds into $\mathrm{PGL}_4(\mathbb{C})$, hence acts on $\mathbb{P}^3$.

In the first case, one checks easily that the subgroup $(\mathbb{Z}/2)^2 \times \mathbb{Z}/3 \subset \mathfrak{A}_4 \times \mathfrak{A}_3 \subset \mathfrak{A}_7$ has no fixed point on X, so that X is not $\mathfrak{A}_7$-linearizable by Proposition 16.1.

In the second case, let us consider the double coverings of complex Lie groups

$$\mathrm{SL}(4) \longrightarrow \mathrm{SO}(6) \longrightarrow \mathrm{PGL}(4)$$

deduced from the isomorphism $\mathbb{C}^6 \cong \wedge^2\mathbb{C}^4$. The standard representation $\mathfrak{A}_7 \subset \mathrm{SO}(6)$, composed with the second arrow, gives the embedding of $\mathfrak{A}_7$ into PGL(4). The inverse image of $\mathfrak{A}_7$ in SL(4) appears as a central extension

$$1 \to \{\pm I\} \longrightarrow \tilde{\mathfrak{A}}_7 \longrightarrow \mathfrak{A}_7 \to 1\,.$$

To apply Lemma 16.2, we need to show that the central element $-I$ of $\tilde{\mathfrak{A}}_7$ is a commutator. Consider the subgroup $\mathfrak{A}_4$ of $\mathfrak{A}_7$ fixing the last three letters. We have a commutative diagram

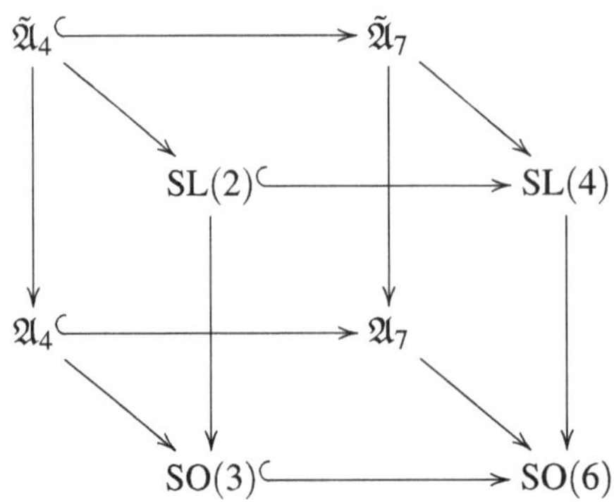

so it suffices to check that $-I$ is a commutator in $\tilde{\mathfrak{A}}_4$. Since SL(2) contains no element of order 2 except $-I$, the Klein subgroup $(\mathbb{Z}/2)^2 \subset \mathfrak{A}_4$ lifts to the quaternion group $Q_8 = \{\pm 1, \pm i, \pm j, \pm k\}$, and we have $(i, j) = -1$. Thus $-I$ is a commutator in $\tilde{\mathfrak{A}}_7$, and Lemma 16.2 shows that $\mathbb{P}^3$ is not $\mathfrak{A}_7$-linearizable. ∎

16.4.3 *The Group* $\mathrm{PSp}_4(\mathbb{F}_3)$

Proposition 16.6. *The essential dimension of* $\mathrm{PSp}_4(\mathbb{F}_3)$ *is* 4.

Proof. The group $\mathrm{Sp}_4(\mathbb{F}_3)$ has a linear representation on the space W of functions on $\mathbb{F}_3^2$, the *Weil representation*, for which we refer to [1], Appendix I. This representation splits as $W = W^+ \oplus W^-$, the spaces of even and odd functions; we have $\dim W^+ = 5$, $\dim W^- = 4$. The central element $(-I)$ of $\mathrm{Sp}_4(\mathbb{F}_3)$ acts on W by ${}^{(-I)}F(x) = F(-x)$, hence it acts trivially on W^+, and as $-\mathrm{Id}$ on W^-. Thus we get a faithful representation of $\mathrm{PSp}_4(\mathbb{F}_3)$ on W^+, hence $\mathrm{ed}(\mathrm{PSp}_4(\mathbb{F}_3)) \le 4$ (Example 16.1.(*b*)).

To prove that we have equality, we observe[1] that $\mathrm{PSp}_4(\mathbb{F}_3)$ contains a subgroup isomorphic to $(\mathbb{Z}/2)^4$. One way to see this is to use the classical isomorphism of

[1] I am indebted to A. Duncan for this observation.

$\mathrm{PSp}_4(\mathbb{F}_3)$ with $\mathrm{SO}_5^+(\mathbb{F}_3)$, the kernel of the spinor norm $\mathrm{SO}_5(\mathbb{F}_3) \to \mathbb{F}_3^* \cong \{\pm 1\}$. Essentially by definition (see [4], §9, no. 5), this group contains the transformations $\sigma_v : x \mapsto -x + 2(x.v)v$ for each length 1 vector v; when v runs over the elements of an orthonormal basis, the σ_v span a subgroup of $\mathrm{SO}_5^+(\mathbb{F}_3)$ isomorphic to $(\mathbb{Z}/2)^4$. Therefore $\mathrm{ed}(\mathrm{PSp}_4(\mathbb{F}_3)) \geq \mathrm{ed}((\mathbb{Z}/2)^4) = 4$ (Corollary 16.1). ■

16.4.4 The Group $\mathrm{SL}_2(\mathbb{F}_8)$

Proposition 16.7. *The essential dimension of* $\mathrm{SL}_2(\mathbb{F}_8)$ *is* ≥ 4.

The group $\mathrm{SL}_2(\mathbb{F}_8)$ has a representation of dimension 7, hence its essential dimension is ≤ 6 – we do not know its precise value.

Proof. The group $\mathrm{SL}_2(\mathbb{F}_8)$ acts on a rational Fano threefold $X \subset \mathbb{P}^8$ in the following way [16]. Let U be an irreducible 9-dimensional representation of $\mathrm{SL}_2(\mathbb{F}_8)$; there exists a non-degenerate invariant quadratic form q on U, unique up to a scalar. Then $\mathrm{SL}_2(\mathbb{F}_8)$ acts on the orthogonal Grassmannian $\mathbb{G}_{\mathrm{iso}}(4, U)$ of 4-dimensional isotropic subspaces of U. This Grassmannian admits a $\mathrm{O}(q)$-equivariant embedding into $\mathbb{P}^{15}$, given by the half-spinor representation [15]. The threefold X is the intersection of $\mathbb{G}_{\mathrm{iso}}(4, U)$ with a subspace $\mathbb{P}^8 \subset \mathbb{P}^{15}$ invariant under $\mathrm{SL}_2(\mathbb{F}_8)$.

Let $N \subset \mathrm{SL}_2(\mathbb{F}_8)$ be the subgroup of matrices $\begin{pmatrix} 1 & a \\ 0 & 1 \end{pmatrix}$, $a \in \mathbb{F}_8$. We will show that N has no fixed point in $\mathbb{G}_{\mathrm{iso}}(4, U)$, and therefore in X.

Let χ_U be the character of the representation U. We have $\chi_U(n) = 1$ for $n \in N$, $n \neq 1$ (see for instance [6], 2.7). It follows that U restricted to N is the sum of the regular representation and the trivial one; in other words, as a N-module we have

$$U = \mathbb{C}_1^2 \oplus \sum_{\lambda \in \hat{N} - \{1\}} \mathbb{C}_\lambda ,$$

where $\mathbb{C}_\lambda$ is the one-dimensional representation associated to the character λ of N. The subspaces $\mathbb{C}_\lambda$ and $\mathbb{C}_\mu$ for $\lambda \neq \mu$ are orthogonal for q; since q is non-degenerate, its restriction to each $\mathbb{C}_\lambda$ $(\lambda \neq 1)$ and to $\mathbb{C}_1^2$ must be non-degenerate.

Now any vector subspace $L \subset U$ stable under N must be the sum of some of the $\mathbb{C}_\lambda$, for $\lambda \neq 1$, and of some subspace of $\mathbb{C}_1^2$; this implies that L cannot be isotropic as soon as $\dim L \geq 2$. Hence N has no fixed point on $\mathbb{G}_{\mathrm{iso}}(4, U)$, and X is not linearizable by Proposition 16.1. ■

This finishes the proof of Proposition 16.4. ■

16.4.5 *About* $\mathbf{PSL_2(\mathbb{F}_{11})}$

According to [16] there are two rationally connected threefolds with an action of $\mathrm{PSL}_2(\mathbb{F}_{11})$, the Klein cubic $X^{\mathrm{k}} \subset \mathbb{P}^4$ given by $\sum_{i\in\mathbb{Z}/5} X_i^2 X_{i+1} = 0$ and a Fano threefold $X^{\mathrm{a}} \subset \mathbb{P}^9$ of degree 14, birational to X^{k}. The group $\mathrm{PSL}_2(\mathbb{F}_{11})$ has order $660 = 2^2.3.5.11$; its abelian subgroups are cyclic, except the 2-Sylow subgroups which are isomorphic to $(\mathbb{Z}/2)^2$. A finite order automorphism of a rationally connected variety has always a fixed point (for instance by the holomorphic Lefschetz formula); one checks easily that a 2-Sylow subgroup of $\mathrm{PSL}_2(\mathbb{F}_{11})$ has a fixed point on both X^{k} and X^{a}. So Proposition 16.1 does not apply, and another approach is needed. In [11] the authors show that the equality $\mathrm{ed}(\mathrm{PSL}_2(\mathbb{F}_{11})) = 3$ would follow from a conjecture of Cassels and Swinnerton-Dyer on the existence of rational points on cubic hypersurfaces.

References

1. A. Adler, S. Ramanan, *Moduli of Abelian Varieties*. Lecture Notes in Mathematics, vol. 1644 (Springer, Berlin, 1996)
2. A. Beauville, De combien de paramètres dépend l'équation générale de degré n? Gaz. Math. **132**, 5–15 (2012)
3. H. Blau, A fixed-point theorem for central elements in quasisimple groups. Proc. Am. Math. Soc. **122**(1), 79–84 (1994)
4. N. Bourbaki, *Algèbre*, chap. 9 (Hermann, Paris, 1959)
5. J. Buhler, Z. Reichstein, On the essential dimension of a finite group. Compos. Math. **106**(2), 159–179 (1997)
6. M. Collins, *Representations and Characters of Finite Groups* (Cambridge University Press, Cambridge/New York, 1990)
7. R. Curtis, *Symmetric Generation of Groups*. Encyclopedia of Mathematics and Its Applications, vol. 111 (Cambridge University Press, Cambridge, 2007)
8. I. Dolgachev, V. Iskovskikh, Finite subgroups of the plane Cremona Group, in *Algebra, Arithmetic, and Geometry: In Honor of Yu. I. Manin. Vol. I*. Progress in Mathematics, vol. 269 (Birkhäuser, Boston, 2009), pp. 443–548
9. A. Duncan, Essential dimensions of A_7 and S_7. Math. Res. Lett. **17**(2), 263–266 (2010)
10. A. Duncan, Finite groups of essential dimension 2. Comment. Math. Helv. **88**, 555–585 (2013)
11. A. Duncan, Z. Reichstein, Versality of algebraic group actions and rational points on twisted varieties. J. Algebr. Geom. (to appear)
12. N. Karpenko, A. Merkurjev, Essential dimension of finite p-groups. Invent. Math. **172**(3), 491–508 (2008)
13. J. Kollár, E. Szabó, Fixed points of group actions and rational maps (Appendix to a paper of Z. Reichstein and B. Youssin). Can. J. Math. **52**(5), 1054–1056 (2000)
14. A. Merkurjev, Essential dimension: a survey. Trans. Groups **18**(2), 415–481 (2013)
15. S. Mukai, Curves and symmetric spaces, I. Am. J. Math. **117**(6), 1627–1644 (1995)
16. Y. Prokhorov, Simple finite subgroups of the Cremona group of rank 3. J. Algebr. Geom. **21**(3), 563–600 (2012)

Chapter 17
Geometric Constructions of Extremal Metrics on Complex Manifolds

Claudio Arezzo

Abstract In this note we review recent progresses on the existence problem of Kähler constant scalar curvature metrics on complex manifolds. The content of this note is an expanded version of author's talk "Constant curvature metrics on algebraic manifolds" at the Giornata Indam at L'Aquila on June 9th 2011.

1991 Math. Subject Classification: 58E11, 32C17

17.1 Prelude, The Real Story

A central problem in differential geometry is to determine, given a compact manifold M, which is the "best" metric we can put on it. Of course large part of the problem is to decide what "best" really means. In fact even the fact that we address such a metric with "the" indicates that we expect right away a difficult property this has to satisfy: *it has to be unique in a suitable sense*, so that the association $\{\text{manifolds}\} \to \{\text{its best metric}\}$ becomes well defined.

On top of this, a geometric-minded person would tend to ask to the metric to have the maximal number of isometries, while from the analytic point of view it would be natural to ask to this mysterious object should be a critical point of a natural functional. Certainly both will want such a metric to exist on a set of manifolds as large as possible (ideally on every manifold!).

Parallel to these requirements even a quick study of the foundations of differential geometry leads to look at the curvature *Riem* of the riemannian manifold and to

C. Arezzo (✉)
Abdus Salam International Center for Theoretical Physics, Trieste, Italy
Universita' di Parma, Parma, Italy
e-mail: arezzo@ictp.it

V. Ancona and E. Strickland (eds.), *Trends in Contemporary Mathematics*,
Springer INdAM Series 8, DOI 10.1007/978-3-319-05254-0_17,

ask whether this can be made *constant*, which being a tensor, is equivalent to say $\nabla Riem = 0$ (where ∇ is the Levi-Civita Connection of g). Such manifolds are called *symmetric spaces*. This turns out to be an extremely strong requirement on the base manifold and in fact, thanks to the fundamental work of Cartan [14] we know that only very special types of manifolds can carry such a structure. It is then natural to try to relax such a request extracting other curvatures form *Riem* by means of mixtures of algebraic and geometric operations.

Right from the birth of riemannian geometry, Riemann himself (even if not aware of Cartan's Theorem) has defined a different more geometric curvature, called *sectional curvature* which measures, roughly speaking, the Gauss curvature of a 2-dimensional surface passing through a given point with a given tangent space. After observing that the knowledge of the sectional curvature at every point in every plane-direction is equivalent to that of *Riem*, he has asked immediately what the constancy of this new object implies (where now it depends both on the point of the base and on the plane in the tangent space) and proved what has rightly passed into history of geometry as the Fundamental Theorem of Riemannian Geometry:

Theorem 17.1. *If the sectional curvature is constant, then the manifold is a quotient, via a finite group of isometries, of a Euclidean, spherical or hyperbolic space.*

Classification of such manifolds has then turned into the (highly nontrivial) problem of classification of finite subgroups of the isometry groups of the classical models, but it is evident from the above theorem that such a condition will be almost never fulfilled.

One can then try a different operation, i.e. *contracting* the Riemann tensor, hence getting a $(2,0)$-tensor, the *Ricci curvature*, *Ric*, and again ask how often this can be made into a constant tensor. The (system of) equations one is then led to study is then $\nabla Ric = 0$, better known in the equivalent form

$$Ric = \lambda g,$$

the *Einstein equation*. The existence of metrics satisfying the Einstein equation is an incredibly difficult problem and this is not a place even to attempt to review its huge literature, so we just restrict ourselves to three beautiful results which highlight to subtleness of the problem:

1. (Hamilton-Perelman) in dimension 3 every simply connected manifold admits an Einstein metric of positive curvature, hence it is a sphere.
2. Not every 4-manifold has an Einstein metric. For example, Hitchin-Thorpe found a severe obstruction in the form

$$\chi(M) \geq \frac{3}{2}\tau(M),$$

   where $\chi(M)$ denotes the Euler characteristic of the manifold and $\tau(M)$ its signature.

3. (Le Brun [31]) there exist homeomorphic not diffeomorphic 4-manifolds M and N, such that M is Einstein and N is not!

Things change quite dramatically if we perform again the contraction which took *Riem* into *Ric*, this time taking *Ric* into a single function R, the *scalar curvature*. The constancy of this function this time does not put any restriction on the base, as we learned from the combined effort of Yamabe-Trudinger-Aubin up to the conclusive fundamental work of Schoen [42]:

Theorem 17.2. *Given any compact riemannian manifold* (M, g) *there exists a smooth function* u *such that the metric* $e^u g$ *has constant scalar curvature.*

It is important to conclude this prelude by recalling that in dimension 2 all the above curvature notions coincide (up to a positive factor) with the classical Gauss curvature for which we know the fundamental

Theorem 17.3. *Every compact surface admits a metric of constant Gauss curvature.*

17.2 Making Everything Complex

The picture quickly described above takes an intriguing twist if we put an extra structure on the base manifold, namely a *complex structure* J. It is natural to assume that J, an endomorphism of the tangent bundle whose square equals minus the identity, is compatible with the metric, i.e. it is an isometry for g and it is parallel, $\nabla J = 0$. In this case we say that (M, g, J) is a Kähler manifold. We will usually think of (M, J) as the datum and seek g with some curvature properties as above among those for which the triple is Kähler.

Even with the restrictions described above, one can still hope that Einstein manifolds form a reasonably large class of manifolds, and certainly constant scalar curvature manifold do! Yet, in the Kähler world this dramatically changes. A first simple indication of this phenomenon comes just from observing that the changes in metrics in Theorem 17.2 (i.e. those of the form $e^u g$) destroy the Kähler structure unless u is constant. A more conclusive observation was done by Futaki [25] who introduced the following character of the Lie algebra of the space of holomorphic vector fields: first observe that on a compact manifold there exists a smooth function (which is unique up to a constant and which depends on g) such that

$$R(g) - \frac{1}{Vol(M, g)} \int_M R(g) dVol_g = \Delta_g h_g$$

and then take a holomorphic vector field X on M to define

$$F(X, g) = \int_M X(h_g) dVol_g \ .$$

Futaki's key observation is that $F(X, g)$ depends not on g but only on its Kähler class, moreover, since clearly if g is Kcsc, F vanishes (since h_g is constant), and yet it is not hard to compute this number in specific cases, as for the blow up of $\mathbb{C}P^2$ in one or two points, to get a nonzero number, such manifolds cannot admit a Kcsc metric (a very nice reference for these results is Tian's monograph [54]).

Given a Kähler structure, Calabi [11, 12] has proposed a notion of "best" metric which puts together all the possible points of view described in our prelude as in a beautiful mosaic. Being easy to observe that the classical Hilbert functional

$$g \to \int_M R(g) dVol_g$$

is constant on a fixed Kähler class, he then looked at the L^2-norm of the scalar curvature

$$\mathcal{C}(g) = \int_M R(g)^2 dVol_g \ .$$

Metrics which are critical points of the Calabi functional have been baptized *extremal*, and they satisfy (on a compact manifold) the following Euler-Lagrange equation: *the* $(1,0)$*-part of the gradient of* $R(g)$ *is a (real) holomorphic vector field.*

Clearly this is a weaker condition than Einstein or Kcsc (corresponding to the zero vector field) and Calabi himself produced the first example of an extremal non Kcsc metric precisely on the above mentioned manifold of $\mathbb{C}P^2$ blown up at one point (in every Kähler class). The analogue question (at least in *some* Kähler class) will be answered in Sect. 17.4.

The difference between extreme and Kcsc is completely encoded in the following important observation by Calabi, which will be very important also in Sect. 17.4:

Theorem 17.4. *Let* (M, g) *be an extremal compact manifold. Then* g *has constant scalar curvature if and only if* $F(\cdot, g) = 0$.

Let's now turn back to our list of requirements for a metric to be called "the best":

1. Extremal metrics are born, as we have just seen, as critical points (minima, in fact) of the simplest possible riemannian functional defined on a fixed Kähler class. Calabi himself addressed the question of whether there could be other functionals, as "natural" as $\mathcal{C}$, defining other notions of canonical metrics. Surprisingly he proved that all riemannian functionals *quadratic in the curvature (not necessarily the scalar curvature)* have the same critical points.
2. Uniqueness: this turned out to be an amazingly challenging problem. It was solved in the Einstein case of non positive curvature by Calabi, as an application of the maximum principle applied to the complex Monge-Ampère equation, and by Bando-Mabuchi [6] in the Einstein positive case. It was then proved by Donaldson [23] in the Kcsc case and *rational* Kähler class and discrete automorphism group, later extended by Mabuchi to the rational extremal case

[38]. It was proved finally in full generality (i.e. extremal and any class) by Chen-Tian [16].

3. Maximal symmetries: the issue of determining the isometry group of an extremal metric was studied immediately by Calabi in his foundational papers. He extended the same statement previously proved in the Einstein and Kcsc cases by Matsushima [39] and Lichnerowicz [36] respectively, in the general extremal case: *the identity component of the group of isometries of (M, g) is a maximal compact subgroup of the identity component of the automorphism group.*
4. Existence: as we said right at the beginning of our paper, we would like our "best metrics" to exists on the largest possible (ideally every) set of manifolds. We have already observed that extremal metrics are certainly more abundant than Einstein and Kcsc ones, yet the just mentioned property of the automorphism group of an extremal manifold shows that not every manifold can be equipped with such a metric. In particular Levine [35] has produced examples of iterated blow ups of $\mathbb{C}P^2$ which do not admit such a metric because their automorphism group contain a copy of $\mathbb{C}$.

   Despite a huge amount of work on this problem it is fair to say that our knowledge on this problem is very scarce and the rest of this paper is devoted to describe some constructions of such special metrics.

   On the general side, let us just mention a beautiful Conjecture by Tian [55] which, roughly speaking, states that given any Kähler manifold one can find a finite number of degenerations of complex structures which make it extremal.

17.3 Existence of Extremal Metrics

In this section we collect some classes of examples of manifolds for which we know, at least in *some* Kähler classes, the existence of an extremal metric. The reader should be warned that this is a huge area of research in which a variety of tools has been implemented ranging from complex analytic results such as Siu-Nadel's multiplier ideal sheaves [44], Tian's α-invariant and twistor theory [27, 29, 30, 33], elliptic and parabolic PDEs (notably Yau's estimates and Ricci and Calabi flows), moment map constructions and symplectic techniques [10]. We do not even attempt an effort towards completeness of the presentation but we try to follow a road which leads to the constructions coming in the next sections.

As we already observed, thanks to Poincarè's Theorem 17.3, the case of Riemann surfaces is completely understood and no obstructions appear. Going in higher dimensions the most classical results are concerned with the Einstein problem of nonpositive curvature. In this case the Kähler class is determined by the first Chern class of the canonical bundle, which is then forced to be positive to be represented by a metric. Under this assumption we have Aubin-Yau's celebrated Theorem:

Theorem 17.5. *If $c_1(M)$ is negative or zero, then M admits a Kähler-Einstein metric, unique in its Kähler class.*

Note that in this case holomorphic vector fields do not appear (or are parallel in the flat case). Passing in positive curvature the problem becomes much more delicate. Already in complex dimension 2, the Calabi-Matsushima-Lichnerowicz condition and the vanishing of the Futaki invariant put two constraints on the type of manifolds and its allowable Kähler class (the first is "class free", while the second is sensitive to the class).

One of the deepest existence results we have up to now, is Tian's proof [52] that in dimension 2 these are necessary and sufficient conditions for the existence of a Kähler-Einstein metric (it follows from the classification of surfaces that in this case the vanishing of the Futaki invariant implies the C-M-L obstruction).

Theorem 17.6. *If M has complex dimension 2 and $c_1(M)$ is positive, then M admits a Kähler-Einstein metric if and only if $F(\cdot, c_1(M)) = 0$. In particular M is the blow up of $\mathbb{C}P^2$ in k points, $3 \leq k \leq 8$ in general position.*

Nothing similar to the above Theorem is known for higher dimensional manifolds even in the lucky situations (as in dimension 3) when we have a complete classification, and it is fair to say that the quest for examples is widely open even in the Einstein case. We have some classes of examples though, e.g. [57].

Theorem 17.7. *A smooth toric manifold with positive first Chern class admits a Kähler-Einstein metric if and only if it has vanishing Futaki invariant.*

One should observe that, thanks to an important result of Mabuchi [37] the condition about the Futaki invariant is easily checked on the associated polytope.

The above mentioned result could lead the reader to think that a plausible Conjecture to put on this problem is the following:

Guess 17.1. *A manifold with positive first Chern class admits a Kähler-Einstein metric if and only if it has vanishing Futaki invariant.*

Unfortunately this turns out to be false. Note that in particular this would imply that any manifold with positive first Chern class and discrete automorphism group has an Einstein metric. In fact, already in dimension 3 where we can look at the Iskovskikh-Mori-Mukai classification of Fano threefolds, an (counter)example appears, the Mukai-Umemura manifold X_{22}. The proof that of this fact is again quite delicate and has been given by Tian [53]. In fact the moduli spaces of such manifolds has been a great source of inspiration in the subject, as Donaldson [24] proved that in this space there is also an element admitting an Einstein metric. Up to then all our knowledge seemed to suggest that the existence (or nonexistence) of Einstein metrics was true (or false) in the whole deformation type.

How to adjust the above guess has been the subject of tremendous work. This (and other conceptually fundamental considerations coming from the known story on bundles, the so-called Hitchin-Kobayashi correspondence) essentially lead to Tian's definition of K-stability in [53] for Einstein metrics, and in [55] in the case of any *rational* Kähler class (see also Donaldson [22]), and finally by Szèkelyhidi [49] for extremal metrics. Very recently Chen-Donaldson-Sun [17–19] and Tian [56] have proved this important connection in the Einstein case:

Theorem 17.8. *A manifold with positive first Chern class admits a Kähler-Einstein metric if and only if it is K-stable.*

The necessity of K-stability was in fact proved earlier by Tian [53], Stoppa [46] and Stoppa-Szèkelyhidi [48] even for extremal metrics.

We have of course not even entered the big delicacies of the above problem, which start right from the definition of K-stability, but for the main concern of this note let us just observe that we are not able to prove the existence of a single Einstein, not to mention extremal, metric by means of K-stability, to the point that we do not even have a proof at the moment of K-stability of $\mathbb{C}P^2$!

We stop for the moment listing cases where some examples of extremal metrics are known to exist to start focusing on some special constructions all inspired by the same general vague question:

Problem 17.1. *Given an extremal manifold (M, g) which geometric operations applied to M preserve the extremal property?*

Of course it is a matter of mathematical taste to order geometric operations by simplicity, but, setting aside the trivial cases of riemannian products, probably all the readers would agree that quotients by finite group actions could be a very good place to start. If the action is assumed to be free, a moment of thought will convince the reader that in this case there is not much to say, since if the group by which we plan to make a quotient has to preserve the complex structure (as it must) it has to be a compact discrete subgroup of the automorphism group, hence it has to stay inside a maximal compact torus, but then by Calabi's structure Theorem mentioned above, it has to be a subgroup of the isometry group hence the quotient will have the same properties as the original manifold.

If the action is not assumed to be free, the same argument as above will produce an extremal metric on the quotient with *conical singularities* along the fixed point set. The existence of such metrics is an extremely interesting property of a manifold which has longly escaped our understanding. A major breakthrough came very recently in the Einstein case and in fact (mixed to a critical analytical step forward, the proof of the so-called *Tian's partial C^0-estimates*) led to the proof of Theorem 17.8.

Very different story going in the opposite direction! Suppose now G is a finite subgroup of $Aut(M)$ and M/G has an extremal metric. What can we deduce on M? In general, very little of course. After all having one such quotient is a very weak condition. Technically the main difficulty come from the simple observation that the pullback of a metric on the quotient is *not* a metric on M, being degenerate along the ramification, and so very hard to use as a "background" metric for all PDEs one would like to solve on M (of course the pullback will be a genuine but incomplete extremal metric on the complement of the ramification but this is not what we look for). But things change drastically if one assumes to have more symmetries, i.e. more than one such ramified Galois covering and this is what we have studied with Ghigi and Pirola in [4]. A first result proved there is the following:

Theorem 17.9. *Let $M, M_1, \ldots, M_k$ be Fano manifolds of the same dimension n. Assume that all the M_i's admit a Kähler-Einstein metric, and that we are given ramified coverings $\pi_i : M \to M_i$ satisfying the following assumptions:*

1. *All the coverings are quotients by some finite group G_i, i.e. $M_i = M/G_i$.*
2. *The G_i's are all contained in some compact subgroup $G \subset Aut(M)$.*
3. *If $R(\pi_i)$ denote the ramification divisor of π_i, then $\cap_{i=1}^{k} R(\pi_i) = \emptyset$.*
4. *The divisors $R(\pi_i)$ are all proportional to the anti canonical divisor of M, i.e. there are rational numbers β_i such that*

$$R(\pi_i) = \beta_i K_M^{-1}$$

Then M admits a Kähler-Einstein metric.

The third condition is of course the critical one, and allows one to manufacture a Kähler metric with many special properties out of the pullbacks of the metrics on the quotients to be used as a starting metric for the continuity method.

Much more delicate (and interesting) is to avoid such a restrictive assumption. In this case one is bound to reproduce some of the analytic results about the continuity method using as a reference form a degenerate metric. This is still, to the author's knowledge, the only place where such a strategy has been implemented successfully to reach smooth metrics, but very deep work in this spirit and different aims has been done using hard pluripotential theory by Bouksom-Eissideaux-Guedj-Zerihai and others [8]. The main result of [4] is then the following

Theorem 17.10. *Let $M, M_1, \ldots, M_k$ be Fano manifolds of the same dimension n. Assume that all the M_i's admit a Kähler-Einstein metric, and that we are given ramified coverings $\pi_i : M \to M_i$ satisfying the following assumptions:*

1. *All the coverings are quotients by some finite group G_i, i.e. $M_i = M/G_i$.*
2. *The G_i's are all contained in some compact subgroup $G \subset Aut(M)$.*
3. *If V_i denotes the reduced divisor associated with the ramification divisors of the π_i's, then the V_i's are all smooth hyper surfaces that intersect transversally in a smooth submanifold V.*
4. *There are rational numbers β_i such that*

$$R(\pi_i) = \beta_i K_M^{-1}$$

and they satisfy

$$\frac{1}{d_1 - 1} + \cdots + \frac{1}{d_k - 1} > \frac{1}{\beta}$$

where $\beta = min\{\beta_i\}$ and d_i are the cardinalities of the G_i's.

Then M admits a Kähler-Einstein metric.

This Theorem (and a minor variation) allowed us to list the following set of examples of Kähler-Einstein manifolds of positive curvature. In particular

1. Hypersurfaces of the form $\{x_0^d + \cdots + x_{k-1}^d + f(x_k, \ldots, x_{n+1}) = 0\} \subset \mathbb{C}P^{n+1}$, where f is a homogeneous polynomial of degree d and $k > n + 2 - d$.
2. Arbitrary intersections of two hyperquadrics.
3. Double covers of $\mathbb{C}P^n$ ramified along a smooth hypersurface of degree $2d$ with $\frac{n+1}{2} < d \leq n$.
4. Double covers of the n-dimensional quadric $Q_n \subset \mathbb{C}P^{n+1}$ with smooth branching locus cut out by a hypersurface of degree $2d$ with $\frac{n}{2} < d < n$

all have Kähler-Einstein metrics.

In a series of works I. Cheltsov has looked for smooth and singular examples using the log-canonical threshold and Tian's α invariant very much in the spirit of previous work by Demailly and Kollár. We refer to [15] for a comprehensive summary of this approach.

Of course one could ask if a similar phenomenon holds for general extremal metrics, but unfortunately nobody has been able to prove anything similar in this generality. The reason for this is that in the Einstein case, the continuity method gives a powerful tool to attack the problem and the variational characterization of Einstein metrics gives us a number of reasonably nice functional whose cohercivity can be studied under coverings. Nothing similar is known for Kcsc or extremal metrics (which being fourth order equation are much more difficult to treat). An intriguing extension of this idea is to study whether, given a manifold which an S^1-action, its Kähler reduction is "closer" to being extremal than the original one. Some preliminary steps in this direction has been given in [1].

Before passing to the next type of geometric operations to be performed on M, we focus for a moment on the question of the behavior of our metrics under deformations of Kähler classes and complex structure. This has been clearly explained in a beautiful paper by LeBrun and Simanca [32]. Since we are going to deform all our structures let us indicate our starting manifold by (M, J_0, g_0, ω_0), where ω_0 represents the real $(1,1)$-form associate to g_0 and J_0. Let us also indicate by ρ_0 the $(1,1)$-form associated to Ric_0 via J_0.

Suppose first J_0 is fixed, $R(g_0)$ is constant and we deform ω_0 (hence g_0) with a harmonic $(1,1)$-form β and $\partial\bar{\partial}\psi$ for some smooth function ψ

$$\omega(t) = \omega_0 + t(\beta + \partial\bar{\partial}\psi) .$$

We then want to compute the scalar curvature of the deformed metric $g(t)$ and expand it in powers of t. This gives

$$R(g(t)) = R(g_0) + t(-\frac{1}{2}\Delta^2\psi + (\rho_0 \cdot \partial\bar{\partial}\psi) + \Delta(\omega_0 \cdot \beta) - 2(\rho_0 \cdot \beta)) + \mathcal{O}(t^2)$$

Having assumed that $R(g_0)$ has constant scalar curvature it is not difficult to deduce that the linearization of the map $(\beta, \psi) \to R(g(t))$ at $(0,0)$ is given by

$$DR_{(0,0)}(\gamma, \phi) = -\frac{1}{2}\Delta^2\phi - (\rho_0 \cdot \partial\bar{\partial}\psi) - 2(\rho_0 \cdot \beta) \, .$$

This computation implies the following important fact

Theorem 17.11. *If* $[\omega_0]$ *is Kcsc and the differential of the Futaki invariant, seen as a map from the Lie algebra of the holomorphic vector fields into the dual of the* $(1,1)$*-real cohomology, is injective at* $[\omega]$*, then all nearby Kähler classes have a Kcsc representative.*

Unfortunately the non-degeneracy condition on the Futaki invariant turns out to be extremely restrictive (in particular never true in the Einstein case), but LeBrun and Simanca managed to circumvent this problem passing through general extremal metrics. Similar infinitesimal analysis allowed them to prove the following remarkable property of extremal metrics (which adds up to all the good ones we listed in Sect. 17.2!):

Theorem 17.12. *If* $[\omega_0]$ *is Kcsc then all nearby Kähler classes have an extremal representative.*

By Calabi's result (Theorem 17.4), this allowed them to construct a huge number a new Kcsc examples, for instance deforming Tian's Einstein complex surfaces.

Now what about moving also J_0? Well, the reader will not be surprised to know that things become extremely complicated once again. This problem has become clear once Burns and De Bartolomeis [9] have found the first examples of ruled surfaces (i.e. projectivizations of vector bundles of rank 2 over Riemann surfaces Σ of genus at least 2) which do not carry any extremal metric in any Kähler class. Since they can be thought as deformations of the trivial product $\Sigma \times \mathbb{C}P^1$, this shows that some key obstruction must be hidden somewhere.

Burns-De Bartolomeis' work has inspired a great deal of work in many directions, especially trying to characterize which vector bundles do not create this problem, very much in the spirit of the classical Narasimhan-Seshadri Theorem, but for our purposes let us just say that the optimal conditions to be imposed on the deformation of complex structure in order to guarantee the possibility to deform the extremal metric have been found by Szèkelyhidi [50] again in terms of a proper stability notion very much in the spirit of K-stability. These are quite difficult to describe, but let us just underline that they all become vacuous when the starting J_0 has no holomorphic vector fields. In fact the following statement can be proved quite easily without such machinery (but of course is far from optimal!):

Theorem 17.13. *If* $[\omega_0]$ *is Kcsc and no holomorphic vector fields, then all nearby Kähler classes for all nearby complex structures have a Kcsc representative.*

What we learned analyzing deformations is that the absence of holomorphic vector fields allow to get optimal results, while their presence puts restrictions on the directions where to move. This is of critical importance in interpreting the results we are about to describe in the next section.

17.4 Blowing Up Points

We now want to describe the study we carried our in collaboration with F. Pacard in the Kcsc case [2, 3], and with him and M. Singer in the extremal one [5] about the effect of blowing up smooth points on an extremal base manifold and, even if no direct relationship with the results just described will be used, we want to think of this operation as a deformation of complex structure and Kähler class. In fact, at least at the beginning, we can attempt a general construction starting form a variety with isolated singular point of quotient type, choosing finitely many points $p_1, \ldots, p_k \in M$ and replacing a small neighborhood of each point p_j, biholomorphic to a neighborhood of the origin in $\mathbb{C}^m/\Gamma_j$, by a (suitably scaled down by a small factor ε) piece of a Kähler manifold (N_j, η_j), biholomorphic to $\mathbb{C}^m/\Gamma_j$ away from a compact subset. This generalized connected sum yields a Kähler manifold or a Kähler orbifold with isolated singularities that we call

$$M \sqcup_{\varepsilon,p_1} N_1 \sqcup \ldots \sqcup_{\varepsilon,p_k} N_k$$

and whose complex structure does not depend on $\varepsilon \neq 0$. In this sense we can guess that this construction shares some similarity with the LeBrun-Simanca smooth deformation theory. In order for the differentiable structures to match on small balls in M and large balls in N_j we need to have a special structure of the "models", i.e. they need to "look like" a euclidean space outside a compact set. Moreover, in order to have at least a zero-th order matching of metrics on the balls of radius ε we need to scale down the metrics on N_j and even to start hoping that the resulting glued metric still has constant scalar curvature, the only possibility is that each (N_j, η_j) is scalar flat (since zero is the only real number which does not blow up when multiplied by ε^{-2}!).

These are clearly necessary conditions for this approach to even have a hope to succeed. Our building bricks are then

(i) (M, ω) is a m-dimensional compact Kähler manifold or orbifold with isolated singularities.
(ii) The scalar curvature of ω is constant.
(iii) Given points $p_1, \ldots, p_n \in M$ which might be either singular or regular points of M, let Γ_j be the finite subgroup of $U(m)$ acting freely on $\mathbb{C}^m - \{0\}$ such that a neighborhood of p_j is biholomorphic to a neighborhood of the origin in $\mathbb{C}^m/\Gamma_j$. Each $\mathbb{C}^m/\Gamma_j$ has an ALE resolution (N_j, η_j) (which might either be a manifold or an orbifold with isolated singularities) endowed with a zero scalar curvature Kähler form η_j. Furthermore, we assume that, away from a compact set, the Kähler form η_j can be expanded as

$$\eta = i\, \partial\, \bar{\partial}\, (\frac{1}{2}\, |u|^2 + \tilde{\varphi}(u)) \tag{17.1}$$

at infinity, where the potential $\tilde{\varphi}$ satisfies

$$\tilde{\varphi}(u) = a\,|u|^{4-2m} + \mathcal{O}(|u|^{3-2m}), \tag{17.2}$$

when $m \geq 3$ and

$$\tilde{\varphi}(u) = a\,\log|u| + \mathcal{O}(|u|^{-1}). \tag{17.3}$$

when $m = 2$.

The main result proved in [2] is then

Theorem 17.14. *Assume that assumptions (i)–(ii) and (iii) are satisfied and there are no nonzero holomorphic vector fields on M. Then, there exists $\varepsilon_0 > 0$ and, for all $\varepsilon \in (0, \varepsilon_0)$, there exists a constant scalar curvature Kähler form $\tilde{\omega}_\varepsilon$ defined on $M \sqcup_{p_1,\varepsilon} N_1 \sqcup_{p_2,\varepsilon} \cdots \sqcup_{p_n,\varepsilon} N_n$.*

As ε tends to 0, the sequence of Kähler forms $\tilde{\omega}_\varepsilon$ converges (in $\mathcal{C}^\infty$ topology) to the Kähler metric ω, away from the points p_j and the sequence of Kähler forms $\varepsilon^{-2}\,\tilde{\omega}_\varepsilon$ converges (in $\mathcal{C}^\infty$ topology) to the Kähler form η_j, on compact subsets of N_j.

The case of the blow up at smooth points is covered by the previous result by taking N_j to be the total space of the line bundle $\mathcal{O}(-1)$ over $\mathbb{P}^{m-1}$ (in this case $\Gamma_j = \{id\}$). The key property (iv) asks for an ALE zero scalar curvature metric η_j on $\mathcal{O}(-1)$ such that $[\eta_j] = -PD[E_j]$ and with appropriate decay at infinity. These Kähler forms have been obtained by Simanca [43]. So in this case we get that, given finitely many points $p_1, \ldots, p_n \in M$ and positive numbers $a_1, \ldots, a_n > 0$, there exists $\varepsilon_0 := \varepsilon_0(M, \omega) > 0$ such that, for all $\varepsilon \in (0, \varepsilon_0)$, the blow up of M at $p_1, \ldots, p_n$ carries a constant scalar curvature Kähler form

$$\omega_\varepsilon \in \pi^*\,[\omega] - \varepsilon^2\,(a_1\,PD[E_1] + \ldots + a_n\,PD[E_n]),$$

where the $PD[E_j]$ are the Poincaré dual of the $(2m-2)$-homology classes of the exceptional divisors of the blow up at p_j.

The problem of deciding to which orbifold points we can apply such a construction is on the contrary very delicate and largely open. The existence of these type of special resolutions is a classical topic in local algebraic geometry. For example, if $m = 2$ or 3, and $\Gamma \subset SU(m)$, then such models do exists, as proved by Kronheimer [28], Roan [40] and Joyce [26], and the metric η_j can be chosen to be even Ricci flat. For general groups $\Gamma \subset U(m)$ very little is known (see e.g. [13]). $SU(2)$-singularities are good enough to prove for example the following extension of Aubin-Yau's Theorem:

Corollary 17.1. *Any compact complex surface of general type admits constant scalar curvature Kähler metrics.*

In order to find a connection with the smooth perturbation theory described in the previous section, we can observe that the parameter ε in the previous construction (the "gluing" parameter) of blowing up, and for this extent also resolution of singularities, give rise to a trivial family of complex manifolds

$M \sqcup_{p_1,\varepsilon} N_1 \sqcup_{p_2,\varepsilon} \cdots \sqcup_{p_n,\varepsilon} N_n$. What is really changing, and quite dramatically, is the deformation of Kähler class which is now "blowing up" at the exceptional divisors in the parameter ε. Yet, even with this huge difference of perspective from the smooth to the singular perturbation, Theorem 17.14 essentially says that the smooth picture still holds.

The appearance of holomorphic vector fields changes things drastically. This is what we studied in [3]. First recall that the Matsushima-Lichnerowicz Theorem asserts that the space of hamiltonian holomorphic vector fields on (M, J, ω) is also the complexification of the real vector space of holomorphic vector fields Ξ which can be written as

$$\Xi = X - i\, J\, X$$

where X is a Killing vector field which vanish somewhere on M. Let us denote by $\mathfrak{h}$, the space of hamiltonian holomorphic vector field and by

$$\xi_\omega : M \longmapsto \mathfrak{h}^*$$

the *moment* map which is defined by requiring that if $\Xi \in \mathfrak{h}$, the function $\zeta_\omega := \langle \xi_\omega(\cdot), \Xi \rangle$ is a (complex valued) Hamiltonian for the vector field Ξ, namely the unique solution of

$$-\bar{\partial}\zeta_\omega = \frac{1}{2}\,\omega(\Xi, -)$$

which is normalized by

$$\int_M \zeta_\omega \, dvol_g = 0$$

With these notations, the result we have obtained in [5] reads:

Theorem 17.15. *Assume that (M, J, ω) is a constant scalar curvature compact Kähler manifold and that $p_1, \ldots, p_n \in M$ and $a_1, \ldots, a_n > 0$ are chosen so that:*

(i) $\xi_\omega(p_1), \ldots, \xi_\omega(p_n)$ span $\mathfrak{h}^*$.

(ii) $\sum_{j=1}^n a_j\, \xi_\omega(p_j) = 0 \in \mathfrak{h}^*$ (balancing condition).

Then, there exist $\varepsilon_0 > 0$ s.t. for all $\varepsilon \in (0, \varepsilon_0)$, there exists on $\tilde{M}$, the blow up of M at $p_1, \ldots, p_n$, a constant scalar curvature Kähler metric g_ε associated to the Kähler form

$$\omega_\varepsilon \in \pi^*[\omega] - \varepsilon^2 \,(a_{1,\varepsilon}^{\frac{1}{m-1}}\, PD[E_1] + \ldots + a_{n,\varepsilon}^{\frac{1}{m-1}}\, PD[E_n]),$$

where the $PD[E_j]$ *are the Poincaré duals of the* $(2m-2)$*-homology classes of the exceptional divisors of the blow up at* p_j *and where*

$$|a_{j,\varepsilon} - a_j| \leq c\, \varepsilon^{\frac{2}{2m+1}} \tag{17.4}$$

if $\mathfrak{h} \neq \{0\}$*. Finally, the sequence of metrics* $(g_\varepsilon)_\varepsilon$ *converges to* g *in* $\mathcal{C}^\infty(M \setminus \{p_1, \ldots, p_n\})$*.*

Therefore, in the presence of nontrivial hamiltonian holomorphic vector fields, the number of points which can be blown up, their position, as well as the possible Kähler classes on the blown up manifold have to satisfy some constraints.

To illustrate this fact, we once more consider the case of $(\mathbb{C}P^m, \omega_{FS})$ to get the:

Theorem 17.16. *There exist* $\varepsilon_0 := \varepsilon_0(m) > 0$ *and for all* $\varepsilon \in (0, \varepsilon_0)$*, there exists a constant (positive) scalar curvature Kähler form* ω_ε *on the blow up of* $\mathbb{C}P^m$ *at*

$$p_1 := [1, 0, \ldots, 0, 0], \ldots, p_{m+1} = [0, \ldots, 0, 1]$$

with

$$\omega_\varepsilon \in \pi^*[\omega_{FS}] - \varepsilon^2 (PD[E_1] + \ldots + PD[E_{m+1}]),$$

where the $PD[E_j]$ *are the Poincaré dual of the* $(2m-2)$*-homology classes of the exceptional divisors of the blow up at* p_j*.*

Observe that all volumes of the exceptional divisors are identical. The above corollary is optimal in the number of points because $\mathbb{C}P^m$ blown up at $n \leq m$ points does not carry any constant scalar curvature Kähler metric since it violates the Matsushima-Lichnerowicz obstruction. Observe also that $\mathbb{C}P^m$ blown up at $p_1, \ldots, p_{m+1}$ still has holomorphic vector fields vanishing somewhere.

It is well known that on $\mathbb{C}P^m$, $m+2$ points forming a projective frame are enough to kill all holomorphic vector fields after blow up, and we can prove that this condition also guarantees the existence of Kähler constant scalar curvature metric.

Theorem 17.17. *Given* $p_1, \ldots, p_n$ *points in* $\mathbb{C}P^m$ *such that* $p_1, \ldots, p_{m+2}$ *form a projective frame, the blow up of* $\mathbb{P}^m$ *at* $p_1, \ldots, p_n$ *carries constant scalar curvature Kähler metrics and no holomorphic vector fields. Moreover* $p_{m+3}, \ldots, p_n$ *can be chosen arbitrarily on* $\mathbb{C}P^m$ *blown up at* $p_1, \ldots, p_{m+2}$*.*

In [5] we have extended the previous analysis to the general extremal case. We have proved a theorem precisely parallel to Theorem 17.15, where the role of the moment map ξ is taken by a 'relative moment map' ξ''. More precisely, denote by K the group of biholomorphic self-maps of M which are also *exact symplectomorphisms* of (M, ω), fix in advance a torus $T \subset K$, and define H to be the centralizer of T in K, denote by $\mathfrak{h}$ the Lie algebra of H, and put $H'' = H/T$ with Lie algebra $\mathfrak{h}'' = \mathfrak{h}/\mathfrak{t}$. If $X \in \mathfrak{h}$, denote by X'' the projection of X to $\mathfrak{h}''$.

By the equivariance of ξ, if $p \in \mathrm{Fix}(T)$, then $\xi(p) \in \mathfrak{h}$, so the 'relative moment map' $\xi''(p)$, the projection of $\xi(p)$ to $\mathfrak{h}''$ is well-defined.

Theorem 17.18. *Let (M, ω) be an extremal Kähler manifold with extremal vector field X_s. Let $T \subset K$ be a torus with $X_s \in \mathfrak{t}$. Suppose that $p_j \in Fix(T)$ and*

(i) There exist $a_j > 0$ such that $\sum_{j=1}^{n} a_j^{m-1}\xi''(p_j) = 0$ (balancing condition).
(ii) $\mathbb{R}\xi''(p_1) + \cdots + \mathbb{R}\xi''(p_n) = \mathfrak{h}''$ (genericity condition).

Then there exists $\varepsilon_0 > 0$, $c > 0$ and $\theta > 0$, such that for all $\varepsilon \in (0, \varepsilon_0)$ there is an extremal Kähler metric ω_ε on $Bl_{p_1,\dots,p_n}(M)$ in the Kähler class

$$\pi^*[\omega] - \varepsilon^2 \sum_{j=1}^{n} \tilde{a}_j PD[E_j],$$

where $\tilde{a}_j$ depends upon ε and $|\tilde{a}_j - a_j| \leq c\varepsilon^\theta$ as $\varepsilon \to 0$.
Furthermore, if

(iii) There is no non-zero element of $\mathfrak{h}''$ which vanishes at $p_1, \dots, p_n$

then we can assume that $\tilde{a}_j = a_j$.

This result has two notable corollaries

Corollary 17.2. *Let M be a compact complex manifold of dimension m acted on by a complex torus T^c so that a dense open subset of M is biholomorphic to T^c. Suppose that ω is a toric extremal Kähler metric on M, so that there exists a compact m-dimensional torus $T \subset T^c$ acting isometrically on (M, ω). Then for any subset $\{p_1, \dots, p_n\} \subset Fix(T)$, $Bl_{p_1,\dots,p_n}(M)$ admits extremal Kähler metrics.*

Corollary 17.3. *Let (M, ω) be a compact extremal Kähler manifold. Then there exist at least two points p_1, p_2 of M such that $Bl_{p_1,p_2}(M)$ admits extremal Kähler metrics.*

In particular $Bl_{p_1,p_2}\mathbb{C}P^2$ has extremal metrics. This has been a key tool used by Chen-LeBrun-Weber [20] to show the remarkable fact that *any compact complex surface with positive first Chern class has an Einstein metric (not necessarily Kähler!).*

As one can see holomorphic vector fields introduce a genuine obstruction in passing from a smooth (where by LeBrun-Simanca's Theorem the extremal property is open) to the singular perturbation. One might wonder why it is so. In the proof of all above theorems they appear naturally since they can be naturally identified with the kernel of the linearized scalar curvature operator, forcing the whole singular perturbation analysis to be carried out on the orthogonal space. From a more geometric point of view the reason is quite subtle and not completely understood. In the Kcsc case, the balancing condition in Theorem 17.15 has been completely explained as an application Theorem 17.4 in this singular perturbation process by

Stoppa [47], Della Vedova [21] and Szèkelyhidi [51], who also strengthened the range of applicability of the above results in removing the genericity condition.

17.5 More Deformations

By what we have just described, we have a good understanding of the following situations:

1. Blowing up and resolving isolated quotient singularities for Kcsc metrics in absence of nontrivial holomorphic vector fields.
2. Blowing up smooth points for both Kcsc and extremal with holomorphic vector fields.

It is then natural to ask the following questions:

Problem 17.2. *What happens when trying to resolve isolated quotient singularities in the presence of nontrivial holomorphic vector fields?*

Problem 17.3. *How this picture changes when one, instead of resolving the singularities (and possibly leaving the world of quotient singularities), wants to perform other natural desingularization processes?*

As for the first question, this has very recently been solved by R. Lena in his PhD thesis [34]. Interestingly, it turns out that the answer strongly depends on the type of the group Γ which creates the singularity. If all the points one is resolving correspond to finite subgroup of $U(m)$ but *not* in $SU(m)$, the same condition as in Theorem 17.15 is required, if some are in $U(m)$ and some in $SU(m)$ these latter points do not interfere with the construction and only the old balancing condition on the first ones is required (this was previously observed by Rollin and Singer [41]), while if only points corresponding to subgroups of $SU(m)$ are there a much more complicated balancing condition appears which involves this time the Laplacian of the moment map.

Problem 17.3 turns out to be extremely challenging and far from being a merely technical extensions of the results described in the previous section. As for now the important breakthrough on this problem has come form the work of Spotti [45] in the Einstein case, and by Biquard-Rollin [7] in the Kcsc case, both performing a "smoothing" of the singularities in complex dimension 2 and without holomorphic vector fields. This approach is somehow dual to the one described above, since in this procedure one can think of the complex structure to move and the Kähler class to stay fixed (they can in fact construct new KE metrics), exactly the opposite of the picture in [2]. Too make statements more precise, let us recall that a *smoothing* of a complex orbifold M_0 is the datum of a flat family of complex varieties over the complex disc $\pi : \mathcal{M} \to \Delta_t \subset \mathbb{C}$, where $M_t := \pi^{-1}(t)$ is a smooth manifold, and it is called *partial* if some of the orbifold singularities persist on M_t. The simplest smoothable orbifold singularity is the quotient $\mathbb{C}^2/\mathbb{Z}_2$, which is biholomorphic to the singular hypersurface in $\mathbb{C}^3$ of equation $V_0 : x^2 + y^2 + z^2 = 0$ (the "node"). In this case a smoothing (its versal family) is simply given by taking $V_s : x^2 + y^2 + z^2 = s$. The crucial fact is that V_s admits an ALE Kähler metric, the Eguchi-Hanson metric (now considered to be Kähler with respect to the complex structure of the smoothing which is different from the complex structure of the resolution). Under a mild technical condition relating the smoothing parameter t of a given partial smoothing $\pi : \mathcal{M} \to \Delta_t \subset \mathbb{C}$ of a nodal compact orbifold to the local parameter s of the versal deformation V_s of the node, in [45] the following theorem is proved:

Theorem 17.19. *Let $\mathcal{M} \to \Delta_t \subset \mathbb{C}$ be a "generic" partial smoothing of a Kähler-Einstein del Pezzo orbifold (M_0, g_0) with nodal singularities and discrete automorphism group. Then, for t sufficiently small, X_t admits a Kähler-Einstein (orbifold) metric g_t, roughly obtained by gluing scaled Eguchi-Hanson spaces to (M_0, g_0), i.e.,*

$$g_t \sim g_0 \# \sum_i \epsilon_i^2(t) g_{EH}^i.$$

In particular (M_t, g_t) converges to (M_0, g_0) in the metric topology for $t \to 0$.

An important highly nontrivial extension of Theorem 17.19 to generic deformations of Kcsc orbifolds with discrete automorphisms has been proved by Biquard and Rollin [7] who also extended the type of singularities to which it can be applied.

All the extensions of Spotti-Biquard-Rollin's results in higher dimension and with holomorphic vector fields would be a major breakthrough in the field. We do not have at moment even a reasonable conjectural statement which this time has to encode two obstructions, one as in the results of Sect. 17.4 coming from the singular deformation of the Kähler class and the second coming from the mentioned obstruction on the deformation of complex structures (note that this time the glued manifolds are NOT biholomorphic when varying the gluing parameter!) found by Szekelyhidi in the smooth case.

References

1. C. Arezzo, A. Della Vedova, G. La Nave, Geometric flows and Kähler reduction. J. Symplectic Geom. (to appear)
2. C. Arezzo, F. Pacard, Blowing up and desingularizing Kähler orbifolds with constant scalar curvature. Acta Math. **196**(2), 179–228 (2006)
3. C. Arezzo, F. Pacard, Blowing up Kähler manifolds with constant scalar curvature. II. Ann. Math. **170**(2), 685–738 (2009)
4. C. Arezzo, A. Ghigi, G.P. Pirola, Symmetries, quotients and Kähler-Einstein metrics. Crelle's J. **157**(1), 1–51 (2006)
5. C. Arezzo, F. Pacard, M. Singer, Extremal metrics on blowups. Duke Math. J. **157**(1), 1–51 (2011)

6. S. Bando, T. Mabuchi, Uniqueness of Einstein Kähler metrics modulo connected group actions, in *Algebraic Geometry*, Sendai, 1985, ed. by T. Oda. Advanced Studies in Pure Mathematics, vol. 10, 1987
7. O. Biquard, Y. Rollin, Smoothing singular extremal Kähler surfaces and minimal Lagrangians (pre-print). arXiv:1211.6957
8. S. Bouksom, P. Eyssidieux, V. Guedj, A. Zerihai, Monge-Ampère equations in big cohomology classes. Acta Math. **205**, 199–262 (2010)
9. D. Burns, P. de Bartolomeis, Stability of vector bundles and extremal metrics. Invent. Math. **92**, 403–407 (1988)
10. E. Calabi, Métriques kählériennes et fibrés holomorphes. Ann. Sci. École Norm. Sup. 4 **12**(2), 269–294 (1979)
11. E. Calabi, Extremal Kähler metrics, in *Seminar on Differential Geometry*, ed. by S.-T. Yau. Annals of Mathematics Studies, vol. 102 (Princeton University Press, Princeton, 1982), pp. 259–290
12. E. Calabi, Extremal Kähler metrics II, in *Differential Geometry and Its Complex Analysis*, ed. by I. Chavel, H.M. Farkas (Springer, Berlin/Heidelberg, 1985)
13. D. Calderbank, M. Singer, Einstein metrics and complex singularities. Invent. Math. **156**(2), 405–443 (2004)
14. È. Cartan, Sur une classe remarquable d'espaces de Riemann, II. Bulletin de la Socií Ḿathḿatique de France **55**, 114–134 (1927)
15. I. Cheltsov, K.A. Shramov, Log canonical thresholds of smooth Fano threefolds, with an appendix by J.P. Demailly. Russ. Math. Surv. **63**(5), 859–958 (2008)
16. X.X. Chen, G. Tian, Geometry of Kähler metrics and holomorphic foliation by discs. Publ. Math. Inst. Hautes Études Sci. **107**, 1–107 (2008)
17. X.X. Chen, S.K. Donaldson, S. Sun, Kähler-Einstein metrics on Fano manifolds, I: approximation of metrics with cone singularities. http://arxiv.org/pdf/1211.4566.pdf
18. X.X. Chen, S.K. Donaldson, S. Sun, Kähler-Einstein metrics on Fano manifolds, II: limits with cone angle less than 2π. http://arxiv.org/pdf/1212.4714.pdf
19. X.X. Chen, S.K. Donaldson, S. Sun, Kähler-Einstein metrics on Fano manifolds, III: limits as cone angle approaches 2π and completion of the main proof. http://arxiv.org/pdf/1302.0282.pdf
20. X.X. Chen, C. LeBrun, B. Weber, On conformally Kähler, Einstein manifolds. J. Am. Math. Soc. **21**(4), 1137–1168 (2008)
21. A. DellaVedova, CM-stability of blow-ups and canonical metrics. arXiv:0810.5584
22. S.K. Donaldson, Scalar curvature and projective embeddings I. J. Differ. Geom. **59**(3), 479–522 (2001)
23. S.K. Donaldson, Lower bounds on the Calabi functional. J. Differ. Geom. **70**, 453–472 (2005)
24. S.K. Donaldson, Kähler geometry on toric manifolds, and some other manifolds with large symmetry, in *Handbook of Geometric Analysis, No. 1*, ed. by L. Lin, P. Li, R.M. Schoen, L. Simon. Advanced Lectures in Mathematics (ALM), vol. 7 (International Press, Somerville, 2008), pp. 29–75
25. A. Futaki, *Kähler-Einstein Metrics and Integral Invariants*. Lecture Notes in Mathematics, vol. 1314 (Springer, Berlin/Heidelberg, 1988)
26. D. Joyce, *Compact Manifolds with Special Holonomy* (Oxford University Press, Oxford/ New York, 2000)
27. J. Kim, M. Pontecorvo, A new method of constructing scalar-flat Kähler surfaces. J. Differ. Geom. **41**(2), 449–477 (1995)
28. P. Kronheimer, The construction of ALE spaces as hyper-Kähler quotients. J. Differ. Geom. **29**(3), 665–683 (1989)
29. C. LeBrun, Scalar-flat Kähler metrics on blown-up ruled surfaces. J. Reine Angew. Math. **420**, 161–177 (1991)
30. C. LeBrun, On the scalar curvature of complex surfaces. Geom. Funct. Anal. **5**, 619–628 (1995)

31. C. LeBrun, Einstein metrics, four-manifolds, and differential topology, in *Surveys in Differential Geometry*, Boston, 2002. Surveys in Differential Geometry, vol. VIII (International Press, Somerville, 2003), pp. 235–255
32. C. LeBrun, S. Simanca, Extremal Kähler metrics and complex deformation theory. Geom. Funct. Anal. **4**, 298–336 (1994)
33. C. LeBrun, M. Singer, Existence and deformation theory for scalar flat Kähler metrics on compact complex surfaces. Invent. Math. **112**, 273–313 (1993)
34. R. Lena, On the desingularization of Kähler orbifolds with constant scalar curvature. PhD thesis, SISSA, 2013
35. M. Levine, A remark on extremal Kähler metrics. J. Differ. Geom. **21**(1), 73–77 (1985)
36. A. Lichnerowicz, Sur les transformations analytiques des variétés kählériennes compactes. C. R. Acad. Sci. Paris **244**, 3011–3013 (1957)
37. T. Mabuchi, Einstein-Kähler forms, Futaki invariants and convex geometry on toric Fano varieties. Osaka J. Math. **24**(4), 705–737 (1987)
38. T. Mabuchi, An energy-theoretic approach to the Hitchin-Kobayashi correspondence for manifolds, I. Invent. Math. **159**, 225–243 (2005)
39. Y. Matsushima, Sur la structure du groupe d'homomorphismes analytiques d'une certaine varité kählériennee. Nagoya Math. J. **11**, 145–150 (1957)
40. S.S. Roan, Minimal resolution of Gorenstein orbifolds. Topology **35**, 489–508 (1971)
41. Y. Rollin, M. Singer, Construction of Kaehler surfaces with constant scalar curvature. J. Eur. Math. Soc. **11**(5), 979–997 (2009)
42. R. Schoen, Conformal deformation of a Riemannian metric to constant scalar curvature. J. Differ. Geom. **20**(2), 479–495 (1984)
43. S. Simanca, Kähler metrics of constant scalar curvature on bundles over CP_{n-1}. Math. Ann. **291**(2), 239–246 (1991)
44. Y.T. Siu, The existence of Kähler-Einstein metrics on manifolds with positive anticanonnical line bundle and a suitably symmetry group. Ann. Math. **127**, 585–627 (1988)
45. C. Spotti, Deformations of Nodal Kähler-Einstein Del Pezzo surfaces with finite automorphism groups. J. Lond. Math. Soc. (2014). doi:10.1112/jlms/jdt076
46. J. Stoppa, K-stability of constant scalar curvature Kähler manifolds. Adv. Math. **221**(4), 1397–1408 (2009)
47. J. Stoppa, Unstable blowups. J. ALgebr. Geom. **19**, 1–17 (2010)
48. J. Stoppa, G. Szèkelyhidi, Relative K-stability of extremal metrics. J. Eur. Math. Soc. **13**(4), 899–909 (2011)
49. G. Szèkelyhidi, Extremal metrics and K-stability. Bull. Lond. Math. Soc. **39**(1), 76–84 (2007)
50. G. Szèkelyhidi, The Kähler-Ricci flow and K-polystability. Am. J. Math. **132**(4), 1077–1090 (2010)
51. G. Szèkelyhidi, On blowing up extremal Kähler manifolds. Duke Math. J. **161**(8), 1411–1453 (2012)
52. G. Tian, On Calabi's conjectures for complex surfaces with positive first Chern class. Invent. Math. **101**(1), 101–172 (1990)
53. G. Tian, On Kähler-Einstein metrics with positive scalar curvature. Invent. Math. **130**, 1–37 (1997)
54. G. Tian, *Canonical Metrics on Kähler Manifolds* (Birkhauser, Boston, 1999)
55. G. Tian, Extremal metrics and geometric stability. Houst. J. Math. **28**(2), 411–431 (2002)
56. G. Tian, K-stability and Kähler-Einstein metrics. http://arxiv.org/pdf/1211.4669v2.pdf
57. X.-J. Wang, X.-H. Zhu, Kähler-Ricci solitons on toric manifolds with positive first Chern class. Adv. Math. **188**(1), 87–103 (2004)

Chapter 18
Deriving Ohm's Law from the Vlasov-Maxwell-Boltzmann System

Laure Saint-Raymond

Abstract Ohm's law states that the current density j at a given location in a plasma is proportional to the electric field E at that location. We propose here a rigorous derivation of this law (and of some extensions of it) starting from a microscopic model consisting of two species of charged particles interacting both via the self-consistent electromagnetic field, and via some collisional processes.

The goal of this paper is to review some recent results [6] about the magneto-hydrodynamic limits of the Vlasov-Maxwell-Boltzmann system. The challenge is to understand the influence of the self-consistent electromagnetic field on the process of hydrodynamic limits [20].

It obviously induces additional (possibly singular) nonlinearities due to the mean field coupling, but it also changes deeply the structure of the transport equation, so that

- Classical renormalized solutions as defined by DiPerna and Lions [9] are not known to exist.
- The hypoelliptic mechanism is much more complicated [4].
- The singular perturbation may also involve some electromagnetic constraints.

The complete asymptotic study therefore requires a number of technical steps, but we will focus here on some key points and refer to [6] for the details.

L. Saint-Raymond (✉)
Université Pierre et Marie Curie & Ecole Normale Supérieure, 45 rue d'Ulm, 75230 Paris Cedex 05, France
e-mail: Laure.Saint-Raymond@ens.fr

V. Ancona and E. Strickland (eds.), *Trends in Contemporary Mathematics*, Springer INdAM Series 8, DOI 10.1007/978-3-319-05254-0_18,

18.1 Scalings and Formal Limits

18.1.1 A Multiscale Problem

The following Vlasov-Maxwell-Boltzmann system describes a plasma consisting of two species of probability densities (f^+, f^-) interacting both via some mean field electromagnetic field (E, B), and via some localized collisions which induce relaxation towards local thermodynamic equilibrium.

$$\begin{cases} \partial_t f^+ + v\cdot\nabla_x f^+ + \frac{q_+}{m_+}(E+v\wedge B)\cdot\nabla_v f^+ = \frac{1}{\tau_+}Q(f^+,f^+) + \frac{1}{\tau_\pm}Q(f^+,f^-), \\ \partial_t f^- + v\cdot\nabla_x f^- - \frac{q_-}{m_-}(E+v\wedge B)\cdot\nabla_v f^- = \frac{1}{\tau_-}Q(f^-,f^-) + \frac{1}{\tau_\pm}Q(f^-,f^+), \\ \mu_0\epsilon_0\partial_t E - \operatorname{rot} B = -\mu_0\left(q_+\int f^+ v dv - q_-\int f^- v dv\right), \qquad \partial_t B + \operatorname{rot} E = 0, \\ \operatorname{div} E = \epsilon_0\left(q_+\int f^+ dv - q_-\int f^- dv\right), \qquad \operatorname{div} B = 0. \end{cases}$$

The qualitative behavior of this system in the fast relaxation limit depends strongly on the relative size of some characteristic parameters measuring typically

- The bulk velocity of each species.
- The thermal speed (also called sound speed).
- The light speed (governing the propagation of electromagnetic waves).
- The typical acceleration due to the Lorentz force.

We will not discuss here other parameters such as the mass or charge ratios of both species that we assume to be of order 1.

According to the precise ordering of these physical parameters, we expect to find different magneto-hydrodynamic regimes listed in a systematic way in [6], classified especially according to the size and mode of propagation of the electromagnetic field.

Our starting point here is the following scaled system

$$\begin{aligned} &\epsilon\partial_t f^\pm + v\cdot\nabla_x f^\pm \pm \delta(\epsilon E + v\wedge B)\cdot\nabla_v f^\pm = \frac{1}{\epsilon}Q(f^\pm,f^\pm) + \frac{\delta^2}{\epsilon}Q(f^\mp,f^\pm) \\ &\partial_t B + \operatorname{rot} E = 0, \quad \partial_t E - \operatorname{rot} B = -\tfrac{\delta}{\epsilon^2}\left(\int f^+ v dv - \int f^- v dv\right) \\ &\operatorname{div} E = \tfrac{\delta}{\epsilon}\left(\int f^+ v dv - \int f^- v dv\right). \end{aligned} \tag{18.1}$$

which is associated to the following entropy inequality

$$\begin{aligned} &\frac{d}{dt}\left(\frac{1}{\epsilon^2}H(f_\epsilon^+|M) + \frac{1}{\epsilon^2}H(f_\epsilon^-|M) + \frac{1}{2}\int(|E_\epsilon|^2 + |B_\epsilon|^2)dx\right) \\ &\qquad + \int \frac{1}{\epsilon^4}\left(D(f_\epsilon^+) + D(f_\epsilon^-)\right) + \frac{2\delta^2}{\epsilon^4}D(f_\epsilon^+, f_\epsilon^-)dxds \le 0 . \end{aligned}$$

defining the relative entropy and the entropy dissipation respectively by

$$H(f|M) = \int \left(f \log \frac{f}{M} - f + M \right) dvdx,$$
$$D(f,g) = -\int (Q(f,g)\log f + Q(g,f)\log g)dv, \qquad D(f) = \tfrac{1}{2}D(f,f).$$

Starting from fluctuations of density of order $O(\epsilon)$ (which is consistent with the scaling of the Strouhal number in the evolution equations), we then obtain the following a priori bounds

$$g_\epsilon^\pm = \frac{1}{\epsilon}\frac{f_\epsilon^\pm - M}{M} = O(1)_{L_t^\infty(L^2(dxMdv))} + O(\epsilon)_{L_t^\infty(L^1(dxMdv))}, \quad (E_\epsilon, B_\epsilon) = O(1)_{L^\infty(L^2(dx))},$$

from which we deduce that, up to extraction of a subsequence,

$$g_\epsilon^\pm \rightharpoonup g^\pm, \quad E_\epsilon \rightharpoonup E, \quad B_\epsilon \rightharpoonup B.$$

A formal asymptotic expansion allows then to characterize these joint limit points.

In this presentation, for the sake of simplicity, we will consider only the subcritical case $\epsilon \ll \delta \ll 1$, but the critical case $\delta = 1$ – even more complex – can be dealt with in a very similar way.

18.1.2 Linear Constraint Equations

Rewriting the system (18.1) in terms of the fluctuations $g_\epsilon^+, g_\epsilon^-$, we get

$$\epsilon\partial_t \begin{pmatrix} g_\epsilon^+ \\ g_\epsilon^- \end{pmatrix} + v\cdot\nabla_x \begin{pmatrix} g_\epsilon^+ \\ g_\epsilon^- \end{pmatrix} + \delta(\epsilon E_\epsilon + v\wedge B_\epsilon)\cdot\nabla_v \begin{pmatrix} g_\epsilon^+ \\ -g_\epsilon^- \end{pmatrix} - \delta E_\epsilon\cdot v \begin{pmatrix} 1+\epsilon g_\epsilon^+ \\ -1-\epsilon g_\epsilon^- \end{pmatrix}$$
$$= -\frac{1}{\epsilon}\begin{pmatrix} \mathscr{L} g_\epsilon^+ + \delta^2 \mathscr{L}(g_\epsilon^+, g_\epsilon^-) \\ \mathscr{L} g_\epsilon^- + \delta^2 \mathscr{L}(g_\epsilon^-, g_\epsilon^+) \end{pmatrix} + \begin{pmatrix} \mathscr{Q}(g_\epsilon^+, g_\epsilon^+ + \delta^2 g_\epsilon^-) \\ \mathscr{Q}(g_\epsilon^-, g_\epsilon^- + \delta^2 g_\epsilon^+) \end{pmatrix}$$

denoting by $\mathscr{L}$ and $\mathscr{Q}$ the usual linearized and quadratic Boltzmann operators (acting only on the v-variable)

$$\mathscr{L}(g,h)(v) = \int M(v_1)(g(v) + h(v_1) - g(v') - h(v_1'))b(v-v_1,\omega)dv_1 d\omega$$
$$\mathscr{Q}(g,h)(v) = \int M(v_1)(g(v)h(v_1) - g(v')h(v_1'))b(v-v_1,\omega)dv_1 d\omega,$$

where b is the collision cross-section giving the statistical repartition of post-collisional velocities in terms of pre-collisional velocities and deflection angle.

At leading order, we find that

$$\mathscr{L} g^+ = \mathscr{L} g^- = 0.$$

Recall that $\mathscr{L}$ is a Fredholm operator on $L^2(M\nu dv)$ with kernel spanned by the collision invariants $1, v$, and $|v|^2$, where the collision frequency is defined by

$$\nu(v) = \int M(v_1)b(v - v_1, \omega)dv_1 d\omega\,.$$

This implies that

$$g^{\pm} = \rho^{\pm} + u^{\pm} \cdot v + \theta^{\pm}\frac{|v|^2 - 3}{2}, \tag{18.2}$$

which expresses the relaxation towards local thermodynamic equilibrium.

More precisely, we expect that for fixed ϵ,

$$g_\epsilon^{\pm} = \rho_\epsilon^{\pm} + u_\epsilon^{\pm} \cdot v + \theta_\epsilon^{\pm}\frac{|v|^2 - 3}{2} + O(\epsilon) + O(\delta^2)$$

denoting by $\rho_\epsilon^{\pm}, u_\epsilon^{\pm}, \theta_\epsilon^{\pm}$ the moments of $g_\epsilon^{\pm}$.

The next order involves the interspecies interactions and leads to the constraint

$$\int \mathscr{L}(g^+ - g^-, g^- - g^+)\begin{pmatrix} 1 \\ v \\ |v|^2 \end{pmatrix} dv = 0\,.$$

Together with the Ansatz (18.2), this shows that

$$u^+ = u^- \equiv u \text{ and } \theta^+ = \theta^- \equiv \theta\,, \tag{18.3}$$

so the bifluid model should have only six degrees of freedom. Note that a careful study of the dissipation actually provides the quantitative estimate

$$u_\epsilon^+ - u_\epsilon^- = O(\epsilon/\delta), \qquad \theta_\epsilon^+ - \theta_\epsilon^- = O(\epsilon/\delta)\,.$$

The incompressibility and Boussinesq constraints

$$\mathrm{div} u = 0, \quad \nabla(\rho^+ + \rho^- + 2\theta) = 0 \tag{18.4}$$

are obtained from the local conservation of mass, and from the local conservation of total momentum (which are equations living in some sense in a space orthogonal to the previous constraints (18.2) and (18.3)).

The asymptotic form of the Maxwell equations is then obviously

$$\partial_t E - \mathrm{rot} B = -j, \qquad \partial_t B + \mathrm{rot} E = 0, \qquad \mathrm{div} B = \mathrm{div} E = 0,$$

denoting by j the limit point of the scaled current $\frac{\delta}{\epsilon}(u_\epsilon^+ - u_\epsilon^-)$.

Note that essentially only weak compactness is required to justify all these linear constraints.

18.1.3 Nonlinear Constraints and Evolution Equations

As usual the most difficult part of the derivation (even at formal level) is to get evolution equations. For fixed ϵ, the conservation laws state

$$
\begin{aligned}
\partial_t u_\epsilon^\pm + \frac{1}{\epsilon}\nabla(\rho_\epsilon^\pm + \theta_\epsilon^\pm) + \frac{1}{\epsilon}\nabla \int M\phi g_\epsilon^\pm dv \mp \frac{\delta}{\epsilon}E_\epsilon \mp (\delta\rho_\epsilon^\pm E_\epsilon + \frac{\delta}{\epsilon}u_\epsilon^\pm \wedge B_\epsilon) \\
= -\frac{\delta^2}{\epsilon^2}\int (\mathscr{L}(g_\epsilon^\pm, g_\epsilon^\mp) + \epsilon \mathscr{Q}(g_\epsilon^\pm, g_\epsilon^\mp))Mvdv \\
\frac{3}{2}\partial_t(\theta_\epsilon^\pm + \rho_\epsilon^\pm) + \frac{5}{2\epsilon}\nabla\cdot u_\epsilon^\pm + \frac{1}{\epsilon}\nabla \int M\psi g_\epsilon^\pm dv \mp \delta u_\epsilon^\pm E_\epsilon \\
= -\frac{\delta^2}{2\epsilon^2}\int (\mathscr{L}(g_\epsilon^\pm, g_\epsilon^\mp) + \epsilon \mathscr{Q}(g_\epsilon^\pm, g_\epsilon^\mp))M|v|^2 dv
\end{aligned}
\tag{18.5}
$$

denoting by ϕ and ψ the kinetic momentum and energy fluxes

$$
\phi = v\otimes v - \frac{1}{3}|v|^2 Id, \quad \psi = \frac{1}{2}v(|v|^2 - 5) \text{ so that } \phi, \psi \in (\mathrm{Ker}\mathscr{L})^\perp .
$$

At this stage, because of the constraints (18.3), the equations are still singular and it is not possible to take limits directly. We therefore consider separately the conservation of total momentum and energy, and the equations for the differences $u_\epsilon^+ - u_\epsilon^-$ and $\theta_\epsilon^+ - \theta_\epsilon^-$ since these quantities are expected to be $O(\epsilon/\delta)$.

To deal with the conservations of total momentum and total energy, we follow the usual strategy [7, 15]: it consists in splitting the flux terms into transport and diffusion components according to the linearized Chapman-Enskog decomposition

$$
\frac{1+\delta^2}{\epsilon}\mathscr{L}(g_\epsilon^+ + g_\epsilon^-) = \mathscr{Q}(g_\epsilon^+, g_\epsilon^+) + \mathscr{Q}(g_\epsilon^-, g_\epsilon^-) - v\cdot\nabla_x(g_\epsilon^+ + g_\epsilon^-) + O(\epsilon) + O(\delta^2)
$$

Straightforward but tedious computations (reported in [6]) then lead to the following asymptotic evolution equations

$$
\begin{aligned}
&\partial_t u + u\cdot\nabla u - \mu\Delta u + \nabla p = \frac{1}{2}j\wedge B, \\
&\partial_t \theta + u\cdot\nabla\theta - \kappa\Delta\theta = 0 .
\end{aligned}
\tag{18.6}
$$

The final step is then to derive the equation relating j to the other macroscopic fields, which is usually referred to as Ohm's law. Combining the equations of mass shows that in the regime under consideration j has to be divergence free. Looking at the most singular term in the equation for $u_\epsilon^+ - u_\epsilon^-$, we get

$$
\begin{aligned}
&\frac{1}{\delta}\nabla(\rho_\epsilon^+ + \theta_\epsilon^+ - \rho_\epsilon^- - \theta_\epsilon^-) - 2E_\epsilon - (u_\epsilon^+ + u_\epsilon^-)\wedge B_\epsilon \\
&= -\frac{\delta}{\epsilon}\int \mathscr{L}(g_\epsilon^+ - g_\epsilon^-, g_\epsilon^- - g_\epsilon^+)Mvdv + O(\epsilon/\delta).
\end{aligned}
$$

Plugging the Ansatz (18.2) and (18.3) in this relation leads to the explicit relation

$$\sigma(\nabla_x p + u \wedge B + E) = j \text{ with } \frac{1}{\sigma} = \frac{1}{6}\int v \cdot \mathscr{L}(v, -v) M dv\,. \tag{18.7}$$

Combining all equations, we end up with the Navier-Stokes-Maxwell system with solenoidal Ohm's law. Note that it differs from the common Ohm law due to the divergence free constraint imposed by the weak scaling of interspecies collisions.

Remark 18.1. In the critical case $\delta = 1$, the quantity

$$\frac{1}{\epsilon}(g_\epsilon^+ - \rho_\epsilon^+ - g_\epsilon^- + \rho_\epsilon^-) \text{ is no more an infinitesimal Maxwellian.}$$

The nonlinear constraint is obtained by inversion of the Fredholm operator

$$\mathbf{L} : g \mapsto \mathscr{L}(g) + \mathscr{L}(g, -g)$$

whose kernel is composed only of constant functions. Ohm's law then becomes

$$j - nu = \sigma(E + u \wedge B - \frac{1}{2}\nabla n) \text{ with } n = \operatorname{div} E,$$

which does not seem to be referenced in the physical literature.

18.2 Main Results

The previous formal asymptotic analysis obviously misses all problems related to the convergence of nonlinear herms, which are known to be rather complicated even in the case of neutral gases with only one species [14, 15, 18]. But even more problematic is the fact that the PDEs under consideration are not known to have solutions (even in the sense of distributions). Indeed renormalization techniques cannot be applied to the Lorentz force [10]. The first step towards a rigorous analysis is therefore to define the objects we will work with.

18.2.1 Existence of Renormalized Solutions with Defect Measure for the Vlasov-Maxwell-Boltzmann System

Following the method of Alexandre and Villani [1], it is possible to establish [5] the existence of very weak solutions of the Vlasov-Maxwell-Boltzmann system in the presence of long-range interactions, i.e. $b \notin L^1_{loc}(dzd\omega)$. The breakdown of the DiPerna-Lions theory on linear transport [8] can be indeed compensated by the regularizing (or at least compactifying) effect due to long-range interactions [2]. This approach provides thus some global existence result for large initial data in the entropy space (defined by the natural entropy bounds).

Theorem 18.1. *Let $b(z,\omega)$ be a collision kernel which derives from an inverse power potential. Then, for any initial condition $\left(f_0^+, f_0^-, E_0, B_0\right)$ satisfying the compatibility conditions on the divergences as well as the bound*

$$H(f_0^+|M) + H(f_0^-|M) + \frac{1}{2}\int (|E_0|^2 + |B_0|^2)dx < +\infty ,$$

there exists a renormalized solution $(f_\epsilon^\pm, E_\epsilon, B_\epsilon)$ to the Vlasov-Maxwell-Boltzmann system (18.1) with a defect measure.

To avoid technicalities, we will not define precisely the notion of renormalized solution with defect measure: essentially the kinetic equations obtained by the DiPerna – Lions renormalization process [9] have to be replaced by inequalities, and the consistency is then obtained from the local conservation of mass.

18.2.2 Existence of Dissipative Solutions for the Navier-Stokes-Maxwell Equations with Ohm's Law

The limiting system obtained formally states

$$\begin{cases} \partial_t u + (u\cdot\nabla u) + \nabla p - \mu\Delta u - \dfrac{1}{2} j \wedge B = 0 \\ j = \sigma \mathbf{P}(E + u\wedge B), \quad \mathrm{div} u = \mathrm{div} E = \mathrm{div} B = 0 \\ \partial_t B + \mathrm{rot} E = 0, \quad \partial_t E - \mathrm{rot} B = -j\, , \end{cases} \tag{18.8}$$

together with a convection-diffusion equation for the temperature, which is essentially linear.

A natural framework to study these equations should be the energy space, i.e. the functional space defined by the (formal) energy conservation.

$$\frac{1}{2}\frac{d}{dt}\left(\|u\|_{L^2}^2 + \frac{1}{2}\|E\|_{L^2}^2 + \frac{1}{2}\|B\|_{L^2}^2 \right) + \mu\|\nabla u\|_{L^2}^2 + \frac{1}{2\sigma}\|j\|_{L^2}^2 = 0\, .$$

We indeed expect solutions in this space to be global. The energy estimate shows actually that all terms in the motion equation and in Ohm's law make sense, but it does not guarantee weak stability of the Lorentz force $j \wedge B$. Furthermore, since the Maxwell equations are hyperbolic, we do not expect to gain regularity (or even compactness) on the electromagnetic field (E, B).

By analogy with the 3D incompressible Euler equations, we have then the following global existence result:

Theorem 18.2. *For any solenoidal initial condition $(u_0, E_0, B_0) \in L^2$, there exists a dissipative solution to (18.8), i.e. a solenoidal vector field $(u, E, B, j) \in C_t^0(w - L^2)$ such that $j = \sigma P(E + u \wedge B)$ and for any smooth $(\bar{u}, \bar{E}, \bar{B}, \bar{j})$ such that* $\mathrm{div}\bar{u} = \mathrm{div}\bar{E} = \mathrm{div}\bar{B} = \mathrm{div}\bar{j} = 0$

$$
\begin{aligned}
&\|2(u-\bar{u})(t)\|_{L^2}^2+\|(E-\bar{E})(t)\|_{L^2}^2+\|(B-\bar{B})(t)\|_{L^2}^2\\
&\le 2\left(\|u_0-\bar{u}_0\|_{L^2}^2+\|E_0-\bar{E}_0\|_{L^2}^2+\|B_0-\bar{B}_0\|_{L^2}^2\right)\exp(\gamma(t))\\
&+\int_0^t\int (j-\bar{j})\cdot(\sigma^{-1}\bar{j}-\bar{E}-\bar{u}\wedge\bar{B})\exp(\gamma(s))\,dxds\\
&-\int_0^t\int\begin{pmatrix}u-\bar{u}\\ E-\bar{E}\\ B-\bar{B}\end{pmatrix}\cdot\begin{pmatrix}2(\partial_t\bar{u}+\bar{u}\cdot\nabla\bar{u}-\mu\Delta\bar{u})-\bar{j}\wedge\bar{B}\\ \partial_t\bar{E}-\mathrm{rot}\bar{B}+\bar{j}\\ \partial_t\bar{B}+rot\bar{E}\end{pmatrix}\exp(\gamma(s))\,dxds
\end{aligned}
$$

where the growth rate $\gamma(t)=C(\mu^{-1}+\sigma)\int_0^t\|\bar{u}(s)\|_{L^\infty}^2ds+C\mu^{-1}\int_0^t\|\bar{j}(s)\|_{L^3}^2ds.$

Dissipative solutions are not known to be weak solutions of the Navier-Stokes-Maxwell equations in conservative form. But they coincide with the unique smooth solution with same initial data as long as the latter does exist.

Note that, as in the case of the 3D incompressible Euler equation, it is possible to prove the local existence of such smooth solutions [17].

18.2.3 Convergence in the Fast Relaxation Limit

Our goal here is to establish the convergence of scaled families of renormalized solutions with defect measure to the Vlasov-Maxwell-Boltzmann system towards dissipative solution to the Navier-Stokes-Maxwell equations with Ohm's law, without any restriction on their size or regularity. As for the incompressible Euler limit of the Boltzmann equation, the proof of convergence relies on some weak-strong stability principle, already used to define dissipative solutions of the limiting system.

Theorem 18.3. *Let* $(f_{\epsilon,0}^{\pm},E_{\epsilon,0},B_{\epsilon,0})$ *be a family of well-prepared initial data*

$$
\begin{aligned}
&g_{\epsilon,0}^{\pm}\rightharpoonup g_0=u_0\cdot v+\theta_0\frac{|v|^2-5}{2}\quad \textit{weakly in } L^1_{loc}(dxMdv),\\
&\frac{1}{\epsilon^2}H(f_{\epsilon,0}^{+}|M)+\frac{1}{\epsilon^2}H(f_{\epsilon,0}^{-}|M)\to\int M(g_0)^2dvdx,\\
&E_{\epsilon,0}\to E_0,\quad B_{\epsilon,0}\to B_0\quad \textit{strongly in } L^2\,.
\end{aligned}
$$

For any ϵ*, let* $(f_\epsilon^{\pm},E_\epsilon,B_\epsilon)$ *be a renormalized solution with defect measure to (18.1) with* $\epsilon<<\delta<<1$*. Denote by* $u_\epsilon=\frac{1}{2}\int(f_\epsilon^{+}+f_\epsilon^{-})vdv$ *and* $\theta_\epsilon=\frac{1}{4}\int(f_\epsilon^{+}+f_\epsilon^{-})(|v|^2-3)dv$ *the bulk velocity and temperature of the plasma.*

Then $(u_\epsilon,\theta_\epsilon,E_\epsilon,B_\epsilon)$ *is weakly compact in* $L^1_{loc}(dtdx)$ *and any of its limit point is a dissipative solution to the Navier-Stokes-Maxwell equations with Ohm's law. In particular it coincides with the unique smooth solution as long as the latter does exist.*

In the case when the limiting system has a smooth solution $(\bar{u}, \bar{E}, \bar{B}, \bar{j})$, as a byproduct of the previous theorem, we actually get some strong convergence with rate γ. This explains why it is much more difficult to deal with ill-prepared initial data (see [20] for a discussion on this topics).

18.3 Main Mathematical Difficulties and Strategy of the Proof

18.3.1 Dealing with Weak Notions of Solutions

As mentioned in the previous section, the first obvious difficulty is to build solutions both to the kinetic and magneto-hydrodynamic systems, as they involve singular terms which are either not stable under weak convergence (electromagnetic coupling in (18.8)), or even not defined by the a priori estimates (collision integrals in (18.1)).

By defining very weak notions of solutions, it is possible to bypass these difficulties, but the counterpart is that formal properties of (18.8) and (18.1) are not known to hold anymore.

When dealing with renormalized solutions in the sense of Alexandre and Villani, one of the major problem is the fact that conservation laws are not satisfied while they are crucial in the study of hydrodynamic limits (see paragraphs 18.1.2 and 18.1.3). We start therefore from some approximate system of conservation laws obtained by renormalization and truncation of large velocities, of the type

$$\begin{aligned}\partial_t \int M\beta_\epsilon(g_\epsilon^\pm)\varphi\chi\left(\frac{|v|^2}{K_\epsilon}\right)dv + \frac{1}{\epsilon}\nabla_x \cdot \int M\beta_\epsilon(g_\epsilon^\pm)\varphi\chi\left(\frac{|v|^2}{K_\epsilon}\right)vdv\\ -\frac{\delta}{\epsilon}E_\epsilon \cdot \int Mv(1+\epsilon g_\epsilon^\pm)\beta_\epsilon'(g_\epsilon^\pm)\varphi\chi\left(\frac{|v|^2}{K_\epsilon}\right)dv\\ -\delta\int \beta_\epsilon(g_\epsilon^\pm)(E_\epsilon + v\wedge B_\epsilon)\cdot\nabla_v\left(M\varphi\chi\left(\frac{|v|^2}{K_\epsilon}\right)\right)dv\\ = \frac{1}{\epsilon^3}\int \beta_\epsilon'(g_\epsilon^\pm)\mathcal{Q}(G_\epsilon^\pm, G_\epsilon^\pm + \delta^2 G_\epsilon^\mp)M\varphi\chi\left(\frac{|v|^2}{K_\epsilon}\right)dv\\ +\epsilon\int\varphi\chi\left(\frac{|v|^2}{K_\epsilon}\right)d\nu_{\epsilon,\beta}^\pm(v),\end{aligned}$$

where β_ϵ is a smooth compactly-supported function such that

$$\beta_\epsilon(z) \equiv z \text{ on } \left[-\frac{1}{2\epsilon}, \frac{1}{2\epsilon}\right],$$

and $\nu_{\beta,\epsilon}^\pm$ are the corresponding defect measures.

The point is that the right-hand side is not zero even for collision invariants φ.

- Because of the symmetry breaking, the first term does not cancel. Proving that these conservation defects vanish in the limit $\epsilon \to 0$ could be done following the strategy of [15], but it requires some equiintegrability on the fluctuations which is not inherited directly from the entropy inequality.
- The second term involves the defect measure coming from the construction of solutions in Theorem 18.1. Proving that it converges to 0 as $\epsilon \to 0$ relies on the following improvement of Theorem 18.1 (see [3, 5, 6] for the precise statement and the proof).

Proposition 18.1. *Let $b(z, \sigma)$ be a collision kernel which derives from an inverse power potential. Then, the renormalized solutions of the Vlasov-Maxwell-Boltzmann system with defect measures built in Theorem 18.1 satisfy the refined entropy inequality*

$$\begin{aligned}&\frac{1}{\epsilon^2}H(f_\epsilon^+|M) + \frac{1}{\epsilon^2}H(f_\epsilon^-|M) + \frac{1}{2}\int (|E_\epsilon|^2 + |B_\epsilon|^2)dx\\ &+\int_0^t\int \frac{1}{\epsilon^4}\left(D(f_\epsilon^+) + D(f_\epsilon^-)\right) + \frac{2\delta^2}{\epsilon^4}D(f_\epsilon^+, f_\epsilon^-)dxds\\ &+\lambda_\epsilon^+([0,t]\times \mathbf{R}^3) + \lambda_\epsilon^-([0,t]\times \mathbf{R}^3) + \lambda_\epsilon^\pm([0,t]\times \mathbf{R}^3) \le C_0\,.\end{aligned}$$

where $\lambda_\epsilon^+, \lambda_\epsilon^-, \lambda_\epsilon^\pm$ control the defect measures $\nu_{\epsilon,\beta}^\pm$ in the renormalized equations.

18.3.2 A Rough Force Term in the Transport Equation

As mentioned in the introduction, the Lorentz force changes deeply the structure of the transport equation, even in the case of a smooth electromagnetic field (in which case renormalization is not a problem).

In [10, 15], the structure of the free transport was used in a crucial way to obtain both:

- The equiintegrability required to prove the convergence of conservation defects (see the previous paragraph) [16].
- And the strong spatial compactness of moments needed to establish the stability of nonlinear terms [11, 12].

These properties are indeed based on the hypoelliptic transfer of L^1 weak compactness from the v-variable to the x-variable [16].

In the presence of source terms involving v-derivatives, L^1 weak compactness is not known to be transferred anymore. Both the proof of Theorem 18.1 and the asymptotic analysis of stability for Ohm's law instead use some transfer of strong compactness [4], which relies on the transport of frequencies in the phase space together with refined tools of harmonic analysis.

Theorem 18.4. *Let the bounded family of nonnegative functions* $\{\phi_\lambda(t,x,v)\}_\lambda \subset L^1(dtdx, L^r(dv))$, *for some* $1 < r < \infty$, *be locally relatively compact in v and such that*

$$(\partial_t + v\cdot\nabla_x)\phi_\lambda = (1-\Delta_{t,x})^{\frac{\beta}{2}}(1-\Delta_v)^{\frac{\alpha}{2}} S_\lambda,$$

for some bounded family $\{S_\lambda(t,x,v)\}_\lambda \subset L^1(dtdx, L^r(dv))$, *where* $\alpha \geq 0$ *and* $0 \leq \beta < 1$.

Then, $\{\phi_\lambda(t,x,v)\}_\lambda$ *is locally relatively compact in* $L^1(dtdxdv)$ *(in all variables).*

18.3.3 A Singular Coupling

The last difficulty which is specific to the electromagnetic forcing we consider here is the singular scaling of the Lorentz force, which is responsible for the lack of weak stability of the limiting system (see paragraph 18.2.2). In particular, we do not expect to be able to prove directly the convergence of the nonlinear term $j_\epsilon \wedge B_\epsilon$ (as there is no a priori strong spatial compactness on j_ϵ and B_ϵ), nor to get the a priori equiintegrability required to establish the convergence of conservation defects.

As for the incompressible Euler limit of the Boltzmann equation, the strategy is therefore to replace such a priori estimates by some loop estimates and the use of Gronwall's lemma. A natural idea (following [13, 19, 22]) would be to get in this way some stability inequality for the scaled modulated entropy

$$\sum_\pm \frac{1}{\epsilon^2} H(f_\epsilon^\pm | f_{app}^\pm) = \sum_\pm \frac{1}{\epsilon^2}\int \left(f_\epsilon^\pm \log\frac{f_\epsilon^\pm}{f_{app}^\pm} - f_\epsilon^\pm + f_{app}^\pm \right) dvdx,$$

which measures in some sense the distance between the fluctuations $g_\epsilon^\pm$ and their expected limits $g_0^\pm \sim \frac{1}{\epsilon}(f_{app}^\pm - M)/M$. Unfortunately this strategy fails:

- Even for weak solutions in the sense of distributions, we would have no estimate on large velocities to control the momentum and energy fluxes.
- For renormalized solutions in the sense of DiPerna and Lions, local conservation laws are not known to hold.
- For solutions in the sense of Alexandre and Villani, even after renormalization, the kinetic equation is relaxed in an inequality.

The main novelty here is to use renormalization techniques together with the relative entropy method. More precisely, we do not use as usual a modulated entropy inequality for renormalized solutions to the kinetic equation. We modulate a renormalized version of the entropy inequality, which requires much less informations on the solutions to the kinetic equation. For any smooth vector field $(\bar{u}, \bar{\theta}, \bar{E}, \bar{B}) \in C^\infty \cap W^{\infty,1}(\mathbf{R}^+ \times \mathbf{R}^3)$ such that $\mathrm{div}\bar{u} = \mathrm{div}\bar{E} = \mathrm{div}\bar{B} = 0$, we define the infinitesimal Maxwellian

$$g = v \cdot \bar{u} + \frac{1}{2}(|v|^2 - 5)\bar{\theta}$$

and the dissipations

$$q = \frac{1}{2}\left(\nabla_x \bar{u} : (\tilde{\phi} + \tilde{\phi}_1 - \tilde{\phi}' - \tilde{\phi}'_1) + \nabla_x \bar{\theta} \cdot (\tilde{\psi} + \tilde{\psi}_1 - \tilde{\psi}' - \tilde{\psi}'_1)\right),$$
$$\eta = \bar{j} \cdot (v - v') \text{ with } \bar{j} = \sigma \mathbf{P}(\bar{E} + \bar{u} \wedge \bar{B}),$$

where $\tilde{\phi}, \tilde{\psi}$ have been already defined as pseudo-inverses of the kinetic momentum and energy fluxes. We then define the modulated renormalized entropy by

$$\begin{aligned}\delta\mathscr{H}(t) =& \frac{1}{2}\int M\left(\beta_\epsilon(g_\epsilon^+)\chi\left(\frac{|v|^2}{K_\epsilon}\right) - g\right)^2 dvdx \\ &+ \frac{1}{2}\int M\left(\beta_\epsilon(g_\epsilon^-)\chi\left(\frac{|v|^2}{K_\epsilon}\right) - g\right)^2 dvdx \\ &+ \frac{1}{2}\|(E_\epsilon - \bar{E})(t)\|_{L^2}^2 + \frac{1}{2}\|(B_\epsilon - \bar{B})(t)\|_{L^2}^2\end{aligned}$$

for some flat truncation β_ϵ, and the modulated renormalized entropy dissipation by

$$\begin{aligned}\delta\mathscr{D}(t) = &\int M\left(\frac{1}{\epsilon^2}\left(\sqrt{(f_\epsilon^+)'_1(f_\epsilon^+)'} - \sqrt{(f_\epsilon^+)_1(f_\epsilon^+)'}\right) - q\right)^2 bd\omega dv_1 dvdx \\ &+ \int M\left(\frac{1}{\epsilon^2}\left(\sqrt{(f_\epsilon^-)'_1(f_\epsilon^-)'} - \sqrt{(f_\epsilon^-)_1(f_\epsilon^-)}\right) - q\right)^2 bd\omega dv_1 dvdx \\ &+ 2\int M\left(\frac{\delta}{\epsilon^2}\left(\sqrt{(f_\epsilon^+)'_1(f_\epsilon^-)'} - \sqrt{(f_\epsilon^+)_1(f_\epsilon^-)}\right) - \eta\right)^2 bd\omega dv_1 dvdx\,.\end{aligned}$$

Proposition 18.2. *Denote by $(f_\epsilon^\pm, E_\epsilon, B_\epsilon)$ any solution to the Vlasov-Maxwell-Boltzmann system (18.1). Then, with the previous notations, we have the following stability inequality*

$$\begin{aligned}&\delta\mathscr{H}(t) + \int_0^t \mathscr{D}(s)ds \le \delta\mathscr{H}(0)\exp(\gamma(t)) + o(1) \\ &- \int_0^t \int \begin{pmatrix} u_\epsilon - \bar{u} \\ \frac{3}{5}\theta_\epsilon - \frac{1}{5}\rho_\epsilon^- - \frac{1}{5}\rho_\epsilon^+ - \bar{\theta} \\ E_\epsilon - E \\ B_\epsilon - \bar{B}\end{pmatrix} \cdot \begin{pmatrix} \partial_t \bar{u} + \bar{u}\cdot\nabla\bar{u} - \mu\Delta\bar{u} - \bar{j}\wedge\bar{B} \\ \partial_t\bar{\theta} + \bar{u}\cdot\bar{\theta} - \kappa\Delta\bar{\theta} \\ \partial_t \bar{E} - rot\bar{B} + \bar{j} \\ \partial_t\bar{B} + rot\bar{E}\end{pmatrix} \exp(\gamma(s))\, dxds\end{aligned} \tag{18.9}$$

Note that (18.9) is very similar to the stability inequality defining dissipative solutions to the Navier-Stokes-Maxwell system with Ohm's law. Combining it with the linear constraints (obtained by weak compactness arguments) and with Ohm's law (obtained by hypoelliptic regularity and compensated compactness) leads to a complete characterization of the limit points (u, θ, E, B).

References

1. R. Alexandre, C. Villani, On the Boltzmann equation for long-range interactions. Commun. Pure Appl. Math. **55**(1), 30–70 (2002)
2. R. Alexandre, L. Desvillettes, C. Villani, B. Wennberg, Entropy dissipation and long-range interactions. Arch. Ration. Mech. Anal. **152**(4), 327–355 (2000)
3. D. Arsénio, From Boltzmann's equation to the incompressible Navier-Stokes-Fourier system with long-range interactions. Arch. Ration. Mech. Anal. **206**(2), 367–488 (2012)
4. D. Arsénio, L. Saint-Raymond, Compactness in kinetic transport equations and hypoellipticity. J. Funct. Anal. **261**, 3044–3098 (2011)
5. D. Arsénio, L. Saint-Raymond, Solutions of the Vlasov-Maxwell-Boltzmann system with long-range interactions. C. R. Acad. Sci. **351**(9–10), 357–360 (2013)
6. D. Arsénio, L. Saint-Raymond, From the Vlasov-Maxwell-Boltzmann system to incompressible viscous electro-magneto-hydrodynamics (preprint in 2014)
7. C. Bardos, F. Golse, C.D. Levermore, Fluid dynamic limits of kinetic equations II: convergence proofs for the Boltzmann equation. Commun. Pure Appl. Math. **46**, 667–753 (1993)
8. R.J. DiPerna, P.-L. Lions, Ordinary differential equations, transport theory and Sobolev spaces. Invent. Math. **98**, 511–547 (1989)
9. R.J. DiPerna, P.-L. Lions, On the Cauchy problem for the Boltzmann equation: global existence and weak stability results. Ann. Math. **130**, 321–366 (1990)
10. R.J. DiPerna, P.-L. Lions, Global weak solutions of Vlasov-Maxwell systems. Commun. Pure Appl. Math. **42**, 729–757 (1989)
11. R.J. DiPerna, P.-L. Lions, Y. Meyer, L^p regularity of velocity averages. Ann. Inst. H. Poincar Anal. Non Linéaire **8**, 271–287 (1991)
12. F. Golse, P.-L. Lions, B. Perthame, R. Sentis, Regularity of the moments of the solution of a transport equation. J. Funct. Anal. **76**, 110–125 (1988)
13. F. Golse, D. Levermore, L. Saint-Raymond, La méthode de l'entropie relative pour les limites hydrodynamiques de modèles cinétiques. Séminaire Equations aux dérivées partielles (Polytechnique) (1999–2000)
14. F. Golse, L. Saint-Raymond, The Navier-Stokes limit of the Boltzmann equation for bounded collision kernels. Invent. Math. **155**(1), 81–161 (2004)
15. F. Golse, L. Saint-Raymond, The Navier-Stokes limit of the Boltzmann equation for hard potentials. J. Math. Pure Appl. **91**(5), 508–552 (2009)
16. F. Golse, L. Saint-Raymond, Velocity averaging in L^1 for the transport equation. C. R. Acad. Sci. **334**, 557–562 (2002)
17. S. Ibrahim, S. Keraani, Global small solutions for the Navier-Stokes-Maxwell system. SIAM J. Math. Anal. **43**(5), 2278–2295 (2011)
18. P.-L. Lions, N. Masmoudi, From the Boltzmann equations to the equations of incompressible fluid mechanics. I, II. Arch. Ration. Mech. Anal. **158**, 173–193, 195–211 (2001)
19. L. Saint-Raymond, Convergence of solutions to the Boltzmann equation in the incompressible Euler limit. Arch. Ration. Mech. Anal. **166**, 47–80 (2003)

20. L. Saint-Raymond, Hydrodynamic limits: some improvements of the relative entropy method. Ann. Inst. H. Poincaré **26**, 705–744 (2009)
21. L. Saint-Raymond, *Hydrodynamic Limits of the Boltzmann Equation*. Lecture Notes in Mathematics, vol. 1971 (Springer, Berlin, 2009)
22. H.T. Yau, Relative entropy and hydrodynamics of Ginzburg-Landau models. Lett. Math. Phys. **22**, 63–80 (1991)

Chapter 19
Kinetic Theory and Gas Dynamics, Some Historical Perspectives

Tai-Ping Liu

Abstract The purpose of the present paper is to present, through historical perspectives, some of the recent developments on the Boltzmann equation in the kinetic theory and multi-dimensional shock waves in gas dynamics.

19.1 Introduction

The basic system for shock waves in gas dynamics is the compressible Euler equations:

$$\begin{aligned}
&\rho_t + \nabla_x \cdot (\rho \boldsymbol{u}) = 0, && \text{continuity equation,}\\
&(\rho \boldsymbol{u})_t + \nabla_x \cdot (\rho \boldsymbol{u} \otimes \boldsymbol{u} + pI) = 0, && \text{momentum equations,}\\
&(\rho E)_t + \nabla_x \cdot (\rho E \boldsymbol{u} + p\boldsymbol{u}) = 0, && \text{energy equation.}
\end{aligned}$$

Here ρ is the density, $\boldsymbol{u}$ velocity, $E = e + |\boldsymbol{u}|^2/2$ total energy, and the system is closed with given constitutive relation with pressure $p = p(\rho, s)$, s entropy. The most important equation for the kinetic theory is the Boltzmann equation

$$\mathsf{f}_t + \boldsymbol{\xi} \cdot \nabla_x \mathsf{f} = \frac{1}{k} Q(\mathsf{f}, \mathsf{f}).$$

We will present some of the recent progresses on multi-dimensional shock waves in gas dynamics and on the Boltzmann equation through historical perspectives.

T.-P. Liu (✉)
Institute of Mathematics, Academia Sinica, Taipei, Taiwan
Stanford University, Stanford, CA, USA
e-mail: liu@math.stanford.edu

V. Ancona and E. Strickland (eds.), *Trends in Contemporary Mathematics*, Springer INdAM Series 8, DOI 10.1007/978-3-319-05254-0_19,

For the historical perspectives, we start with the panel discussion on Wednesday morning August 17, 1949 chaired by von Neumann with participants including Liepmann, von Karman, Burgers and Heisenberg:

DISCUSSION ON THE EXISTENCE AND UNIQUENESS OR MULTIPLICITY OF SOLUTIONS OF THE AERODYNAMICAL EQUATIONS

It is included in the collected work of von Neumann and reprinted as

Discussion on the existence and uniqueness or multiplicity of solutions of the aerodynamical equations, Bull. Amer. Math. Soc. 47 (2010), 145–154.

One of our main points is to highlight the importance of physical reasoning in the mathematical analysis.

19.2 Gas Dynamics Equations

Due to great complexity of shock reflections, the Euler equations are often simplified, [3]. For smooth flows, if the initial value is assumed to be isentropic, then the flow stays as isentropic, $p = p(\rho)$, and the Euler equations are simplified to

$$\begin{aligned} &\rho_t + \nabla_x \cdot (\rho \boldsymbol{u}) = 0, && \text{continuity equation,} \\ &(\rho \boldsymbol{u})_t + \nabla_x \cdot (\rho \boldsymbol{u} \otimes \boldsymbol{u} + p(\rho) I) = 0, && \text{momentum equations.} \end{aligned}$$

The stationary Isentropic Euler equations are

$$\begin{aligned} &\nabla_x \cdot (\rho \boldsymbol{u}) = 0, && \text{continuity equation,} \\ &\nabla_x \cdot (\rho \boldsymbol{u} \otimes \boldsymbol{u} + p(\rho) I) = 0, && \text{momentum equations.} \end{aligned}$$

In the case, there is a self-similarity with self-similar variables, $\boldsymbol{\xi} = \boldsymbol{x}/t$, we can simplified the isentorpic Euler equations with the psuedo velocity $\boldsymbol{v} = \boldsymbol{u} - \boldsymbol{\xi}$:

$$\begin{aligned} &\nabla_\xi \cdot (\rho \boldsymbol{v}) = -n\rho, && \text{continuity equation,} \\ &\nabla_\xi \cdot (\rho \boldsymbol{v} \otimes \boldsymbol{v} + p(\rho) I) = -(n+1)\rho \boldsymbol{v}, && \text{momentum equations.} \end{aligned}$$

The Euler equations possess two kind of discontinuities, the shock waves and vortex sheets. The study of vortex is the main concern for the theory of incompressible fluid equations. To focus on the shock waves, the isentropic Euler equations are further simplified by assuming the flows to be irrotational with the velocity as gradient of the potential ϕ,

$$\boldsymbol{u} = \nabla_x \phi.$$

This and the Euler equations yields the Bernoulli equation

$$\phi_t + \frac{1}{2}|\nabla_x \phi|^2 + \Pi(\rho) = A \text{ (constant)},$$
$$\Pi'(\rho) = \frac{p'(\rho)}{\rho}, \ \sqrt{p'(\rho)} = c \text{ (sound speed)}.$$

The Bernoulli equation and the continuity equation yields the potential flow equation:

$$\phi_{tt} + 2\nabla_x \phi \cdot \nabla_x(\phi_t) + (\nabla_x \phi)^t \nabla_x^2 \phi \nabla_x \phi - c^2 \Delta \phi = 0.$$

Much have been done for the stationary potential flow equation:

$$(\nabla_x \phi)^t \nabla_x^2 \phi \nabla_x \phi - c^2 \Delta \phi = 0.$$

This equation is of mixed types. It is elliptic for subsonic flows, $|\nabla_x \phi| = |\boldsymbol{u}| < c$; and hyperbolic for supersonic flows, $|\nabla_x \phi| = |\boldsymbol{u}| > c$. Another simplification is for self-similar flows:

$$\phi(\boldsymbol{x}, t) = t\psi(\boldsymbol{\xi}), \ \chi(\boldsymbol{\xi}) = \psi(\boldsymbol{\xi}) - \frac{1}{2}|\boldsymbol{\xi}|^2,$$
$$\nabla_{\xi} \psi = \nabla_x \phi = \boldsymbol{u}, \text{ velocity.}$$
$$\nabla_{\xi} \chi = \boldsymbol{u} - \boldsymbol{\xi}, \text{ pseudo-velocity,}$$

and the self-similar potential flow equation is:

$$(\nabla_{\xi} \psi - \boldsymbol{\xi})^t \nabla_{\xi}^2 \psi (\nabla_{\xi} \psi - \boldsymbol{\xi}) - c^2 \Delta \psi = 0,$$
$$(\nabla_{\xi} \psi - \boldsymbol{\xi})^t \nabla_{\xi}^2 \psi (\nabla_{\xi} \psi - \boldsymbol{\xi}) - c^2 \Delta \psi = 2c^2 + |\nabla_{\xi} \chi|^2.$$

It is elliptic for psudo-subsonic flows, $|\nabla_{\xi} \chi| = |\boldsymbol{u} - \boldsymbol{\xi}| < c$, and hyperbolic for psudo-supersonic flows, $|\boldsymbol{u} - \boldsymbol{\xi}| > c$.

For two-dimensional potential flow, $\boldsymbol{x} = (x, y)$, $\boldsymbol{\xi} = (\xi, \eta)$, $\boldsymbol{u} = (u, v) = (\phi_x, \phi_y)$, we have

$$\phi_{tt} + 2\phi_x \phi_{xt} + 2\phi_y \phi_{yt} + [(\phi_x)^2 - c^2]\phi_{xx} + 2\phi_x \phi_y \phi_{xy} + [(\phi_y)^2 - c^2]\phi_{yy} = 0,$$

and the stationary flow equation becomes

$$[(\phi_x)^2 - c^2]\phi_{xx} + 2\phi_x \phi_y \phi_{xy} + [(\phi_y)^2 - c^2]\phi_{yy} = 0,$$

and the self-similar flow equation, $(\xi, \eta) = (x/t, y/t)$, $\chi_{\xi} = u - \xi$, $\chi_{\eta} = v - \eta$, becomes

$$((\chi_{\xi})^2 - c^2]\chi_{\xi\xi} + 2\chi_{\xi}\chi_{\eta}\chi_{\xi\eta} + ((\chi_{\eta})^2 - c^2]\chi_{\eta\eta} = 2c^2 + (\chi_{\xi})^2 + (\chi_{\eta})^2.$$

Notice that the stationary and self-similar equations differ only in the lower order term. Thus they are of the same type. As it turns out, the presence of the lower order term $2c^2 + (\chi_\xi)^2 + (\chi_\eta)^2$ in the self-similar equation has important implication.

19.3 Boltzmann Equation

The kinetic theory starts with the sistribution function $f(\boldsymbol{x}, \boldsymbol{\xi}, t)$, $\boldsymbol{x} \in \mathbb{R}^3$ space variable, $\boldsymbol{\xi} \in \mathbb{R}^3$ microscopic velocity. The macroscopic variables in fluid dynamics are obtained as averages:

$$\begin{cases} \rho(\boldsymbol{x},t) \equiv \int_{\mathbb{R}^3} \mathsf{f}(\boldsymbol{x},t,\boldsymbol{\xi})d\boldsymbol{\xi}, & \text{density,} \\ \rho\boldsymbol{v}(\boldsymbol{x},t) \equiv \int_{\mathbb{R}^3} \boldsymbol{\xi}\mathsf{f}(\boldsymbol{x},t,\boldsymbol{\xi})d\boldsymbol{\xi}, & \text{momentum,} \\ \boldsymbol{v} = (v^1, v^2, v^3), & \text{fluid velocity,} \\ \rho e(\boldsymbol{x},t) \equiv \int_{\mathbb{R}^3} \frac{|\boldsymbol{\xi}-\boldsymbol{v}|^2}{2}\mathsf{f}(\boldsymbol{x},t,\boldsymbol{\xi})d\boldsymbol{\xi}, & \text{internal energy,} \\ \rho E(\boldsymbol{x},t) \equiv \int_{\mathbb{R}^3} \frac{|\boldsymbol{\xi}|^2}{2}\mathsf{f}(\boldsymbol{x},t,\boldsymbol{\xi})d\boldsymbol{\xi} = \rho e + \frac{1}{2}\rho|\boldsymbol{v}|^2, & \text{total energy,} \\ p^{ij}(\boldsymbol{x},t) \equiv \int_{\mathbb{R}^3} (\xi^i - v^i)(\xi^j - v^j)\mathsf{f}(\boldsymbol{x},t,\boldsymbol{\xi})d\boldsymbol{\xi}, & \\ \quad \boldsymbol{P} = (p^{ij})_{1\le i,j\le 3}, & \text{stress tensor,} \\ q^i(\boldsymbol{x},t) \equiv \int_{\mathbb{R}^3} (\xi^i - v^i)\frac{1}{2}|\boldsymbol{v}-\boldsymbol{\xi}|^2\mathsf{f}(\boldsymbol{x},t,\boldsymbol{\xi})d\boldsymbol{\xi}, & \text{heat flux.} \end{cases}$$

An important reason for considering the kinetic formulation is that one can pose physically more realistic boundary condition. The following are some of the typical boundary conditions around a boundary with given velocity $\mathbf{0}$ and temperature θ_w:

$$\begin{aligned} &\mathsf{f}(\boldsymbol{x},t,\boldsymbol{\xi}) = \mathsf{f}(\boldsymbol{x},t,\boldsymbol{\xi} - 2(\boldsymbol{\xi}\cdot\boldsymbol{n})\boldsymbol{n}), && \text{specular reflection boundary condition;} \\ &\mathsf{f}(\boldsymbol{x},t,\boldsymbol{\xi}) = \mathsf{M}_{(\rho_0,\mathbf{0},\theta_w)}(\boldsymbol{\xi})\boldsymbol{\xi}\cdot\boldsymbol{n} > 0, && \text{diffuse reflection boundary condition;} \\ &\mathsf{f}(\boldsymbol{x},t,\boldsymbol{\xi}) = \mathsf{M}_{(\rho_w,\mathbf{0},\theta_w)}(\boldsymbol{\xi})\boldsymbol{\xi}\cdot\boldsymbol{n} > 0, && \text{complete condensation boundary condition;} \end{aligned}$$

for $\boldsymbol{x}$ on the boundary, $\boldsymbol{n}$ the unit outer normal to the boundary. ρ_w given and ρ_0 determined by zero mass flux at the boundary. The Maxwell type boundary conditions are interpolation of diffuse and specular boundary conditions.

The most important equation in the kinetic theory is the Boltzmann equation

$$\mathsf{f}_t + \boldsymbol{\xi}\cdot\nabla_x\mathsf{f} = \frac{1}{k}Q(\mathsf{f},\mathsf{f}).$$

The left hand side is the transport term and the right hand side is the collision operator:

$$\boldsymbol{Q}(\mathsf{f},\mathsf{f}) \equiv \int_{\substack{\mathbb{R}^3\times S^2 \\ (\boldsymbol{\xi}-\boldsymbol{\xi}_*)\cdot\Omega\ge 0}} (\mathsf{f}(\boldsymbol{\xi}')\mathsf{f}(\boldsymbol{\xi}'_*) - \mathsf{f}(\boldsymbol{\xi}_*)\mathsf{f}(\boldsymbol{\xi}))B(\boldsymbol{\xi}-\boldsymbol{\xi}_*,\Omega)d\boldsymbol{\xi}_*d\Omega;$$

$$\begin{cases} \xi' = \xi - [(\xi - \xi_*) \cdot \Omega]\, \Omega, \\ \xi'_* = \xi_* + [(\xi - \xi_*) \cdot \Omega]\, \Omega. \end{cases}$$

The cross section $B(\xi - \xi_*, \Omega)$ depends on the inter-molecular forces between the particles. For the hard sphere models

$$B(\xi - \xi_*, \Omega) = (\xi - \xi_*) \cdot \Omega.$$

The collision operator redistribute the particle velocity, but preserve the mass, momentum and energy:

$$\int_{\mathbb{R}^3} \begin{pmatrix} 1 \\ \xi \\ \frac{1}{2}|\xi|^2 \end{pmatrix} Q(\mathsf{g}, \mathsf{h}) d\xi = 0, \quad \begin{pmatrix} \text{mass} \\ \text{momentum} \\ \text{energy} \end{pmatrix}.$$

This yields the conservation laws by integrating the Boltzmann equation times $1, \xi$ and $|\xi|^2/2$:

$$\begin{cases} \partial_t \rho + \partial_{\boldsymbol{x}} \cdot (\rho \boldsymbol{v}) = 0, & \text{conservation of mass,} \\ \partial_t (\rho \boldsymbol{v}) + \partial_{\boldsymbol{x}} \cdot (\rho \boldsymbol{v} \otimes \boldsymbol{v} + \boldsymbol{P}) = 0, & \text{conservation of momentum,} \\ \partial_t (\rho E) + \partial_{\boldsymbol{x}} \cdot (\rho \boldsymbol{v} E + \boldsymbol{P} \boldsymbol{v} + \boldsymbol{q}) = 0, & \text{conservation of energy.} \end{cases}$$

These are 5 equations, 1 for the continuity equation, 3 for the momentum equations and 1 for the energy equation. However, there are 14 unknowns, 1 for the density ρ, 3 for the fluid velocity $\boldsymbol{v}$, 1 for the energy E, 6 for the stress $\boldsymbol{P}$, and 3 for the heat flux $\boldsymbol{q}$. Thus the conservation laws are under-determined. The fluid dynamics systems are obtained under some simplifications:

$$p_E^{ij} = p\delta_{ij},\ \boldsymbol{q}_E = 0, \qquad \text{Euler stress and heat flux;}$$

$$\begin{cases} p_{NS}^{ij} = p\delta_{ij} - \mu[\frac{\partial v^i}{\partial x^j} + \frac{\partial v^j}{\partial x^i} - \frac{2}{3}\sum_{k=1}^{3} \frac{\partial v^k}{\partial x^k}\delta_{ij}] - \mu_B \sum_{k=1}^{3} \frac{\partial v^k}{\partial x^k}\delta_{ij}, & \text{Navier-Stokes stress;} \\ \boldsymbol{q}_{NS} = \kappa \nabla_{\boldsymbol{x}} \theta, & \text{Navier-Stokes heat flux.} \end{cases}$$

Another basic element for the theory of the Boltzmann equation is the H-Theorem, by noticing that

$$\int_{\mathbb{R}^3} \log \mathsf{f} Q(\mathsf{f}, \mathsf{f}) d\xi = \frac{1}{4} \int_{\mathbb{R}^3} \int_{\mathbb{R}^3} \int_{S_+^2} \log \frac{\mathsf{f}\mathsf{f}_*}{\mathsf{f}'\mathsf{f}'_*} [\mathsf{f}'\mathsf{f}'_* - \mathsf{f}\mathsf{f}_*] B d\Omega d\xi_* d\xi \le 0,$$

$= 0$ if and only if $f = \mathsf{M}$ the Maxwellian, or the gas is in thermo-equilibrium $Q(\mathsf{M}, \mathsf{M}) = 0$:

$$\mathsf{f}(\boldsymbol{x},t,\boldsymbol{\xi}) = \frac{\rho(\boldsymbol{x},t)}{(2\pi R\theta(\boldsymbol{x},t))^{3/2}} e^{-\frac{|\xi-v(x,t)|^2}{2R\theta(x,t)}} \equiv \mathsf{M}_{[(\rho,v,\theta)]},$$

θ, the temperature, $p = R\rho\theta$. Integrating the Boltzmann equation times $\log f$ yields an inequality:

$$H_t + \nabla_x \mathbf{H} \leq 0,\ H \equiv \int f \log f \boldsymbol{\xi},\ \mathbf{H} \equiv \int f \boldsymbol{\xi} f \log f \boldsymbol{\xi}.$$

The H-Theorem shows that the Boltzmann equation is irreversible. It also indicates that a Boltzmann solution has the tendency to approach the 5-dimensional manifold of thermo-equilibrium states, in the space of functions in the microscopic variable $\boldsymbol{\xi}$.

19.4 Non-uniqueness

We now start with the historical perspective through the discussion in von Neumann Panel. The main topic is the following question raised by von Neumann in the opening address:

> von Neumann:
> Occasionally the simplest hydrodynamical problems have several solutions, some of which are very difficult to exclude on mathematical grounds only. For instance, a very simple hydrodynamical problem is that of the supersonic flow of a gas through a concave corner, which obviously leads to the appearance of a shock wave. In general, there are two different solutions with shock waves, and it is perfectly well known from experimentation that only one of the two, the weaker shock wave, occurs in nature. But I think that all stability arguments to prove that it must be so, are of very dubious quality.

For supersonic flight, to reduce the drag and the temperature, the tip of the flying object is made pointed. Prandtl uses shock polar analysis and found analytically that there are two possible shock reflections at the tip. Both satisfies the basic entropy condition from general theory of hyperbolic conservation laws, [3,5,11], in that the entropy increases across the shock. Both also satisfy the requirement that after the shock the flow is parallel to the flying object. The gas is subsonic behind the strong one and supersonic behind the weak one. As von Neumann just pointed out, it is the weak one that is observed in experiments. This non-uniqueness issue is puzzling.

Liepmann, of Liepmann and Roshko [14], pointed out that, from the theory of partial differential equations, the proper boundary condition depends on the type of the equation. In this case, the type of the stationary equations in turns depends on the solution.

> Liepmann:
> I would like to add a remark about the question of the two shock waves. I think that the experiments cannot be safely cited to settle whether only the solution with the weaker shock appears in nature, because the theoretical case refers to an infinite wall (or to the flow along the two sides of an infinite wedge), which case cannot be realized in practice. With the stronger one of the two shock waves you have subsonic flow behind the shock wave,

> which means that behind the shock wave you have a region where the theory of the elliptic differential equation applies and where the field is influenced by the boundary conditions at a finite or an infinite distance downstream. In the case of the other shock wave the velocity remains supersonic, so that you have conditions such as those obtained with hyperbolic equations. Thus one cannot exclude a priori that conditions downstream may influence the flow and thus may lead to a predilection for one type of shock wave about the other type.

The stationary equation is usually posted as a boundary value problem with boundary data given at infinity. Non-uniqueness of stationary solutions with given boundary condition is common in the theory of incompressible flows and also in elasticity. It is expected, though exact analysis of this for compressible flows has been shown only for the quasi-one dimensional nozzle flows, [15, 16]. It is clear that the resolution of the Prandtl non-uniqueness paradox cannot be done within the class of stationary solutions, see also [19]. On the other hand, the self-similar flow is equivalent to the initial value problem with self-similar initial data and boundary condition, and is expected to be unique. We will address the Prandtl paradox from this point of view.

19.5 Stationary Versus Time-Dependent

von Karman, the leader in fluid dynamics expressed a very basic point that one should consider only stationary solutions which can be realized as the large time state of a plausible time dependent process.

> von Karman:
> I would like to say something about this question of uniqueness of solutions. I don't think that there is any reason that if you put a problem in a form which has no physical meaning, there shall not be two solutions. And I think the case of stationary motion as such belongs to this category, because it can occur only as a limiting case. Any physical process starts from somewhere and goes to somewhere. In the case of the two shock waves, if instead of considering a stationary motion you consider an accelerated motion, you will first get a detached shock wave ahead of the obstacle (when the Mach number has just passed through unity). Then, with increasing velocity the solution will approach the correct solution for the steady case, I should think, without any difficulty. Such a case comes near to what you can actually realize in an experiment. Is that not correct?

von Neumann responded to this and von Karman further clarified his point.

> von Neumann:
> I may not have chosen that example which fits best to your argument. It has, of course, to be admitted that to postulate stationarity is to postulate a general trait of the solution one wants, which may hold only approximately in the physical situation that can actually be realized. However, it is not necessary to take the stationary flow through a corner. The following problem also has two solutions. If you take a plane shock which hits a wall and you consider the reflection of the shock from the wall, then under a wide variety of conditions (in fact, in most cases) there are two solutions. In this case stationarity has not been postulated.
> von Karman:
> I only mean the following thing. I suppose we start from a certain state of rest of the gas, which must be a solution of our equations. Then we change the conditions gradually and

follow the system step by step. I believe that in such a case you will always get a solution and only one solution. There is no proof that there is only one, but I believe it to be so. For, after all, a gas is a molecular system, which follows the general equations of classical mechanics. But if you take first an infinite cone, or an infinite wedge both of which are situations which can never be realized and furthermore you ask for a stationary solution; in such a case there is no reason why there should be only one solution. Since the equations are non-linear, you can often, without violating continuity, pass from one solution to another one by following an envelope, and in such a case you can scarcely find a mathematical reason why one solution should be preferred to the other. But if you start from an actually existing (observed) state and then determine the next phase, I believe you will find only one completely determined result.
Concerning Dr. von Neumann's example of the reflection of waves from a wall, I do not know the answer, but I believe that no case in which infinitely extending waves or walls are involved is really defined physically.

Burgers echoed this point:

Burgers:
Dr. von Neumann mentioned a case of nonstationary theory where you have also two solutions: a shock wave hitting a wall. But in the picture you gave (Figure 3) the wall was infinite, so that here again one must ask: How does the situation arise, when you have an actual, finite wall? It may be that you could treat the problem for an actual situation, in which a shock wave travelling in unlimited space reaches the edge of a wall (see Figure 4), you might obtain a definite solution.

von Neumann pointed out that the uniqueness of time dependent solution had not been shown and is still a hard question.

von Neumann:
In that case you assume that the state at the time $t = 0$ is given and you ask whether there is or is not a unique continuation of the solution at later times. The answer to this question in its full generality is not known; there seem to be a great many mathematical difficulties.

To realize von Karman's idea, one would need to study the solutions when the flying object is accelerated gradually and study its time-asymptotic state. This is a very hard problem, as many shock waves will be formed and to understand the interaction of multi-dimensional shocks with each other and with the flying object is a formidable task, way beyond the present techniques in mathematical analysis. Nevertheless, von Karman's point of the uniqueness of the time evolutionary solution should be taken for granted even though there is no proof of that, as also pointed out by von Neumann.

Before addressing this issue further, we digress to study what separates the stationary potential equation and the self-similar equation. As we have pointed out, the main different between the stationary potential equation and the self-similar is the presence of the lower order term $2c^2+(\chi_\xi)^2+(\chi_\eta)^2$ in the self-similar equation. This difference is exploited in the paper [7] showing, for any spatial dimension, the following

Ellipticity Principle: A pseudo-supersonic bubble cannot grow inside a pseudo-subsonic region through continuous variation of parameters.

It is well-known that a supersonic bubble around an airfoil usually begin to grow as the upstream Mach number increases from 0.6 to 0.7. As the theory for hyperbolic systems differs in major way from that for the elliptic equations, the Ellipticity Principle is a property of fundamental importance.

We now go back to the suggestion of von Karman. To make mathematical analysis possible and to address the main goal of studying the time-asymptotic states, we suppose that, instead of gradual acceleration, a wedge is instantaneously accelerated to a supersonic speed. Then the initial value is a constant state and is therefore self-similar. The boundary and boundary condition are also self-similar when considering the wedge. The potential flow equation clearly obeys self-similarity transformation $\boldsymbol{x} \to \alpha\boldsymbol{x}$, $t \to \alpha t$ for any positive α. Thus the solution is self-similar. For the Prandtl paradox, we then study the self-similar potential equation and show in [8] that for 2 spatial dimensions the following theorem.

Theorem 19.1. *For sufficiently pointed wedge, the self-similar solution contains a weak, supersonic shock attached to the edge of the wedge.*

The proof of the theorem uses the global method of Leray-Schauder degree and makes use of the Ellipticity Principle. Besides the weak shock at the tip of the wedge, the self-similar solution also contains a shock parallel to the wedge and a pseudo subsonic region in between. As time increases, the weak shock dominates and Prandtl paradox is resolved in this way.

It would be important to study the nonlinear stability of the weak shock reflection. This issue is subtle, as numerical computations give strong indication that both strong and weak shock reflections are stable upon compact supported perturbation. As also hinted by Liepmann, we expect the strong shock reflection to be unstable if the downstream state is also perturbed.

19.6 Kinetic Theory and Gas Dynamics

We now turn to the relation between the kinetic theory and the gas dynamics. As mentioned before, the H-Theorem indicates that a Boltzmann solution has the tendency to approach the 5-dimensional manifold of thermo-equilibrium states. For space homogeneous case

$$\mathsf{f}_t = \frac{1}{k} Q(\mathsf{f}, \mathsf{f}),$$

the H-Theorem implies that a solution f approaches the Maxwellian $\mathsf{M}_\mathsf{f} = \mathsf{M}_{[(\rho,\boldsymbol{v},\theta)]}$, with the time-invariant macroscopic variables calculated from f, e.g. $\rho = \int \mathsf{f} d\boldsymbol{\xi}$. Crucial to such a study is the strength of the H-Theorem in measuring the distance $|\mathsf{f} - \mathsf{M}_\mathsf{f}|$ in terms of $-\int \log \mathsf{f} Q(\mathsf{f}, \mathsf{f}) d\boldsymbol{\xi}$, the so-called Cercignani conjecture, [6]. For space inhomogeneous case, there are rich phenomena of fluid-like flows around the 5-dimensional equilibrium manifold.

Making the physically unrealistic thermo-equilibrium assumption, $\mathsf{f} = \mathsf{M}_\mathsf{f} = \mathsf{M}_{[(\rho(x,t),\boldsymbol{v}(x,t),\theta(x,t))]}$,, the conservation laws become the Euler equations in the gas dynamics. Nevertheless, the Euler equations can be directly related to the Boltzmann equation in the limit of $k \to 0$. This is because, for the Boltzmann equation

$$\mathsf{f}_t + \boldsymbol{\xi} \cdot \nabla_x \mathsf{f} = \frac{1}{k} Q(\mathsf{f}, \mathsf{f}),$$

as the Knudsen number $k \to 0$, one would expect the collision $Q(\mathsf{f}, \mathsf{f}) \to 0$. The H-Theorem would imply that the solution approaches local Maxwellians $\mathsf{f} \to M_{\mathsf{f}}$. Thus the Boltzmann equation is reduced to the Euler equations in the gas dynamics. One would conclude that as $k \to 0$, the Boltzmann solutions would approach the solutions of the Euler equations in gas dynamics. However, this can be true only outside the shock and initial layers. Thus this convergence is not expected to be everywhere, but almost everywhere:

OPEN PROBLEM: Show that the Boltzmann solution tends local Maxwellians almost everywhere and its macroscopic variables tend to a weak solution of the Euler equations in gas dynamics, as $k \to 0$.

This is a part of the Hilbert Sixth Problem in relating the kinetic theory with the fluid dynamics. This has been shown only when the limiting Euler solutions contain finite number of non-interacting shocks, [25, 26]. The analysis makes use of the stability theory of nonlinear waves for conservation laws, e.g. [22, 24]. There is the general existence theory of Glimm [9] for one spatial dimension hyperbolic conservation laws, including the Euler equations in the gas dynamics. A problem parallel to the zero mean free path problem for the Boltzmann equation is to consider the zero dissipation limit of viscous conservation laws. This has been done for artificial viscosity, [1],

$$\mathsf{u}_t + \mathsf{f}(\mathsf{u})_x = \epsilon \mathsf{u}_{xx}, \ \epsilon \to 0,$$

generalizing the existence theory of Glimm. The zero dissipation theory has not been generalized to the case of physical viscosity such as the case of Navier-Stokes equations in gas dynamics. Write the Boltzmann solution as $\mathsf{f} = \mathsf{M}_\mathsf{f} + \mathsf{G}_\mathsf{f}$. The Navier-Stokes equations are derived through simplification

$$\nabla \mathsf{G}_\mathsf{f} \ll \mathsf{G}_\mathsf{f} \ll \mathsf{M}_\mathsf{f}.$$

This is the Chapman-Enskog expansion assuming that the flow is near equilibrium and slow varying. As the Boltzmann equation is dissipative, the Navier-Stokes equations in gas dynamics can be justified time-asymptotically. Thus this is so time-asymptotically for the perturbation of a global Maxwellian, [20], or for the perturbation of rarefaction waves, [18].

Depending on the Mach, Raynolds and Knudsen numbers, other fluid dynamics equations, e.g. Stokes, Navier-Stokes, and other fluid dynamics equations are suitable to describe the gas outside the boundary layer. For some weak perturbation of a global Maxwellian, the zero mean free path, time-asymptotic limit would yield the incompressible fluid equations. By sophisticated entropy and other functional analytic methods, the Diperna-Lions weak solutions of the Boltzmann equation are shown to converge to the weak solutions of the incompressible Navier-Stokes equations, [10].

19.7 Boundary

The kinetic theory differs from the fluid dynamics mainly on the formulation of the boundary condition and the boundary effects. In the panel discussion, Heisenberg raised the important point that gas dynamics equations may not be valid near vacuum and around boundary. Near vacuum, there are not enough collision to drive the gas toward equilibrium and the fluid dynamics equations may not hold. Even though the kinetic boundary layer may be very thin, the study of the kinetic equation is needed there to provide proper boundary condition for the gas dynamics equations.

Heisenberg:

I have one question in connection with these applications of the hydrodynamical equations. Should one assume from the beginning that these equations actually could be used to such a large extent? if we take the case of the gas expanding into a vacuum, the density at the front is so low that the mean free path becomes larger than the distance to the assumed front. Should one not start from the kinetic picture and say that at the front the molecules will sort themselves out according to their velocities? Then the physical front would be formed by a selection of those molecules which had the highest velocities and did not suffer a collision for a long time. One should expect that there, especially, we have a velocity distribution different from the normal one, and therefore we should not apply the ordinary concepts like temperature and so on. I do not know how big the actual difference is, but I have tried to estimate it. One feels at least that there is a rather large region in which ordinary hydrodynamics cannot be applied, simply because the concepts of temperature and so on would be rather useless.

von Neumann:

Therefore, while it is certainly not rigorously true, don't you think it is sensible, first of all, to apply hydrodynamic theory, and get a solution? If you then discuss in what portions of the field the mean free path is small compared to the distances over which all essential changes occur (one of the most important portions is that where the distance from the boundary is small), it is reasonable to assume that the hydrodynamical equations may at least be used in such regions. When one has to deal with the boundary regions, the Maxwell-Boltzmann theory should be called upon. Now what I have to say is that if one accepts this, and if one estimates how large these extraordinary regions are, in the cases which are of interest in the present context, they turn out to be fairly small. Properly speaking, in the case of the Riemann expansion into vacuum, the region where you have to be careful is quite large but it involves very little substance and very little energy. Hence, in many cases, the correction of the hydrodynamical solution in that region need not be discussed.

Heisenberg:

I certainly agree chiefly with what you say. I only would like to observe that the failures of hydrodynamical solutions determine the boundary conditions. The boundary conditions react back on the solutions of the hydrodynamic equations, and since these boundary conditions cannot be determined from hydrodynamics and require a detailed study of molecular processes, the two things are interconnected. With you, I believe that on the whole we can talk about hydrodynamical equations and their solutions, but the selection of the solutions to be used depends on the boundary conditions and to this extent we get these non-hydrodynamical parts of the field into our problem.

von Neumann:

The boundary layer theory for a fluid of low viscosity certainly furnishes a monumental warning. The naive and yet prima facie seemingly reasonable procedure would be to apply the ordinary equations of the ideal fluid and then to expect that viscosity will somehow take care of itself in a narrow region along the wall. We have learned that this procedure may lead to great errors; a complete theory of the boundary layer may give you completely different conditions also for the flow in the bulk of the field. It is possible that the same discipline will be necessary for the boundary with a vacuum. All I would like to say now is that there is yet no evidence for this.

There have been very substantial progresses since the Panel Discussion on the boundary issue, cf. [23]. A boundary layer contains discontinuous, Knudsen, and viscous layers. An interesting phenomenon is the striking bifurcations for transonic evaporation/condensation. There is now an exact mathematical theory for this, [21]. In gas dynamics, there are several well-known bifurcation phenomena, e.g. regular versus Mach reflections, attached versus detached shocks etc. Experimental evidences seem to indicate no clean criterion for the transitions. Boundary effects may be strong for the consideration. For recent results on exact mathematical analysis on the shock reflections see [8] and [2].

About the vacuum, the gas dynamics equations may not hold as discussed above. Nevertheless, von Neumann made the good point that it is worthwhile to study the vacuum with gas dynamics equations. There have been progresses on this, e.g. [4, 17].

19.8 Conclusions

We conclude by few remarks on the derivation of the Boltzmann equation from Newtonian particle systems. This is an issue of historical and philosophical importance, as the Newtonian system is time reversible and the Boltzmann equation, with its H-Theorem, is irreversible. The fluid dynamics and the classical thermodynamics can be derived from the Boltzmann equation. There are three basic elements in this derivation: First, it is to describe the gas dynamics phenomena starting from the physics on the molecular level. Second, the mean free path, in the gas dynamics scale, is fixed in the limit of infinite molecules. Then the molecular chaos hypothesis, also in the gas dynamics scale, is valid in the limit for later time when assumed at the initial time. The derivation usually starts with the Liouville equation, [12], $\mathfrak{f}^N(\boldsymbol{x}^1, \cdots, \boldsymbol{x}^N, \boldsymbol{\xi}^1, \cdots, \boldsymbol{\xi}^N, t)$ distribution function for the j-th particle at $(\boldsymbol{x}^j, \boldsymbol{\xi}^j)$, $\boldsymbol{F} J$ force on the j-th particle:

$$\partial_t \mathfrak{f}^N + \sum_{j=1}^{N} \boldsymbol{\xi}^j \partial_{\boldsymbol{x}^j} \mathfrak{f}^N + \sum_{j=1}^{N} \boldsymbol{F}^j \cdot \partial_{\boldsymbol{x}^j} \mathfrak{f}^N = 0.$$

Landford [13] shows that the molecular chaos hypothesis propagates for $1/5$ mean free time. There are several important open problems. One is to extend Landford's result to later times. Another is to derive the Boltzmann equation, in the spirit of Grad [12] for infinite range inter-molecular potentials. It would be desirable to start the derivation with the deterministic, classical Newtonian particle system, and not the probabilistic Liouville equation.

Acknowledgements Part of the article was presented in INdAm Day Lecture on June 7, 2012 in Genova. The author would like to thank Professor Tommaso Ruggeri for the opportunity. The research is supported in part by Investigator Award of Academia Sinica and National Science Council Grant 96-2628-M-001-011.

References

1. S. Bianchini, A. Bressan, Vanishing viscosity solutions of nonlinear hyperbolic systems. Ann. Math. **161**, 223–242 (2005)
2. G.-Q. Chen, M. Feldman, Global solutions of shock reflection by large-angle wedges for potential flow. Ann. Math. **171**(2), 1067–1182 (2010)
3. R. Courant, K.O. Friedrichs, *Supersonic Flow and Shock Waves* (Wiley-Interscience, New York, 1948)
4. D. Coutandd, S. Shkoller, Well-posedness in smooth function spaces for the moving-boundary 3-D compressible Euler equations in physical vacuum. Arch. Ration. Mech. Anal. **206**, 515–616 (2012)
5. C. Dafermos, *Hyperbolic Conservation Laws in Continuum Physics* (Springer, Berlin/New York, 1999)
6. L. Desvillettes, C. Villani, On the trend to global equilibrium for spatially inhomogeneous kinetic systems: the Boltzmann equation. Invet. Math. **159**, 245–316 (2005)
7. V. Elling, T.-P. Liu, The ellipticity principle for self-similar potential flows. J. Hyperbolic Differ. Equ. **2**(4), 909–917 (2005)
8. V. Elling, T.-P. Liu, Supersonic flow onto a solid wedge. Commun. Pure Appl. math. **61**(10), 1347–1448 (2008)
9. J. Glimm, Solutions in the large for nonlinear hyperbolic systems of equations. Commun. Pure Appl. Math. **41**, 697–715 (1965)
10. F. Golse, L. Saint-Raymond, The Navier-Stokes limit of the Boltzmann equation for bounded collision kernels. Invent. Math. **155**, 81–161 (2004)

11. P.D. Lax, Hyperbolic systems of conservation laws, II. Commun. Pure Appl. Math. **10**, 537–566 (1957)
12. H. Grad, Principle of the kinetic theory of gases, in *Thermodynamics of Gases*, ed. by S. Flügge (Springer, Berlin/Heidelberg, 1958)
13. O. Lanford, The evolution of large classical system, in *Dynamical Systems, Theory and Applications*, ed. by J. Moser. Lecture Notes in Physics (Springer, Heidleberg, 1975), pp. 1–111
14. H.W. Liepmann, A. Roshko, *Elements of Gasdynamics* (Wiley, New York, 1957)
15. T.-P. Liu, Transonic gas flow in a duct of varying area. Arch. Ration. Mech. Anal. **80**, 1–18 (1982)
16. T.-P. Liu, Nonlinear stability and instability of transonic flows through a nozzle. Commun. Math. Phys. **83**, 243–260 (1982)
17. T.-P. Liu, Z. Xin, T. Yang, Vacuum states for compressible flow. Discret. Contin. Dyn. Syst. **4**, 1–32 (1998)
18. T.-P. Liu, T. Yang, S.-H. Yu, H.-J. Zhao, Nonlinear stability of rarefaction waves for the Boltzmann equation. Arch. Ration. Mech. Anal. **181**, 333–371 (2006)
19. T.-P. Liu, Multi-diemnstional gas flow: some historical perspectives. Bull. Inst. Math. Acad. Sin. (N.S.) **6**, 269–291 (2011)
20. T.-P. Liu, S.-H. Yu, Solving Boltzmann equation, Part I: Green's function. Bull. Inst. Math. Acad. Sin. (N.S.) **6**(1), 115–243 (2011)
21. T.-P. Liu, S.-H. Yu, Invariant manifolds for steady Boltzmann flows and applications. Arch. Ration. Mech. Anal. **207**, 869–997 (2013)
22. T.-P. Liu, Y. Zeng, Shock Waves in Conservation laws with physical viscosity, memoirs. Am. Math. Soc. (to appear)
23. Y. Sone, *Molecular Gas Dynamics: Theory, Techniques, and Applications* (Birkhauser, Boston, 2007)
24. S.-H. Yu, Nonlinear wave propagation over a Boltzmann shock profile. J. Am. Math. Soc. **23**, 1041–1118 (2010)
25. S.-H. Yu, Hydrodynamics limits with shock waves of the Boltzmann equaiton. Commun. Pure Appl. Math. **58**(3), 409–443 (2005)
26. S.-H. Yu, Initial and shock layers for Boltzmann equation. Arch. Ration. Mech. Anal. **211**, 1–60 (2014)

Chapter 20
Recent Advances in Nonlinear Potential Theory

Giuseppe Mingione

Abstract Recent developments in Nonlinear Potential Theory show the possibility of estimating solutions to nonlinear degenerate equations via potentials. We give a brief description of these results.

20.1 Pointwise Estimates

The classical potential theory deals with the fine properties – including regularity – of harmonic functions and, more in general, of solutions to linear elliptic equations. In this case, a central tool is given by Riesz potentials, defined for $\alpha > 0$ as

$$I_\alpha(\mu)(x) := \int_{\mathbb{R}^n} \frac{d\mu(y)}{|x-y|^{n-\alpha}} ,$$

where μ is a Borel (signed) measure defined on $\mathbb{R}^n$, whenever $x \in \mathbb{R}^n$. When dealing with estimates in bounded domains, it is also useful to deal with their "truncated versions", which are defined by

$$\mathbf{I}^\mu_\beta(x,R) := \int_0^R \frac{|\mu|(B(x,\varrho))}{\varrho^{n-\beta}}\,\frac{d\varrho}{\varrho}, \qquad \beta > 0,$$

whenever $x \in \mathbb{R}^n$ and $0 < R \le \infty$. The classical representation formulas via convolution with the fundamental solution

G. Mingione (✉)
Dipartimento di Matematica e Informatica, Università di Parma, Parco Area delle Scienze 53/a, Campus, 43124 Parma, Italy
e-mail: giuseppe.mingione@math.unipr.it

V. Ancona and E. Strickland (eds.), *Trends in Contemporary Mathematics*, Springer INdAM Series 8, DOI 10.1007/978-3-319-05254-0_20,

$$u(x) = \int_{\mathbb{R}^n} G(x,y)\, d\mu(y)\,, \qquad G(x,y) := |x-y|^{2-n} \tag{20.1}$$

allow to reconstruct pointwise properties of solutions to equations as

$$-\Delta u = \mu \qquad \text{in } \mathbb{R}^n \tag{20.2}$$

via potentials; here we confine to the case $n > 2$ for simplicity. For instance the following inequalities hold:

$$|u(x)| \lesssim I_2(|\mu|)(x)| \qquad \text{and} \qquad |Du(x)| \lesssim I_1(|\mu|)(x)\,. \tag{20.3}$$

The formulas in (20.3) together with (20.1), allow to address issues as

- Study of the optimal regularity properties of solutions to (20.2) with respect to the regularity of the datum μ; for example in various function spaces.
- Study of the fine properties of solutions: sets of Lebesgue points of solutions and gradients, removability of singularities etc.
- Various convergence properties of sequences of solutions

The main advantage of this approach is that, ultimately, it allows to unrelate the solution u from the equation, and to perform the analysis only looking at the potential of μ. Needless to say, via suitable localisation arguments, the same approach extends to other, more general linear equations for a formula as (20.1) holds. Now, although all this seems at a first sight to be peculiar of the linear situation, being linked to the existence of fundamental solutions, recent developments from the last years have shown that a similar approach can be pursued in several nonlinear situations too. This is the main object of this note, which is an extended written version of the lecture delivered by the author and Indam Day on June 7, 2012, in Genoa. A larger and more comprehensive presentation, with proofs, can be found in the Guide [32].

20.1.1 Some Notation

Constants generically denoted by c are always larger or equal than one; relevant dependence on parameters is indicated using parenthesis. We denote by $B(x,r) \equiv B_r(x) := \{\tilde{x} \in \mathbb{R}^n : |\tilde{x} - x| < r\}$ the open ball with center x and radius $r > 0$. When not important we shall omit denoting the center as follows: $B_r \equiv B(x,r)$. Moreover, with B being a generic ball with radius r, we will denote by σB the ball concentric to B having radius σr, $\sigma > 0$. With $\mathcal{O} \subset \mathbb{R}^n$ being a measurable subset with positive measure, and with $g\colon \mathcal{O} \to \mathbb{R}^k$, $k \geq 1$, being a measurable map, we shall denote by

$$(g)_{\mathcal{O}} \equiv \fint_{\mathcal{O}} g\, d\tilde{x} := \frac{1}{|\mathcal{O}|} \int_{\mathcal{O}} g(x)\, dx$$

its integral average; here $|\mathcal{O}|$ denotes the Lebesgue measure of $\mathcal{O}$. In the following, μ will always denote a Borel measure with finite total mass, which is initially defined on a certain open subset $\Omega \subset \mathbb{R}^n$. Since this will not affect the rest, with no loss of generality, all such measures will be considered as defined in the whole $\mathbb{R}^n$ so that $|\mu|(\mathbb{R}^n) < \infty$.

20.2 Nonlinear Potential Estimates

We shall treat nonlinear elliptic equations of divergence form of the type

$$- \operatorname{div} a(Du) = \mu \qquad \text{in } \Omega \tag{20.4}$$

and sometimes also having measurable coefficients as

$$- \operatorname{div} a(x, Du) = \mu\,. \tag{20.5}$$

Here $\Omega \subset \mathbb{R}^n$ denotes an open subset, and we shall always consider the case $n \geq 2$. When considering equations as in (20.4) we shall assume the following growth and ellipticity assumptions on the C^1-regular vector field $a\colon \mathbb{R}^n \to \mathbb{R}^n$:

$$\begin{cases} |a(z)| + |\partial a(z)|(|z|^2 + s^2)^{1/2} \leq L(|z|^2 + s^2)^{(p-1)/2} \\ \nu(|z|^2 + s^2)^{(p-2)/2}|\lambda|^2 \leq \langle \partial a(z)\lambda, \lambda\rangle \\ p \geq 2 \end{cases} \tag{20.6}$$

whenever $z, \lambda \in \mathbb{R}^n$, where $s \geq 0$ and $0 < \nu < L$. The restriction to the case $p \geq 2$ is done for sake of simplicity; we shall make a few remarks on the case $p < 2$ later. When instead dealing with equations as in (20.5) we assume the weaker growth and monotonicity conditions on the Carathéodory vector field $a\colon \Omega \times \mathbb{R}^n \to \mathbb{R}^n$:

$$\begin{cases} |a(x, z)| \leq L(|z|^2 + s^2)^{(p-1)/2} \\ \nu(|z_1|^2 + |z_2|^2 + s^2)^{(p-2)/2}|z_1 - z_2|^2 \leq \langle a(x, z_1) - a(x, z_2), z_1 - z_2\rangle \\ p \geq 2 \end{cases} \tag{20.7}$$

satisfied whenever $z_1, z_2, x \in \Omega$. By taking $s = 0$, both assumptions (20.6) and (20.7) are satisfied by the p-Laplacean equation

$$- \Delta_p u := -\operatorname{div}(|Du|^{p-2}Du) = \mu \tag{20.8}$$

which is, when $p > 2$, the most prominent model example for us; see [35].

The assumptions in (20.6) and (20.7) settle the Sobolev space $W^{1,p}(\Omega)$ as the natural one in which considering our equations; the situation is anyway not exactly so. It is indeed easy to see that if $u \in W^{1,p}(\Omega)$ is a local solution to (20.4) then μ belongs to the dual space of $W^{1,p}(\Omega)$, while on the other hand it is possible to consider distributional solutions to (20.4) also in the case μ is a more general Borel measure. These solutions do not in general belong to $W^{1,p}(\Omega)$ and are therefore called very weak solutions. In turn, very weak solutions are too general to be considered, and therefore, also keeping in mind the various methods for solving the existence problems, one is led to consider SOLA (Solutions Obtained as Limits of Approximations), according to the work of [4]. Here are the precise definitions, referring to [39] for more details.

Definition 20.1. A function $u \in W^{1,1}_{\mathrm{loc}}(\Omega)$ is called a very weak solution to Eq. (20.5) in Ω if solves (20.5) in the sense of distributions, and if $a(x, Du) \in L^1_{\mathrm{loc}}(\Omega, \mathbb{R}^n)$. A function $u \in W^{1,1}_{\mathrm{loc}}(\Omega)$ is a SOLA to Eq. (20.5) under assumptions (20.7), iff is a very weak solution and there exists a sequence of local energy solutions $\{u_k\} \subset W^{1,p}_{\mathrm{loc}}(\Omega)$ to the equations $-\mathrm{div}\, a(x, Du_k) = \mu_k$, such that $u_k \to u$ locally in $W^{1,p-1}(\Omega)$.

Our aim here is to present sharp nonlinear analogs of the estimates (20.3). These lie at the core of what is nowadays called Nonlinear Potential Theory, a field whose name stems from the seminal papers of Havin and Maz'ya [16, 17]. In these papers the authors laid the fundamentals of this field and, in particular, introduced and studied two nonlinear potentials (see also [2]). The first is called Havin-Maz'ya potential and is defined via a nonlinear iteration of Riesz potentials:

$$\mathbf{V}_{\beta,p}(|\mu|)(x) := I_\beta \left\{ \left[I_\beta(|\mu|) \right]^{1/(p-1)} \right\}(x).$$

The second potential, which is nowadays called Wolff potential, is in fact connected to the 1970 breakthrough paper of Maz'ya [36] about the validity of the Wiener criterion for the p-Laplacean; it is defined as follows:

$$\mathbf{W}^{\mu}_{\beta,p}(x, R) := \int_0^R \left(\frac{|\mu|(B(x,\varrho))}{\varrho^{n-\beta p}} \right)^{1/(p-1)} \frac{d\varrho}{\varrho}, \qquad \beta > 0$$

whenever $x \in \mathbb{R}^n$ and $0 < R \leq \infty$. Notice that Wolff potentials have been obtained from Riesz potentials by inserting the scaling exponent of Eq. (20.8) under the sign of integral. The three classes of potentials – Riesz, Havin-Maz'ya and Wolff – coincide when $p = 2$. A fundamental result of Havin and Maz'ya [17] asserts that Wolff potentials can be controlled by Riesz potentials $\mathbf{W}^{\mu}_{\beta,p}(x, \infty) \lesssim \mathbf{V}_{\beta,p}(|\mu|)(x)$, in the case $p > 2 - 1/n$ and $p\beta < n$; a related integral inequality for the case $1 < p \leq 2 - 1/n$ is contained in the paper by Hedberg and Wolff [18]. The result of Havin and Maz'ya allows to reduce the study of Wolff potentials to that of Riesz potentials in many important situations. For instance, when studying the mapping properties of Wolff potentials in various rearrangement invariant function spaces [6]. Wolff potentials are nowadays a common tool in order to face several basic questions of Nonlinear Potential Theory; see for instance [19, 40].

The first result towards a nonlinear extension of formulas (20.3) appeared in the fundamental papers of Kilpeläinen and Malý [20, 21], where the authors proved a pointwise potential estimate for solutions in terms of Wolff potentials. Subsequently, different proofs and approaches have been found. We summarise some of the results available in the following:

Theorem 20.1. *Let $u \in W^{1,p-1}(\Omega)$ be a SOLA to the equation with measurable coefficients* (20.5), *under the assumptions* (20.7). *Then*

- [14,21,22,42] *There exists a constant $c \equiv c(n,p,\nu,L)$ such that the inequality*

$$|u(x)| \leq c\mathbf{W}^{\mu}_{1,p}(x,R) + c\fint_{B(x,R)} (|u| + Rs)\, d\tilde{x} \tag{20.9}$$

  holds whenever $B(x,R) \subset \Omega$ and the right hand side is finite.
- [14,21,22,42] *Moreover, if*

$$\lim_{R\to 0} \mathbf{W}^{\mu}_{1,p}(x,R) = 0 \ \textit{ locally uniformly in } \Omega \textit{ w.r.t. } x$$

  then u is continuous in Ω.
- [32] *Finally, the condition $\mathbf{W}^{\mu}_{1,p}(x,R) < \infty$ implies that the following limit exists and therefore defines the precise representative of u at the point x:*

$$\lim_{\varrho\to 0} (u)_{B(x,\varrho)} =: u(x)\,. \tag{20.10}$$

  In particular, the set of non-Lebesgue points of u has p-capacity zero.

Notice that in Theorem 20.1, by writing $u(x)$ for the precise representative of u at x, the set of points x for which (20.9) holds in the sense that $u(x)$ exists is in fact the one for which $\mathbf{W}^{\mu}_{1,p}(x,R) < \infty$ and (20.9) itself becomes nontrivial. This follows from (20.10). We explicitly remark that the full strength of the previous result is already in the case $p = 2$. Indeed, the real point here is passing from linear to nonlinear equations, because fundamental solutions and representation formulas are not available in the nonlinear setting. We also remark that estimate (20.9) is optimal in that when $\mu, u \geq 0$ then it holds that (see [21])

$$\mathbf{W}^{\mu}_{1,p}(x,R) \lesssim u(x) \lesssim \mathbf{W}^{\mu}_{1,p}(x,2R) + \inf_{B(x,R)} u\,.$$

The validity of a nonlinear analog of the second inequality in (20.3) has remained a controversial open problem since the paper of Kilpeläinen and Malý [21]. The first result has been obtained in [38]. We report a global version aimed at highlighting the similarities with the linear case.

Theorem 20.2 ([38]). *Let $u \in W^{1,1}(\mathbb{R}^n)$ be a SOLA to Eq.* (20.4) *considered in $\mathbb{R}^n$, under the assumptions* (20.6) *with $p = 2$. Then there exists a constant $c \equiv c(n, \nu, L)$ such that the following estimate holds for a.e. $x \in \mathbb{R}^n$:*

$$|Du(x)| \leq c \int_{\mathbb{R}^n} \frac{d|\mu|(y)}{|x-y|^{n-1}}\,.$$

The degenerate case $p > 2$ is intriguing and reserves surprises. The standard orthodoxy in Nonlinear Potential Theory prescribes to replace Riesz potentials with Wolff potentials whenever one is dealing with the p-Laplacean, and indeed a first gradient potential estimate using Wolff potentials has been obtained in [14]. On the other hand let us observe that Eq. (20.8) can be formally decomposed as

$$\begin{cases} -\operatorname{div} v = 0 \\ v = |Du|^{p-2}Du\,. \end{cases}$$

In other words, Eq. (20.8) can be read both as a nonlinear equation in the gradient and as a linear equation with respect to the nonlinear vector field of the gradient v. This paves the way to an estimate that involves linear Riesz potentials, and that improves the ones involving Wolff potentials, as those in [14]. It indeed holds the following, surprising:

Theorem 20.3. *Let $u \in W^{1,p-1}(\Omega)$ be a SOLA to Eq.* (20.4) *under the assumptions* (20.6). *Then*

- [23, 27] *There exists a constant $c \equiv c(n, p, \nu, L)$ such that the inequality*

$$|Du(x)|^{p-1} \leq c\mathbf{I}_1^{\mu}(x,R) + c\left(\fint_{B(x,R)} (|Du|+s)\,d\tilde{x}\right)^{p-1} \tag{20.11}$$

  holds whenever $B(x,R) \subset \Omega$ and the right hand side is finite.
- [23, 27] *Moreover, if*

$$\lim_{R\to 0} \mathbf{I}_1^{\mu}(x,R) = 0 \quad \textit{locally uniformly in } \Omega \textit{ w.r.t. } x \tag{20.12}$$

  then Du is continuous in Ω.
- [32] *Finally, the condition $\mathbf{I}_1^{\mu}(x,R) < \infty$ implies that the following limit exists and therefore defines the precise representative of Du at the point x:*

$$\lim_{\varrho\to 0} (Du)_{B(x,\varrho)} =: Du(x)\,.$$

  In particular, the Hausdorff dimension of the set of non-Lebesgue points of Du does not exceed $n-1$.

Estimate (20.11) still holds when $2 - 1/n < p < 2$ [13], but in this case it does not improve the analogous estimates via Wolff potentials previously proved in [14]. The main outcome of Theorem 20.3 is that the gradient theory of general quasilinear, possibly degenerate equations with measure data, is now reduced to the one of the Poisson equation, up to the C^0-level. In particular, essentially all the classical estimates known for the model equation (20.8), plus more difficult borderline cases, now follow with a unified approach via Riesz potentials as in the case of Poisson equation (20.2). It is important to observe that Theorem 20.3 cannot hold for solutions to equations with measurable coefficients as in (20.5), since in this case the gradients of solutions are only known to be in L^q for some $q > p$ (Gerhing's lemma). In this situation it is possible to prove a few nonlocal estimates for the level sets of the gradient using the level sets of the Riesz potential, as shown in [37]. In particular, in this last paper we have given a nonlinear analog of the basic theorems of Adams [1] valid for the Poisson equation. Remarkable extensions of estimate (20.11) have been recently provided by Baroni [3].

It is worth mentioning a few relevant corollaries of the continuity criterion in (20.12). It is possible to obtain a full nonlinear analog of a famous theorem of Stein [41] which is in fact the limit case of Sobolev embedding theorem. This claims that if $v \in W^{1,1}$ is a Sobolev function defined in $\mathbb{R}^n$ with $n \geq 2$, then $Dv \in L(n,1)$ implies that v is continuous. We recall that the Lorentz space $L(n,1)$ (over a subset Ω) is defined as the set of measurable maps $g\colon \Omega \to \mathbb{R}^k$ such that

$$\int_0^\infty |\{x \in \Omega : |g(x)| > \lambda\}|^{1/n} \, d\lambda < \infty .$$

Another way to state Stein's theorem concerns the regularity of solutions $u\colon \Omega \to \mathbb{R}^m$ to the Laplacean system and amounts to observe that $\Delta u \in L(n,1)$ implies the continuity of Du. This follows by the previous result and classical Calderón-Zygmund theory. The condition $\mu \in L(n,1)$ allows to satisfy condition (20.12) and therefore we conclude with the following:

Theorem 20.4 (Nonlinear Stein theorem [27]). *Let $u \in W^{1,p}(\Omega)$ be a solution to Eq.* (20.4)*, under the assumptions* (20.6) *and such that $\mu \in L(n,1)$ locally in Ω. Then Du is continuous in Ω.*

Without appealing to potentials, but by using different means, the result of the previous theorem also holds for systems.

Theorem 20.5 (Vectorial nonlinear Stein theorem [33]). *Let $u \in W^{1,p}(\Omega, \mathbb{R}^m)$, $m \geq 1$, be a vector valued solution to the p-Laplacean system $-\Delta_p u = F$, with $p > 1$. Assume that the components of the vector field $F\colon \Omega \to \mathbb{R}^m$ locally belong to the space $L(n,1)$. Then Du is continuous in Ω.*

We refer to [7, 8] for a global Lipschitz continuity result involving the Lorentz space $L(n,1)$ and the p-Laplacean system, while a Lipschitz local result has been obtained in [12].

20.3 Universal Potential Estimates

Theorems 20.1–20.3 allow to estimate the size of solutions and their gradients via potentials. The idea is now to use potentials to estimate also derivatives of intermediate order i.e. fractional derivatives. This will in turn allow to give bounds for the oscillation of solutions via potentials. Before going on, we give a suitable definition of fractional derivatives, using the ideas of DeVore and Sharpley [9].

Definition 20.2 (Calderón spaces [9]). Let $\alpha \in (0,1]$, $q \geq 1$, and let $\Omega \subset \mathbb{R}^n$ be an open subset. A measurable function v, finite a.e. in Ω, belongs to the Calderón space $C^{\alpha}_q(\Omega)$ if and only if there exists a nonnegative function $m \in L^q(\Omega)$ such that

$$|v(x) - v(y)| \leq [m(x) + m(y)]|x - y|^{\alpha} \tag{20.13}$$

holds for almost every couple $(x, y) \in \Omega \times \Omega$.

Calderón spaces C^{α}_q are strictly related to classical fractional Sobolev spaces $W^{\alpha,q}$, and in this setting $m(\cdot)$ represents a α-fractional derivative, in L^q-sense, of v. The advantage is that the typical nonlocal character of fractional derivatives is reduced to a minimal status: only two points are considered in (20.13). There is always a canonical choice for the function $m(\cdot)$ in (20.13); for this we need another definition.

Definition 20.3 (Fractional sharp maximal operator). Let $\beta \in [0,1]$, $x \in \Omega$ and $R \leq \operatorname{dist}(x, \partial\Omega)$, and let $f \in L^1(\Omega)$; the function defined by

$$M^{\#}_{\beta,R}(f)(x) := \sup_{0<r\leq R} r^{-\beta} \fint_{B(x,r)} |f - (f)_{B(x,r)}| \, d\tilde{x}$$

is called the restricted (centered) sharp fractional maximal function of f.

The connection with the fractional derivatives is then given by the following result, that relies on the original methods of Campanato [5]:

Proposition 20.1 ([9,32]). *Let $f \in L^1(B_{8R/5})$; for every $\alpha \in (0,1]$ the inequality*

$$|f(x) - f(y)| \leq \frac{c}{\alpha} \left[M^{\#}_{\alpha,R}(f)(x) + M^{\#}_{\alpha,R}(f)(y) \right] |x - y|^{\alpha} \tag{20.14}$$

holds whenever $x, y \in B_{2R/5}$, for a constant c depending only on n. More precisely, x and y are Lebesgue points of f whenever $M^{\#}_{\alpha,R}(f)(x)$ and $M^{\#}_{\alpha,R}(f)(y)$ are finite, respectively. Therefore, whenever the right hand side in (20.14) *is finite, the values of f are defined as follows:*

$$f(x) := \lim_{\varrho \to 0} (f)_{B(x,\varrho)} \qquad \text{and} \qquad f(y) := \lim_{\varrho \to 0} (f)_{B(y,\varrho)} \,.$$

The previous result tells that, in order to give a bound for the fractional derivatives of a function v – in the sense of Definition 20.2 – it is sufficient to give a bound for the fractional maximal operator of v. When considering solutions to nonlinear equations we indeed have

Theorem 20.6 (Uniform maximal-potential estimates [32]). *Let $u \in W^{1,p-1}(\Omega)$ be a SOLA to Eq.* (20.4) *under assumptions* (20.6). *Then for every ball $B(x,R) \subset \Omega$ the following estimate:*

$$M^{\#}_{\alpha,R}(u)(x) \leq c\left[\mathbf{I}^{\mu}_{p-\alpha(p-1)}(x,R)\right]^{1/(p-1)} + cR^{1-\alpha}\fint_{B_R}(|Du|+s)\,d\tilde{x} \qquad (20.15)$$

holds uniformly in $\alpha \in [0,1]$, with $c \equiv c(n,p,\nu,L)$.

Applying estimate (20.14) together with (20.15) yields a pointwise estimate on the oscillations of solutions to (20.4), that is

Theorem 20.7 (Uniform Riesz potential estimate [32]). *Let $u \in W^{1,p-1}(\Omega)$ be a SOLA to Eq.* (20.4) *under assumptions* (20.6). *Let $B_R \subset \Omega$ be such that $x, y \in B_{R/4}$; then*

$$\begin{aligned}|u(x)-u(y)| \leq c&\left[\mathbf{I}^{\mu}_{p-\alpha(p-1)}(x,R)+\mathbf{I}^{\mu}_{p-\alpha(p-1)}(y,R)\right]^{1/(p-1)}|x-y|^{\alpha}\\ &+c\fint_{B_R}(|u|+Rs)\,d\tilde{x}\cdot\left(\frac{|x-y|}{R}\right)^{\alpha} \qquad (20.16)\end{aligned}$$

holds provided the right hand side is finite and $0 < \alpha \leq 1$. Moreover, whenever $\tilde{\alpha} \in (0,1]$ is fixed, the dependence of the constant c is uniform for $\alpha \in [\tilde{\alpha},1]$ as c depends only n, p, ν, L and $\tilde{\alpha}$.

We note that the previous estimate gives back (20.11) when $\alpha = 1$, and extends it in the whole range of differentiability $\alpha \in (0,1]$. In view of Definition 20.2 estimate (20.16) can be interpreted, with a strong abuse of notation, as

$$|\partial^{\alpha}u(x)|^{p-1} \lesssim I_{p-\alpha(p-1)}(|\mu|)(x)\,, \qquad 0<\alpha\leq 1\,,$$

a formula that, needless to say, has only a symbolic meaning. The case $\alpha = 0$ is not included in Theorem 20.7, and it cannot, as when $\alpha = 0$ Wolff potentials come into the play. As a matter of fact the validity of (20.16) would ultimately contradict the optimality of (20.9). On the other hand it is also possible to produce a formula to estimate oscillations of solutions via Wolff potentials when α is not very large. Actually, we are going to give a nonlinear potentials formulation of the classical DeGiorgi's theory for solutions to (20.5) in the homogeneous case $\mu = 0$. This theory provides the existence of a *universal Hölder continuity exponent* $\alpha_m \in (0,1)$, depending only on n, p, ν, L, but not on the solutions or of the vector field $a(\cdot)$, such that

$$u \in C^{0,\alpha}_{\text{loc}}(\Omega) \qquad \text{for every } \alpha < \alpha_m$$

and

$$|u(x)-u(y)| \leq c \fint_{B_R} (|u|+Rs)\, d\tilde{x} \cdot \left(\frac{|x-y|}{R}\right)^{\alpha} .$$

The previous estimate holds whenever $x, y \in B_{R/2}$ and $B_R \subset \Omega$, for a constant depending only on n, p, ν, L and α. For the case $\mu \neq 0$ we then have the following:

Theorem 20.8 (De Giorgi's theory via potentials [25]). *Let $u \in W^{1,p-1}(\Omega)$ be a SOLA to the equation with measurable coefficients* (20.5) *under assumptions* (20.7). *Fix $\tilde{\alpha} < \alpha_m$, then the inequality*

$$|u(x)-u(y)| \leq c \left[\mathbf{W}^{\mu}_{1-\alpha(p-1)/p,p}(x,R) + \mathbf{W}^{\mu}_{1-\alpha(p-1)/p,p}(y,R)\right] |x-y|^{\alpha}$$
$$+c \fint_{B_R} (|u|+Rs)\, d\tilde{x} \cdot \left(\frac{|x-y|}{R}\right)^{\alpha} \qquad (20.17)$$

holds whenever $B_R \subset \Omega$ and $x, y \in B_{R/2}$ and $\alpha \in [0, \tilde{\alpha}]$, provided the right hand side is finite. The constant c depends only on n, p, ν, L and $\tilde{\alpha}$.

In view of the available regularity theory for the gradient of solutions to equations as (20.4) we wonder if a similar result holds for the oscillations of the gradient. For this let us recall the basic information about the maximal regularity of solutions to homogeneous equations as in (20.4). This theory goes back to the fundamental contribution of Ural'tseva [43], who proved the Hölder continuity of the gradient of solutions to (20.8) with $\mu = 0$; subsequently, different proofs have been given in [10, 15, 34]. The outcome is the existence of another positive exponent $\alpha_M \in (0,1)$, depending only on n, p, ν and L, such that

$$Du \in C^{0,\alpha}_{\mathrm{loc}}(\Omega) \qquad \text{for every } \alpha < \alpha_M$$

holds and

$$|Du(x)-Du(y)| \leq c \fint_{B_R} (|Du|+s)\, d\tilde{x} \cdot \left(\frac{|x-y|}{R}\right)^{\alpha} .$$

We then have

Theorem 20.9 (Ural'tseva theory via potentials [25]). *Let $u \in W^{1,p-1}(\Omega)$ be a SOLA to Eq.* (20.4) *under assumptions* (20.6). *Fix $\tilde{\alpha} < \min\{1/(p-1), \alpha_M\}$, then the inequality*

$$|Du(x)-Du(y)| \leq c \left[\mathbf{W}^{\mu}_{1-\frac{(1+\alpha)(p-1)}{p},p}(x,R) + \mathbf{W}^{\mu}_{1-\frac{(1+\alpha)(p-1)}{p},p}(y,R)\right] |x-y|^{\alpha}$$
$$+c \fint_{B_R} |Du-(Du)_{B_R}|\, d\tilde{x} \cdot \left(\frac{|x-y|}{R}\right)^{\alpha} \qquad (20.18)$$

holds whenever $B_R \subset \Omega$ and $x, y \in B_{R/2}$, and $\alpha \in [0, \tilde{\alpha}]$, provided the right hand side is finite. The constant c depends only on n, p, ν, L and $\tilde{\alpha}$.

Remark 20.1 (Comparison with the linear case). When $p = 2$, Wolff and Riesz potentials coincide and we have results that hold uniformly in the range $\alpha \in [0, 1]$. In this case Theorems 20.7 and 20.8 provide a complete analog of the estimates available in the linear case (20.2) via fundamental solutions (20.1). Indeed, by using the elementary inequality

$$\left||x-\tilde{x}|^{2-n} - |y-\tilde{x}|^{2-n}\right| \leq c\left||x-\tilde{x}|^{2-n-\alpha} + |y-\tilde{x}|^{2-n-\alpha}\right||x-y|^{\alpha}, \quad (20.19)$$

which is valid whenever $x, y, \tilde{x} \in \mathbb{R}^n$, and (20.1), we get

$$|u(x) - u(y)| \leq c\,[I_{2-\alpha}(|\mu|)(x) + I_{2-\alpha}(|\mu|)(y)]\,|x-y|^{\alpha}, \quad 0 \leq \alpha \leq 1.$$

This is exactly the global version of estimates (20.16) and (20.17) when $p = 2$ (it is sufficient to let $R \to \infty$ there and to assume global $W^{1,1}$-regularity on u). Differentiating (20.1) under the sign of integral and again applying (20.19) yields

$$|Du(x) - Du(y)| \leq c\,[I_{1-\alpha}(|\mu|)(x) + I_{1-\alpha}(|\mu|)(y)]\,|x-y|^{\alpha}, \quad 0 \leq \alpha < 1$$

which is again the global analog of (20.18) for $p = 2$ (but for the upper bound on α).

20.4 Nonlinear Parabolic Equations

Here we briefly summarise the nonlinear potential estimates that are available in the parabolic setting; again we restrict to the case $p \geq 2$ for simplicity, while we refer to [28] for the subquadratic case. Again for ease of presentation, we deal with local energy solutions and a priori estimates, without treating SOLA of parabolic equations. The case of general measure data problems is treated in [30, 31] to which we refer the interested readers.

Dealing with parabolic problems is more difficult, and requires additional new ideas. In particular, the potential theoretic approach developed for the elliptic case has to meet the fundamental concept of intrinsic geometry introduced by DiBenedetto [11]. The basic idea is to rebalance the lack of scaling (for $p \neq 2$) of equations as

$$u_t - \operatorname{div}(|Du|^{p-2}Du) = 0 \qquad \text{in } \Omega \times (-T, 0) \subset \mathbb{R}^{n+1} \quad (20.20)$$

by using certain special cylinders adapted to the solution itself, indeed called *intrinsic parabolic cylinders*. This means that, instead of using standard parabolic cylinders $Q_r(x_0,t_0) := B(x_0,r)\times(t_0-r^2,t_0)$, one uses *cylinders whose time-length is stretched accordingly to the size of the gradient on the cylinder itself.* These are of type

$$Q_r^\lambda(x_0,t_0) := B(x_0,r)\times(t_0-\lambda^{2-p}r^2,t_0)\,,\qquad \lambda>0\,, \tag{20.21}$$

on which it simultaneously happens that a condition of the type

$$\fint_{Q_r^\lambda(x_0,t_0)} |Du|\,dx\,dt \lesssim \lambda \tag{20.22}$$

is satisfied. The terminology "intrinsic geometry" stems exactly from this point. When considered on intrinsic cylinders, estimates become homogeneous and therefore they become suitable to be used in the iteration procedures which are typical of regularity theory. To see this fact, let us make a comparison. If we consider standard parabolic cylinders, then a priori estimates exhibit an anisotropicity linked to the one of the equation, that is

$$\sup_{Q_{r/2}(x_0,t_0)} |Du| \le c(n,p)\fint_{Q_r(x_0,t_0)} (|Du|+s+1)^{p-1}\,dx\,dt\,. \tag{20.23}$$

When instead considering intrinsic cylinders with (20.21) and (20.22) being in force, *estimates become dimensionally homogeneous*:

$$\left(\fint_{Q_r^\lambda(x_0,t_0)} (|Du|+s)^{p-1}\,dx\,dt\right)^{1/(p-1)} \lesssim \lambda \Longrightarrow |Du(x_0,t_0)| \le \lambda\,. \tag{20.24}$$

Both (20.23) and (20.24) are basic results of DiBenedetto and Friedman, for which we refer to [11]. The basic idea introduced in [28,30,31] is now to consider "intrinsic potentials" linked to the local intrinsic geometry, i.e. caloric Riesz potentials of the type

$$\mathbf{I}^{\mu}_{\beta,\lambda}(x_0,t_0;r) := \int_0^r \frac{|\mu|(Q_\varrho^\lambda(x_0,t_0))}{\varrho^{N-\beta}}\,\frac{d\varrho}{\varrho}\,. \tag{20.25}$$

Here $N := n+2$ is the usual parabolic dimension and $\lambda>0$ is a parameter to be chosen in the formulation of the results. Note that for $\lambda=1$ the one in (20.25) gives back the usual caloric Riesz potential. We now come to the results. Our emphasis will be on a priori estimates and for simplicity we shall treat the case of energy distributional solutions to (20.20), that is, functions u such that

$$u\in C^0(-T,0;L^2(\Omega))\cap L^p(-T,0;W^{1,p}(\Omega))\,.$$

The results hold for general equations of the type

$$u_t - \operatorname{div} a(Du) = 0 \qquad \text{in } \Omega \times (-T, 0) \subset \mathbb{R}^{n+1} \tag{20.26}$$

under assumptions (20.6) on the vector field $a(\cdot)$.

Theorem 20.10 (Intrinsic Riesz potential bound [31]). *Let u be a solution to (20.26) with $p \geq 2$. There exist a constant $c > 1$ depending only on n, p, ν, L, such that the following implication holds:*

$$c\mathbf{I}^{\mu}_{1,\lambda}(x_0, t_0; r) + c \left(\fint_{Q^{\lambda}_r(x_0,t_0)} (|Du| + s)^{p-1} \, dx \, dt \right)^{1/(p-1)} \leq \lambda \tag{20.27}$$
$$\Longrightarrow |Du(x_0, t_0)| \leq \lambda$$

whenever $Q^{\lambda}_r(x_0, t_0) \subset \Omega_T$ and (x_0, t_0) is Lebesgue point of Du.

As expected, (20.27) extends (20.24) to the case $\mu \neq 0$. As a matter of fact Theorem 20.10 implies a gradient linear potential estimate involving standard Riesz potentials that again reduces to (20.23) when $\mu = 0$. We indeed have the following:

Theorem 20.11 (Riesz potential bound in classic form [31]). *Let u be a solution to (20.26) with $p \geq 2$. There exists a constant c, depending only on n, p, ν, L, such that*

$$|Du(x_0, t_0)| \leq c\mathbf{I}^{\mu}_1(x_0, t_0; r) + c \fint_{Q_r(x_0,t_0)} (|Du| + s + 1)^{p-1} \, dx \, dt$$

holds whenever $Q_r(x_0, t_0) \subset \Omega_T$ is a standard parabolic cylinder and (x_0, t_0) is Lebesgue point of Du.

The proof of Theorem 20.10 opens the way to an optimal continuity criterion for the gradient that involves classical (caloric) Riesz potentials and that, as such, is again independent of p.

Theorem 20.12 (Gradient continuity via linear potentials [31]). *Let u be a solution to (20.26) with $p \geq 2$. If*

$$\lim_{r \to 0} \mathbf{I}^{\mu}_1(x, t; r) = 0$$

locally uniformly w.r.t. (x, t), then Du is continuous in Ω_T.

An important corollary involves Lorentz spaces; as in the elliptic case this result can be proved both for general equations as in (20.26) and for the evolutionary p-Laplacean system. We give the formulation directly in this case; when the right hand side is time-independent then we recover the full content of Theorem 20.5.

Theorem 20.13 (Parabolic nonlinear Stein theorem [29]). *Let $u \in W^{1,p}(\Omega, \mathbb{R}^m)$, $m \geq 1$, be a vector valued solution to the evolutionary p-Laplacean system $u_t - \triangle_p u = F$ with $p > 2n/(n+2)$. Assume that one of the two conditions are satisfied:*

- *The components of the vector field $F\colon \Omega \to \mathbb{R}^m$ locally belong to the space $L(N,1)$.*
- *The components of the vector field $F\colon \Omega \to \mathbb{R}^m$ are time independent and locally belong to the space $L(n,1)$.*

Then Du is continuous in Ω.

In the previous theorem, the lower bound $p > 2n/(n+2)$ is not a technical assumption, but is essential and cannot be avoided. A related Lipschitz regularity result is in [24].

The results above are based on a very delicate interplay between the new approaches necessary to derive potential estimates, and the classical approaches to prove regularity in the parabolic case when $\mu \equiv 0$. In particular, we make a new presentation of the basic regularity results of DiBenedetto [11], making them suitable to be applied in this new context. As a sample, we propose Theorem 20.14 below. It deals with homogeneous equations as

$$v_t - \operatorname{div} a(Dv) = 0 \tag{20.28}$$

but it works for the evolutionary p-Laplacean system as well, as in fact shown in [26]. It has basically two important features: on one hand it incorporates the basic elements of the gradient regularity theory of equations as (20.28). On the other one, it allows to derive a completely homogeneous decay estimate, which is totally similar to the standard elliptic one, provided certain intrinsic geometry conditions are satisfied.

Theorem 20.14 (Elliptic type excess decay [30]). *Let v be a solution to* (20.28) *in a cylinder $Q \equiv Q_r^\lambda(x_0,t_0)$ of the type in* (20.21). *Consider numbers*

$$A, B \geq 1 \qquad \textit{and} \qquad \varepsilon \in (0,1)\,.$$

Then there exists a constant $\sigma \in (0,1/4)$ depending only on $n, p, \nu, L, A, B, \varepsilon$ such that if

$$\frac{\lambda}{B} \leq \sup_{\sigma Q} |Dv| \leq s + \sup_{\frac{1}{4}Q} |Dv| \leq A\lambda$$

holds, then

$$\fint_{\tau Q} |Dv - (Dv)_{\tau Q}|\, dx\, dt \leq \varepsilon \fint_{\frac{1}{4}Q} |Dv - (Dv)_{\frac{1}{4}Q}|\, dx\, dt$$

holds too, whenever $\tau \in (0, \sigma]$. *Here we are denoting*

$$\tau Q_r^\lambda(x_0, t_0) := B(x_0, \tau r) \times (t_0 - \lambda^{2-p}(\tau r)^2, t_0) .$$

References

1. D.R. Adams, A note on Riesz potentials. Duke Math. J. **42**, 765–778 (1975)
2. D.R. Adams, N.G. Meyers, Thinnes and Wiener criteria for nonlinear potentials. Indiana Univ. Math. J. **22**, 169–197 (1972)
3. P. Baroni, Linear potential estimates for a class of elliptic equations. Calc. Var. & PDE, to appear
4. L. Boccardo, T. Gallouët, Nonlinear elliptic and parabolic equations involving measure data. J. Funct. Anal. **87**, 149–169 (1989)
5. S. Campanato, Proprietà di hölderianità di alcune classi di funzioni. Ann. Scuola Norm. Sup. Pisa (III) **17**, 175–188 (1963)
6. A. Cianchi, Nonlinear potentials, local solutions to elliptic equations and rearrangements. Ann. Scuola Norm. Sup. Cl. Sci. (V) **10**, 335–361 (2011)
7. A. Cianchi, V. Maz'ya, Global Lipschitz regularity for a class of quasilinear elliptic equations. Commun. PDE **36**, 100–133 (2011)
8. A. Cianchi, V. Maz'ya, Global boundedness of the gradient for a class of nonlinear elliptic systems. Arch. Ration. Mech. Anal. **212**, 29–177 (2012)
9. R.A. DeVore, R.C. Sharpley, Maximal functions measuring smoothness. Mem. Am. Math. Soc. **47**(293) (1984)
10. E. DiBenedetto, $C^{1+\alpha}$ local regularity of weak solutions of degenerate elliptic equations. Nonlinear Anal. **7**, 827–850 (1983)
11. E. DiBenedetto, *Degenerate Parabolic Equations*. Universitext (Springer, New York, 1993)
12. F. Duzaar, G. Mingione, Local Lipschitz regularity for degenerate elliptic systems. Ann. Inst. H. Poincaré Anal. Non Linèaire **27**, 1361–1396 (2010)
13. F. Duzaar, G. Mingione, Gradient estimates via linear and nonlinear potentials. J. Funct. Anal. **259**, 2961–2998 (2010)
14. F. Duzaar, G. Mingione, Gradient estimates via non-linear potentials. Am. J. Math. **133**, 1093–1149 (2011)
15. L.C. Evans, A new proof of local $C^{1,\alpha}$ regularity for solutions of certain degenerate elliptic P.D.E. J. Differ. Equ. **45**, 356–373 (1982)
16. M. Havin, V.G. Maz'ja, A nonlinear analogue of the Newtonian potential, and metric properties of (p, l)-capacity. Dokl. Akad. Nauk SSSR **194**, 770–773 (1970)
17. M. Havin, V.G. Maz'ja, Nonlinear potential theory. Russ. Math. Surv. **27**, 71–148 (1972)
18. L. Hedberg, Th.H. Wolff, Thin sets in nonlinear potential theory. Ann. Inst. Fourier (Grenoble) **33**, 161–187 (1983)
19. J. Heinonen, T. Kilpeläinen, O. Martio, *Nonlinear Potential Theory of Degenerate Elliptic Equations*. Oxford Mathematical Monographs (Clarendon Press/Oxford University Press, New York, 1993)
20. T. Kilpeläinen, J. Malý, Degenerate elliptic equations with measure data and nonlinear potentials. Ann. Scuola Norm. Sup. Pisa Cl. Sci. (IV) **19**, 591–613 (1992)
21. T. Kilpeläinen, J. Malý, The Wiener test and potential estimates for quasilinear elliptic equations. Acta Math. **172**, 137–161 (1994)
22. R. Korte, T. Kuusi, A note on the Wolff potential estimate for solutions to elliptic equations involving measures. Adv. Calc. Var. **3**, 99–113 (2010)
23. T. Kuusi, G. Mingione, A surprising linear type estimate for nonlinear elliptic equations. Compt. Rend. Acad. Sci. Ser. i Math. **349**, 889–892 (2011)

24. T. Kuusi, G. Mingione, Potential estimates and gradient boundedness for nonlinear parabolic systems. Rev. Mat. Iber. **28**, 535–576 (2012)
25. T. Kuusi, G. Mingione, Universal potential estimates. J. Funct. Anal. **262**, 4205–4269 (2012)
26. T. Kuusi, G. Mingione, New perturbation methods for nonlinear parabolic problems. J. Math. Pures Appl. (IX) **98**, 390–427 (2012)
27. T. Kuusi, G. Mingione, Linear potentials in nonlinear potential theory. Arch. Ration. Mech. Anal. **207**, 215–246 (2013)
28. T. Kuusi, G. Mingione, Gradient regularity for nonlinear parabolic equations. Ann Scuola Norm. Sup. Pisa Cl. Sci. (V) **12**, 755–822 (2013)
29. T. Kuusi, G. Mingione, Borderline gradient continuity for nonlinear parabolic systems. Mathematische Ann. doi:10.1007/s00208-014-1055-1
30. T. Kuusi, G. Mingione, The Wolff gradient bound for degenerate parabolic equations. J. Eur. Math. Soc. **16**, 835–892 (2014)
31. T. Kuusi, G. Mingione, Riesz potentials and nonlinear parabolic equations. Arch. Ration. Mech. Anal. **212**, 727–780 (2014)
32. T. Kuusi, G. Mingione, Guide to nonlinear potential estimates. Bull. Math. Sci. **4**, 1–82 (2014)
33. T. Kuusi, G. Mingione, A nonlinear Stein theorem. Calc. Var. PDE. doi:10.1007/s00526-013-0666-9
34. J.L. Lewis, Regularity of the derivatives of solutions to certain degenerate elliptic equations. Indiana Univ. Math. J. **32**, 849–858 (1983)
35. P. Lindqvist, Notes on the p-Laplace equation, University of Jyväskylä, Report 102 (2006)
36. V.G. Maz'ja, The continuity at a boundary point of the solutions of quasi-linear elliptic equations. Vestnik Leningrad. Univ. (Russian) **25**, 42–55 (1970)
37. G. Mingione, Gradient estimates below the duality exponent. Math. Ann. **346**, 571–627 (2010)
38. G. Mingione, Gradient potential estimates. J. Eur. Math. Soc. **13**, 459–486 (2011)
39. G. Mingione, Nonlinear measure data problems. Milan J. Math. **79**, 429–496 (2011)
40. N.C. Phuc, I.E. Verbitsky, Quasilinear and Hessian equations of Lane-Emden type. Ann. Math. (II) **168**, 859–914 (2008)
41. E.M. Stein, Editor's note: the differentiability of functions in $\mathbb{R}^n$. Ann. Math. (II) **113**, 383–385 (1981)
42. N.S. Trudinger, X.J. Wang, On the weak continuity of elliptic operators and applications to potential theory. Am. J. Math. **124**, 369–410 (2002)
43. N.N. Ural'tseva, Degenerate quasilinear elliptic systems. Zap. Na. Sem. Leningrad. Otdel. Mat. Inst. Steklov. (LOMI) **7**, 184–222 (1968)

Chapter 21
Partial Regularity Results in Optimal Transportation

G. De Philippis and A. Figalli

Abstract This note describes some recent results on the regularity of optimal transport maps. As we shall see, in general optimal maps are not globally smooth, but they are so outside a "singular set" of measure zero.

21.1 The Optimal Transportation Problem

The optimal transportation problem, whose origin dates back to Monge [19], aims to find a way to transport a distribution of mass from one place to another by minimizing the transportation cost. Mathematically, the problem can be formulated as follows: given two probability measures μ and ν (representing respectively the initial and final configuration of the mass that we want to transport) defined on the measurable spaces X and Y, one says that a map $T : X \to Y$ *transports* μ onto ν if $T_\sharp\mu = \nu$, i.e.,

$$\nu(A) = \mu\left(T^{-1}(A)\right) \qquad \forall\, A \subset Y \text{ measurable.}$$

Then, given a cost function $c : X \times Y \to \mathbb{R}$ (so that $c(x, y)$ represents the cost to transport a unit of mass from x to y), one wants to minimize the transportation cost among all possible transport maps.

G. De Philippis (✉)
Hausdorff Center for Mathematics, Universität Bonn, Endenicher Allee 60, 53115 Bonn, Germany
e-mail: guido.de.philippis@hcm.uni-bonn.de

A. Figalli
Department of Mathematics, The University of Texas at Austin, 2515 Speedway Stop C1200, Austin, TX 78712-1202, USA
e-mail: figalli@math.utexas.edu

V. Ancona and E. Strickland (eds.), *Trends in Contemporary Mathematics*,
Springer INdAM Series 8, DOI 10.1007/978-3-319-05254-0_21,

Since transporting a unitary mass from x to $T(x)$ costs $c(x, T(x))$, the cost to transport the whole mass μ is simply given by $\int_X c(x, T(x))\,d\mu(x)$. Hence the optimal transportation problem consists in solving the minimization problem

$$\min_{T_{\#}\mu=\nu} \left\{ \int_X c(x, T(x))\,d\mu(x) \right\}. \tag{21.1}$$

When $T : X \to Y$ minimizes the transportation cost we call it an *optimal transport map*.

Even in Euclidean spaces with the cost c given by the Euclidean distance $|x-y|$ or its square $|x-y|^2$, the problem of the existence of an optimal transport map is far from being trivial. Moreover, it is easy to build examples where the Monge problem is ill-posed simply because there is no transport map: this happens for instance when μ is a Dirac mass while ν is not. This means that one needs some restrictions on the measures μ and ν.

We notice that when $X, Y \subset \mathbb{R}^n$, $\mu(dx) = f(x)dx$, and $\nu(dy) = g(y)dy$, if $T : X \to Y$ is a sufficiently smooth transport map one can rewrite the transport condition $T_{\#}\mu = \nu$ as a Jacobian equation. Indeed, if $\chi : \mathbb{R}^n \to \mathbb{R}$ denotes a test function, the condition $T_{\#}\mu = \nu$ gives

$$\int_{\mathbb{R}^n} \chi(T(x)) f(x)\,dx = \int_{\mathbb{R}^n} \chi(y) g(y)\,dy.$$

Now, assuming in addition that T is a diffeomorphism, we can set $y = T(x)$ and use the change of variable formula to obtain that the second integral is equal to

$$\int_{\mathbb{R}^n} \chi(T(x)) g(T(x)) \big|\det(\nabla T(x))\big|\,dx.$$

By the arbitrariness of χ, this gives the Jacobian equation

$$f(x) = g(T(x))\,\big|\det(\nabla T(x))\big|. \tag{21.2}$$

21.1.1 The Quadratic Cost on $\mathbb{R}^n$

In [2, 3], Brenier considered the case $X = Y = \mathbb{R}^n$ and $c(x, y) = |x-y|^2/2$, and proved the following theorem (which was also obtained independently by Cuesta-Albertos and Matrán [8] and by Rachev and Rüschendorf [20]).

Theorem 21.1. *Let μ and ν be two compactly supported probability measures on $\mathbb{R}^n$. If μ is absolutely continuous with respect to the Lebesgue measure, then:*

(i) There exists a unique solution $\hat{T}$ to the optimal transport problem with cost $c(x, y) = |x-y|^2/2$.

(ii) The optimal map $\hat{T}$ is characterized by the structure $\hat{T}(x) = \nabla u(x)$ for some convex function $u : \mathbb{R}^n \to \mathbb{R}$, which is called the "potential" associated to $\hat{T}$.

Let us point out, for further use, that the minimization problem for the cost $|x-y|^2/2$ is equivalent to the minimization problem for the cost $-x \cdot y$. Indeed, for any transport map T we have

$$\int_{\mathbb{R}^n} \frac{|T(x)|^2}{2}\, d\mu(x) = \int_{\mathbb{R}^n} \frac{|y|^2}{2}\, d\nu(y)$$

(this is a direct consequence of the condition $T_{\#}\mu = \nu$), hence

$$\begin{aligned}\int_{\mathbb{R}^n} \frac{|x-T(x)|^2}{2}\, d\mu(x) &= \int_{\mathbb{R}^n} \frac{|x|^2}{2}\, d\mu(x) + \int_{\mathbb{R}^n} \frac{|T(x)|^2}{2}\, d\mu(x) \\ &\quad + \int_{\mathbb{R}^n} \big(-x \cdot T(x)\big)\, d\mu(x) \\ &= \int_{\mathbb{R}^n} \frac{|x|^2}{2}\, d\mu(x) + \int_{\mathbb{R}^n} \frac{|y|^2}{2}\, d\nu(y) \\ &\quad + \int_{\mathbb{R}^n} \big(-x \cdot T(x)\big)\, d\mu(x),\end{aligned}$$

and since the first two integrals in the right hand side are independent of T we see that the two problems

$$\min_{T_{\#}\mu=\nu} \int_{\mathbb{R}^n} \frac{|x-T(x)|^2}{2}\, d\mu(x) \qquad \text{and} \qquad \min_{T_{\#}\mu=\nu} \int_{\mathbb{R}^n} \big(-x \cdot T(x)\big)\, d\mu(x)$$

are equivalent (that is, they have the same minimizers).

Having found a solution to (21.1), a natural question is the one concerning its *regularity*:
Assuming that X and Y are two bounded smooth open sets in $\mathbb{R}^n$, let $\mu(dx) = f(x)dx$ and $\nu(y) = g(y)dy$ be two probability measures with smooth densities f and g such that $X = \{f > 0\}$ and $Y = \{g > 0\}$. Then, is it true that the optimal map $\hat{T}$ (or equivalently the "potential" u) is smooth?

As observed by Caffarelli [6], one cannot expect any general regularity result for u without making some geometric assumptions on the support of the target measure. Indeed, let $n = 2$ and suppose that $X = B_1$ is the unit ball centered at the origin and $Y = \big(B_1^+ + e_1\big) \cup \big(B_1^- - e_1\big)$ is the union of two half-balls (here (e_1, e_2) denote the canonical basis of $\mathbb{R}^2$), where

$$B_1^+ := \big(B_1 \cap \{x_1 > 0\}\big), \qquad B_1^- := \big(B_1 \cap \{x_1 < 0\}\big).$$

Then, if $f = \frac{1}{|X|}\mathbf{1}_X$ and $g = \frac{1}{|Y|}\mathbf{1}_Y$, it is easily seen that the optimal map $\hat{T}$ is given by

$$\hat{T}(x) := \begin{cases} x + e_1 \text{ if } x_1 > 0, \\ x - e_1 \text{ if } x_1 < 0, \end{cases}$$

which corresponds to the gradient of the convex function $u(x) = |x|^2/2 + |x_1|$.

Thus (as one could also show by an easy topological argument) in order to hope for a regularity result for u we need at least to assume the connectedness of Y. However, not even this is sufficient. Indeed, starting from the above construction and considering a sequence of domains X'_ε where one adds a small strip of width $\varepsilon > 0$ to glue together $\left(B_1^+ + e_1\right) \cup \left(B_1^- - e_1\right)$, one can also show that for $\varepsilon > 0$ small enough the optimal map will still be discontinuous (see [6] or [25, Theorem 12.3] for more details).

In order to understand what is happening in the previous example, let us try to write down what is the equation satisfied by a potential u. Since $\hat{T} = \nabla u$, the Jacobian equation (21.2) gives that u formally solves the Monge-Ampère equation

$$\det(D^2u(x)) = \frac{f(x)}{g(\nabla u(x))} \qquad f\, dx\text{-a.e.} \tag{21.3}$$

coupled with the "boundary condition"

$$\nabla u(X) = Y \tag{21.4}$$

which corresponds to the fact that $\hat{T}$ transports $f(x)dx$ onto $g(y)dy$ (recall that $X = \{f > 0\}$ and $Y = \{g > 0\}$). So, one may in principle hope to apply the regularity theory for Monge-Ampère in order to show that u is actually smooth.

However this is just a formal computation, and what one can rigorously show is the following: the transport condition $(\nabla u)_\# f = g$ means that

$$\int_{(\nabla u)^{-1}(A)} f = \int_A g \qquad \forall\, A \subset Y.$$

From this fact it is possible to prove (see for instance [24, Lemma 4.6]) that

$$\int_E f = \int_{\partial u(E)} g \qquad \forall\, E \subset X,$$

where ∂u denotes the subdifferential of u:

$$\partial u(x) := \left\{p \in \mathbb{R}^n \,:\, u(z) \geq u(x) + p\cdot(z-x) \quad \forall\, z \in \mathbb{R}^n\right\}, \qquad \partial u(E) := \bigcup_{x\in E} \partial u(x).$$

Hence, since $Y = \{g > 0\}$ we get

$$\int_E f = \int_{\partial u(E)\cap Y} g \qquad \forall\, E \subset X,$$

and, if $\lambda \leq f, g \leq 1/\lambda$ on X and Y respectively, we deduce that[1]

$$\lambda^2|E| \leq |\partial u(E) \cap Y| \leq |E|/\lambda^2 \qquad \forall\, E \subset X. \tag{21.5}$$

Hence, we can see this as a "weak" form of the Monge-Ampère equation.

On the other hand, the "right" notion of weak solution of (21.3) (i.e., a notion of solution which allows one to obtain a satisfactory regularity theory) is the one of *Alexandrov solution* [1]: we say that a convex function $u : X \to \mathbb{R}$ is an Alexandrov solution of

$$\lambda^2 \leq \det D^2 u \leq 1/\lambda^2 \tag{21.6}$$

if

$$\lambda^2|E| \leq |\partial u(E)| \leq |E|/\lambda^2 \qquad \forall\, E \subset X. \tag{21.7}$$

Note that (21.7) is the condition which would follow from (21.3) when u is smooth and $\lambda \leq f, g \leq 1/\lambda$: indeed, if $u \in C^2(X)$ then $\partial u = \nabla u$, hence by the Area Formula

$$|\partial u(E)| = |\nabla u(E)| = \int_E \det(D^2 u),$$

and (21.7) follows from (21.3).

The difference between (21.5) and (21.7) is at the base of the previous counterexample. Indeed, (21.7) provides enough control on the behavior of ∂u to show that if u is strictly convex[2] then $\partial u(x)$ is a singleton for all $x \in X$. By convexity of u, this implies that u is continuously differentiable in X, and actually one can also show that ∇u is Holder continuous (see [4, 5] and [10, Section 2.4]). On the other hand, (21.5) only gives information on the behavior of the intersection $\partial u(E) \cap Y$ for $E \subset X$.

In the counterexample above with $X = B_1$ and $Y = \big(B_1^+ + e_1\big) \cup \big(B_1^- - e_1\big)$, the potential u was given by $u(x) = |x|^2/2 + |x_1|$. Hence, for any x of the form $x = (0, x_2)$ the set ∂u is multivalued, namely $\partial u(x) = [-1, 1] \times \{x_2\}$. Thus

$$\partial u(\{0\} \times [-1, 1]) = [-1, 1]^2.$$

[1]Here and in the sequel, $|E|$ denotes the Lebesgue measure of a set E.

[2]By an example of Pogorelov this turns out to be a necessary condition, see for instance [24, Section 4.1.3].

This would not be possible if u satisfied (21.7) since ∂u has to map sets of measure zero onto sets of measure zero. However, since the intersection of $[-1,1]^2$ with $Y = \{g > 0\}$ has measure zero, one is not able to detect the singularity of u using (21.5).

Hence, in order to avoid this kind of counterexamples one should make sure that the target Y always covers the image of X through the subdifferential map ∂u. A way to ensure this is that Y is convex. Indeed (see for instance [6] or [10, Theorem 3.3]) if Y is convex then $\partial u(X) \subset \overline{Y}$ and (21.5) becomes (21.7). This information allows one to prove regularity [6, 7]. More precisely the following holds:

Theorem 21.2. *Let $X, Y \subset \mathbb{R}^n$ be two bounded open sets, let $f : X \to \mathbb{R}^+$ and $g : Y \to \mathbb{R}^+$ be two probability densities bounded away from zero and infinity on X and Y respectively, and denote by $\hat{T} = \nabla u : X \to Y$ the unique optimal transport map sending f onto g for the cost $|x-y|^2/2$. Assume that Y is convex. Then:*

(a) $\hat{T} \in C^{0,\alpha}_{\mathrm{loc}}(X)$.
(b) If in addition $f \in C^{k,\beta}_{\mathrm{loc}}(X)$ *and* $g \in C^{k,\beta}_{\mathrm{loc}}(Y)$ *for some* $\beta \in (0,1)$, *then* $\hat{T} \in C^{k+1,\beta}_{\mathrm{loc}}(X)$.
(c) Furthermore, if $f \in C^{k,\beta}(\overline{X})$, $g \in C^{k,\beta}(\overline{Y})$, *and both X and Y are smooth and uniformly convex, then* $\hat{T} : \overline{X} \to \overline{Y}$ *is a global diffeomorphism of class* $C^{k+1,\beta}$.

Even if this result is very satisfactory, one still would like to understand how "bad" can be the set where u is not regular when one removes the convexity assumption on the target. As shown in [14] (see also [12] for a more precise description of the singular set in two dimensions), in this case one can prove that the optimal transport map is actually smooth outside a closed set of measure zero. More precisely, the following holds:

Theorem 21.3. *Let $X, Y \subset \mathbb{R}^n$ be two bounded open sets, let $f : X \to \mathbb{R}^+$ and $g : Y \to \mathbb{R}^+$ be two probability densities bounded away from zero and infinity on X and Y respectively, and denote by $\hat{T} = \nabla u : X \to Y$ the unique optimal transport map sending f onto g for the cost $|x-y|^2/2$. Then there exist two relatively closed sets $\Sigma_X \subset X$ and $\Sigma_Y \subset Y$, with $|\Sigma_X| = |\Sigma_Y| = 0$, such that $\hat{T} : X \setminus \Sigma_X \to Y \setminus \Sigma_Y$ is a homeomorphism of class $C^{0,\alpha}_{\mathrm{loc}}$ for some $\alpha > 0$. In addition, if $c \in C^{k+2,\beta}_{\mathrm{loc}}(X \times Y)$, $f \in C^{k,\beta}_{\mathrm{loc}}(X)$, and $g \in C^{k,\beta}_{\mathrm{loc}}(Y)$ for some $k \geq 0$ and $\beta \in (0,1)$, then $\hat{T} : X \setminus \Sigma_X \to Y \setminus \Sigma_Y$ is a diffeomorphism of class $C^{k+1,\beta}_{\mathrm{loc}}$.*

Proof (Sketch of the proof). As explained above, when Y is not convex there could be points $x \in X$ such that $\partial u(x) \not\subset Y$. Let us define[3]

$$\mathrm{Reg}_X := \{x \in X : \partial u(x) \subset Y\} \qquad \Sigma_X := X \setminus \mathrm{Reg}_X.$$

[3]Actually, in [12, 14] the regular set is defined in a slightly different way and it is in principle smaller. However, the advantage of that definition is that it allows for a more refined analysis of the singular set (see [12]).

By the continuity property of the subdifferential of a convex function, it is immediate to see that Reg_X is open. Moreover it follows from the condition $(\nabla u)_\sharp(f\,dx) = g\,dy$ that $\nabla u(x) \in Y$ for a.e. $x \in X$, thus $|\Sigma_X| = 0$. Hence

$$\lambda^2|E| \leq |\partial u(E)| \leq |E|/\lambda^2 \qquad \forall\, E \subset \mathrm{Reg}_X$$

provided $\lambda \leq f, g \leq 1/\lambda$. A (non-trivial) adaptation of Caffarelli's techniques permits to prove that u is smooth inside Reg_X (the main issue here is to show that u is strictly convex).

21.1.2 The Case of a General Cost

After Theorem 21.1 many researchers started to work on the problem of showing existence (and regularity) of optimal maps in the case of more general costs, both in an Euclidean setting and in the case of Riemannian manifolds. Since, at least locally, any Riemannian manifold looks like $\mathbb{R}^n$, here we shall only focus on the Euclidean case (see [13] and [10] for more results).

Let us introduce first some conditions on the cost function $c : X \times Y \to \mathbb{R}$, where $X, Y \subset \mathbb{R}^n$:

(C0) The cost function $c : X \times Y \to \mathbb{R}$ is of class C^2 with $\|c\|_{C^2(X\times Y)} < \infty$.
(C1) For any $x \in X$, the map $Y \ni y \mapsto -D_x c(x, y) \in \mathbb{R}^n$ is injective.
(C2) For any $y \in Y$, the map $X \ni x \mapsto -D_y c(x, y) \in \mathbb{R}^n$ is injective.
(C3) $\det(D_{xy}c)(x, y) \neq 0$ for all $(x, y) \in X \times Y$.

We also introduce the concept of c-convex functions which generalizes the one of convex functions that appeared in the case $c(x, y) = -x \cdot y$ (see Theorem 21.1 and recall that, by the discussion immediately after that theorem, the costs $-x \cdot y$ and $|x - y|^2/2$ are equivalent): a function $u : X \to \mathbb{R} \cup \{+\infty\}$ is *c-convex* if it can be written as

$$u(x) = \sup_{y\in Y}\{-c(x, y) + \lambda_y\}$$

for some family of constants $\lambda_y \in \mathbb{R}$.

The following is a basic result in optimal transport theory.

Theorem 21.4. *Let $c : X \times Y \to \mathbb{R}$ satisfy **(C0)–(C1)**. Given two probability densities f and g supported on X and Y respectively, there exists a c-convex "potential" $u : X \to \mathbb{R}$ such that the map $\hat{T} : X \to Y$ implicitly defined by*

$$D_x c(x, \hat{T}(x)) + \nabla u(x) = 0 \tag{21.8}$$

is the unique optimal transport map sending f onto g.

Since c satisfies **(C1)** we can define the *c-exponential map*:

$$\text{for any } x \in X, y \in Y, p \in \mathbb{R}^n, \qquad \text{c-exp}_x(p) = y \quad \Leftrightarrow \quad p = -D_x c(x, y).$$

This allows us to rewrite (21.8) as $\hat{T}(x) = \text{c-exp}_x(\nabla u(x))$.

Let us try again to understand which PDE is satisfied by a potential u. Assuming that u is smooth, we see that the c-convexity of u implies that

$$D^2 u(x) + D_{xx} c\big(x, \text{c-exp}_x(\nabla u(x))\big) \geq 0. \tag{21.9}$$

Moreover, using **(C2)** one can show that $\hat{T}$ is injective and that $\hat{T}^{-1}$ is the optimal map between ν and μ for the symmetrized cost $c^*(x, y) = c(y, x)$.

Hence, differentiating (21.8) with respect to x and using (21.2) and (21.9), we obtain

$$\begin{aligned} &\det\Big(D^2 u(x) + D_{xx} c\big(x, \text{c-exp}_x(\nabla u(x))\big)\Big) \\ &\quad = \big|\det\big(D_{xy} c\big(x, \text{c-exp}_x(\nabla u(x))\big)\big)\big| \frac{f(x)}{g(\text{c-exp}_x(\nabla u(x)))}. \end{aligned} \tag{21.10}$$

Hence, at least formally, u solves a Monge-Ampère type equation of the form

$$\det\big(D^2 u - \mathscr{A}(x, \nabla u)\big) = h(x, \nabla u) \tag{21.11}$$

with

$$\mathscr{A}(x, p) := -D_{xx} c\big(x, \text{c-exp}_x(p)\big) \tag{21.12}$$

and

$$h(x, \nabla u) := \big|\det\big(D_{xy} c\big(x, \text{c-exp}_x(\nabla u(x))\big)\big)\big| \frac{f(x)}{g(\text{c-exp}_x(\nabla u(x)))}.$$

Observe that, in the case $c(x, y) = -x \cdot y$, $\mathscr{A} \equiv 0$ and (21.11) reduces to the classical Monge-Ampère equation. As we showed in the previous section, in order to get regularity of u one needs to assume the convexity of the target domain. The issue is in some sense the following: the Monge-Ampère equation enjoys some a-priori regularity estimates (these are the so called Pogorelov estimates, see for instance [16, Section 17.6]) which allows one to obtain regularity of solutions provided one has suitable boundary conditions. In the case $c(x, y) = -x \cdot y$ the boundary condition was $\nabla u(X) = Y$ and convexity of Y was enough to ensure regularity.

Now, for the general case we have to face two difficulties: in addition to identify some suitable notion of convexity on the domains to handle the boundary conditions, one also needs some analogous of the Pogorelov estimates for the general class of equations (21.11).

The breakthrough for the regularity of solutions to this class of equations came with the paper of Ma, Trudinger and Wang [18], where the authors found a mysterious condition on the cost functions that turned out to be sufficient to prove the regularity of u. More precisely, the condition to be imposed on the cost (that we call here "MTW condition") is the following:

$$D^2_{p_\eta p_\eta}\mathscr{A}(x,p)[\xi,\xi] \leq 0 \qquad \forall\, x,p,\ \forall\,\xi \perp \eta. \tag{21.13}$$

Since $\mathscr{A}$ depends on first and second order derivatives of the cost (see (21.12)), the MTW condition is a fourth-order condition on c.

Under this condition, Ma, Trudinger, and Wang could prove the following result [18,22,23] (see also [21]), that generalizes Theorem 21.2(c):

Theorem 21.5. *Let $c : X \times Y \to \mathbb{R}$ satisfy* **(C0)–(C3)**. *Assume that the MTW condition holds, and that f and g are smooth and bounded away from zero and infinity on their respective supports X and Y. Also, suppose that:*

- *X and Y are smooth.*
- *$D_x c(x,Y)$ is uniformly convex for all $x \in X$.*
- *$D_y c(X,y)$ is uniformly convex for all $y \in Y$.*

Then $u \in C^\infty(\overline{X})$ and the map $T_u : \overline{X} \to \overline{Y}$ defined as $T_u(x) := c\text{-}exp_x(\nabla u(x))$ is a smooth diffeomorphism.

In [17] Loeper started a systematic study of the MTW condition and its relation to the geometry of optimal transport maps. Among other things, he was able to prove that the MTW condition (21.13) is essentially equivalent to the following statement (see [17], [25, Chapter 12] for a more precise discussion):
For any c-convex function u, its c-subdifferntial

$$\partial_c u(x) := \big\{y \in \overline{Y} : u(z) + c(z,y) \geq u(x) + c(x,y) \qquad \forall\, z \in X\big\} \tag{21.14}$$

is connected *for every $x \in X$.*

Note that, when $c(x,y) = -x \cdot y$, c-convex function are just convex and $\partial_c u(x)$ reduces to $\partial u(x)$ (which is convex, thus connected).

Connectedness of the c-subdifferential turns out to be a necessary condition for the regularity of optimal maps, see [17], [25, Theorem 12.7], [15]. Hence, in view of its equivalence with the MTW condition, Loeper proved the following: if there exist $x, p, \xi \perp \eta$ such that the MTW condition fail, then one can construct smooth positive probability densities f and g (whose supports satisfy the appropriate global convexity assumptions) such that the optimal map between $\mu = f dx$ and $\nu = g dy$ is discontinuous. Moreover, Loeper also found a nice connection with geometry: if $c = d^2/2$ with d a Riemannian distance, then

$$D^2_{p_k p_\ell} A_{ij}(x,0)\xi^i\xi^j\eta^k\eta^\ell = -\frac{2}{3}\mathrm{Sect}_x([\xi,\eta]) \qquad \forall\,\xi,\eta \in T_x M,\ \xi \perp \eta,$$

where $\mathrm{Sect}_x([\xi,\eta])$ denotes the sectional curvature of the 2-plane generated by ξ and η. Since (as we just mentioned above) the MTW condition is necessary for regularity, one gets the following[4]:

Corollary 21.1. *Let $c = d^2/2$ on a smooth Riemannian manifold M, and assume that* $\mathrm{Sect}_x < 0$ *at some point along some 2-plane in T_xM. Then one can construct $f, g \in C^\infty(M)$ with $f, g > 0$ such that $\hat{T} \notin C^0$.*

Let us also mention that the MTW condition is quite restrictive and it is satisfied only in very particular cases. These include the costs:

- $|x-y|^2/2$ (or equivalently $-x\cdot y$).
- $-\log|x-y|$.
- $\sqrt{a^2-|x-y|^2}$ with $a>0$.
- $\sqrt{a^2+|x-y|^2}$ with $a>0$.
- $|x-y|^p$ with $-2<p<1$.

And the case $c = d^2/2$ on the following manifolds:

- $\mathbb{R}^n$ and $\mathbb{T}^n$.
- $\mathbb{S}^n$, its quotients (like $\mathbb{RP}^n$), and its submersions (like $\mathbb{CP}^n$ or $\mathbb{HP}^n$).
- Products of any of the examples listed above (for instance, $\mathbb{S}^{n_1}\times\ldots\times\mathbb{S}^{n_k}\times\mathbb{R}^\ell$ or $\mathbb{S}^{n_1}\times\mathbb{CP}^{n_2}\times\mathbb{T}^{n_3}$).
- Smooth perturbations of $\mathbb{S}^n$.

Because the MTW condition is usually false, a natural question is: Can one prove a partial regularity result for general cost functions?

Notice that in the case $-x\cdot y$ one exploits the fact that $\partial u(x) \subset \Lambda$ a.e. (see the sketch of the proof of Theorem 21.3), which means (very roughly) that at most points we are as in the case of a convex target, and hence a locally regularity theory is in principle available. However, in the general case, besides the global obstruction given by the geometry of the source and target domains there is also the local obstruction given by the failure of the MTW condition. For instance, by Corollary 21.1, if M has negative sectional curvature then MTW fails *at every point*! This means that there is no hope to say that, as in the quadratic case, we are in a "good" situation at almost every point. Still, in [9] we have been able to prove the following result:

Theorem 21.6. *Let $X, Y \subset \mathbb{R}^n$ be two bounded open sets, and let $f : X \to \mathbb{R}^+$ and $g : Y \to \mathbb{R}^+$ be two continuous probability densities bounded away from zero and infinity on X and Y respectively. Assume that the cost $c : X\times Y \to \mathbb{R}$ satisfies* **(C0)–(C3)**, *and denote by $\hat{T} : X \to Y$ the unique optimal transport map sending f onto g. Then there exist two relatively closed sets $\Sigma_X \subset X$ and $\Sigma_Y \subset Y$, with*

[4] Although we did not state them here, many existence and uniqueness result for optimal transport maps on Riemannian manifolds are known (see for instance [11]), and they include for instance the case $c(x,y) = d(x,y)^2/2$.

$|\Sigma_X| = |\Sigma_Y| = 0$, such that $\hat{T} : X \setminus \Sigma_X \to Y \setminus \Sigma_Y$ is a homeomorphism of class $C^{0,\beta}_{\mathrm{loc}}$ for any $\beta < 1$. In addition, if $c \in C^{k+2,\alpha}_{\mathrm{loc}}(X \times Y)$, $f \in C^{k,\alpha}_{\mathrm{loc}}(X)$, and $g \in C^{k,\alpha}_{\mathrm{loc}}(Y)$ for some $k \geq 0$ and $\alpha \in (0,1)$, then $\hat{T} : X \setminus \Sigma_X \to Y \setminus \Sigma_Y$ is a diffeomorphism of class $C^{k+1,\alpha}_{\mathrm{loc}}$.

This result can be suitably localized to obtain a regularity result for the squared distance function on Riemannian manifolds:

Theorem 21.7. *Let M be a smooth Riemannian manifold, and let $f, g : M \to \mathbb{R}^+$ be two continuous probability densities, locally bounded away from zero and infinity on M. Let $\hat{T} : M \to M$ denote the optimal transport map for the cost $c = d^2/2$ sending f onto g, d being the Riemannian distance on M. Then there exist two closed sets $\Sigma_X, \Sigma_Y \subset M$, with $|\Sigma_X| = |\Sigma_Y| = 0$, such that $\hat{T} : M \setminus \Sigma_X \to M \setminus \Sigma_Y$ is a homeomorphism of class $C^{0,\beta}_{\mathrm{loc}}$ for any $\beta < 1$. In addition, if both f and g are of class $C^{k,\alpha}$, then $\hat{T} : M \setminus \Sigma_X \to M \setminus \Sigma_Y$ is a diffeomorphism of class $C^{k+1,\alpha}_{\mathrm{loc}}$.*

Proof (Idea of the proof of Theorem 21.6). Let x_0 be a point where the potential u associated to $\hat{T}$ is twice differentiable (since u is c-convex, one can use Alexandrov's Theorem to show that u is twice differentiable at almost every point), set $y_0 := \hat{T}(x_0)$, and assume without loss of generality that $x_0 = y_0 = 0$. Then u looks like a parabola near zero, and up to subtracting a linear function we have

$$u(x) = Mx \cdot x + o(|x|^2).$$

We now observe that the cost $c(x, y)$ is equivalent to

$$\hat{c}(x, y) := c(x, y) - c(x, 0) - c(0, y) + c(0, 0).$$

Indeed, as in the quadratic case, for any transport map T we have

$$\begin{aligned}\int_{\mathbb{R}^n} \hat{c}(x, T(x))\, d\mu(x) &= \int_{\mathbb{R}^n} c(x, T(x))\, d\mu(x) - \int_{\mathbb{R}^n} c(x, 0)\, d\mu(x) \\ &\quad - \int_{\mathbb{R}^n} c(0, T(x))\, d\mu(x) + \int_{\mathbb{R}^n} c(0, 0)\, d\mu(x) \\ &= \int_{\mathbb{R}^n} c(x, T(x))\, d\mu(x) - \int_{\mathbb{R}^n} c(x, 0)\, d\mu(x) \\ &\quad - \int_{\mathbb{R}^n} c(0, y)\, d\nu(y) + c(0, 0),\end{aligned}$$

and the last three terms are independent of T.

So, without loss of generality we can assume that $\hat{c} = c$, and by Taylor's expansion we get

$$c(x, y) = Lx \cdot y + O(|x|^2|y| + |y|^2|x|).$$

Hence, up to applying the linear transformations $x \mapsto M^{1/2}x$ and $y \mapsto -M^{-1/2}L^*y$, we can assume that

$$u(x) = \frac{|x|^2}{2} + o(|x|^2)$$

and

$$c(x, y) = -x \cdot y + O(|x|^2|y| + |y|^2|x|)$$

near $(0, 0)$.

In addition, since f and g are continuous,

$$f(x) = f(0) + \omega(|x|), \qquad g(y) = g(0) + \omega(|y|),$$

for some modulus of continuity $\omega : \mathbb{R}^+ \to \mathbb{R}^+$.

By compactness we prove the following key result:

Lemma 21.1. *Assume*

$$\left|u(x) - \frac{|x|^2}{2}\right| \le \eta \quad \textit{in } B_1,$$

$$\|c(x, y) + x \cdot y\|_{C^2(B_1 \times B_1)} \le \delta,$$

$$\|f - 1\|_{L^\infty(B_1)} + \|g - 1\|_{L^\infty(B_1)} \le \delta.$$

Then, provided $\eta > 0$ is universally small, there exists a modulus of continuity $\hat{\omega}$ such that

$$|u - \phi| \le \hat{\omega}(\delta) \quad \textit{in } B_{1/2},$$

where $\nabla\phi$ is an optimal transport map for the quadratic cost between two constant densities. In addition

$$\|\phi\|_{C^3(B_{1/2})} \le C.$$

We apply the lemma as follows: first we rescale u, c, f, g once:

$$\psi(x) \mapsto u_1(x) := \frac{u(hx)}{h^2}, \qquad c(x, y) \mapsto c_1(x, y) := \frac{1}{h^2}c(hx, hy),$$

$$f(x) \mapsto f_1(x) := f(hx), \qquad g(x) \mapsto g_1(x) := g(hx)$$

for some $h \ll 1$.

Since

$$u(x) = \frac{|x|^2}{2} + o(|x|^2),$$

for h small we have

$$\left| u_1(x) - \frac{|x|^2}{2} \right| \leq \eta \quad \text{in } B_1.$$

Thus we can apply Lemma 21.1 with $\delta = \min\{\omega(h), Ch\}$ to obtain

$$|u_1 - \phi| \leq \overline{\omega}(h) \quad \text{in } B_{1/2}, \qquad \|\phi\|_{C^3(B_{1/2})} \leq C,$$

for some modulus of continuity $\overline{\omega}$.

Let $P(x) := \frac{1}{2} D^2\phi(0)x \cdot x$. Then

$$|\phi(x) - \phi(0) - \nabla\phi(0) \cdot x - P(x)| \leq Cr^3 \quad \text{in } B_r$$

for any $r \in (0, 1/2)$, therefore

$$|u_1 - \phi(0) - \nabla\phi(0) \cdot x - P(x)| \leq \overline{\omega}(h) + Cr^3 \quad \text{in } B_r.$$

We are now in position to iterate the rescaling argument: set

$$u_2(x) := \frac{u_1(rx) - \phi(0) - \nabla\phi(0) \cdot x}{r^2} \qquad c_2(x, y) := \frac{c_1(rx, ry) - \phi(0) - \nabla\phi(0) \cdot x}{r^2},$$

$$f_2(x) := f_1(rx), \quad g_2(x) := g_1(rx)$$

Then, since $P(rx)/r^2 = P(x)$ we obtain

$$|u_2(x) - P(x)| \leq \frac{\overline{\omega}(h)}{r^2} + Cr \leq \eta \qquad \text{in } B_1$$

provided we choose first $r = r(\eta) \ll 1$ and then $h = h(r, \eta) \ll 1$. Let us observe that $P(x)$ is not exactly $|x|^2/2$ (as we would need to iterate the argument again), but we can show that it is of the form $Ax \cdot x$ for some symmetric matrix A satisfying $\lambda \mathrm{Id} \leq A \leq \Lambda \mathrm{Id}$ for some universal constants $0 < \lambda \leq \Lambda < \infty$. This is actually enough for us to keep iterating this argument and show that, for any $\alpha < 1$, there exists $C > 0$ such that

$$|u(x) - u(0) - \nabla u(0) \cdot x| \leq C|x|^{1+\alpha}.$$

Since this argument can be reapplied at any point near 0, we get $u \in C^{1,\alpha}$ in a neighborhood of 0.

This is the main step of the proof since it allows us to get rid of the local obstruction given by the failure of the MTW condition. Indeed, since u is C^1 near 0, recalling (21.14) it is easy to see that (for x in a neighborhood of 0)

$$\partial_c u(x) = \{\text{c-exp}(\nabla u(x))\},$$

in particular $\partial_c u(x)$ is connected. Relying on this, we can show that u enjoys a comparison principle, and this allows us to use a second approximation argument with solutions of the classical Monge-Ampère equation to conclude that u is $C^{2,\sigma'}$ in a smaller neighborhood for some $\sigma' > 0$. Then higher regularity follows from Schauder's theory.

These results imply that $\hat{T}$ is of class $C^{0,\beta}$ in neighborhood of $\bar{x}$ (resp. $\hat{T}$ is of class $C^{k+1,\alpha}$ if $c \in C^{k+2,\alpha}_{\text{loc}}$ and $f, g \in C^{k,\alpha}_{\text{loc}}$). Being our assumptions completely symmetric in x and y, we can apply the same argument to the optimal map T^* sending g onto f (here optimal means with respect to the cost $c^*(x, y) = c(y, x)$). Since $T^* = \hat{T}^{-1}$, it follows that $\hat{T}$ is a global homeomorphism of class $C^{0,\beta}_{\text{loc}}$ (resp. $\hat{T}$ is a global diffeomorphism of class $C^{k+1,\alpha}_{\text{loc}}$) outside a closed set of measure zero, concluding the proof.

References

1. A.D. Alexandrov, Existence and uniqueness of a convex surface with a given integral curvature. C. R. (Doklady) Acad. Sci. URSS (N. S.) **35**, 131–134 (1942)
2. Y. Brenier, Décomposition polaire et réarrangement monotone des champs de vecteurs (French). C. R. Acad. Sci. Paris Sér. I Math. **305**(19), 805–808 (1987)
3. Y. Brenier, Polar factorization and monotone rearrangement of vector-valued functions. Commun. Pure Appl. Math. **44**(4), 375–417 (1991)
4. L.A. Caffarelli, A localization property of viscosity solutions to the Monge-Ampère equation and their strict convexity. Ann. Math. (2) **131**(1), 129–134 (1990)
5. L.A. Caffarelli, Some regularity properties of solutions of Monge-Ampère equation. Commun. Pure Appl. Math. **44**(8–9), 965–969 (1991)
6. L.A. Caffarelli, The regularity of mappings with a convex potential. J. Am. Math. Soc. **5**(1), 99–104 (1992)
7. L.A. Caffarelli, Boundary regularity of maps with convex potentials. II. Ann. Math. (2) **144**(3), 453–496 (1996)
8. J.A. Cuesta-Albertos, C. Matrán, Notes on the Wasserstein metric in Hilbert spaces. Ann. Probab. **17**(3), 1264–1276 (1989)
9. G. De Philippis, A. Figalli, Partial regularity for optimal transport maps (2012, preprint)
10. G. De Philippis, A. Figalli, The Monge-Ampère equation and its link to optimal transportation. Bull. Amer. Math. Soc. (to appear)
11. A. Fathi, A. Figalli, Optimal transportation on non-compact manifolds. Isr. J. Math. **175**, 1–59 (2010)
12. A. Figalli, Regularity properties of optimal maps between nonconvex domains in the plane. Commun. Partial Differ. Equ. **35**(3), 465–479 (2010)
13. A. Figalli, Regularity of optimal transport maps [after Ma-Trudinger-Wang and Loeper]. Séminaire Bourbaki. Vol. 2008/2009. Exposés 997–1011. Astérisque **332**, Exp. No. 1009 (2010)

14. A. Figalli, Y.-H. Kim, Partial regularity of Brenier solutions of the Monge-Ampère equation. Discret. Contin. Dyn. Syst. **28**(2), 559–565 (2010)
15. A. Figalli, L. Rifford, C. Villani, Necessary and sufficient conditions for continuity of optimal transport maps on Riemannian manifolds. Tohoku Math. J. (2) **63**(4), 855–876 (2011)
16. D. Gilbarg, N.S. Trudinger, *Elliptic Partial Differential Equations of Second Order* (Springer, New York, 1983)
17. G. Loeper, On the regularity of solutions of optimal transportation problems. Acta Math. **202**(2), 241–283 (2009)
18. X.-N. Ma, N.S. Trudinger, X.-J. Wang, Regularity of potential functions of the optimal transportation problem. Arch. Ration. Mech. Anal. **177**(2), 151–183 (2005)
19. G. Monge, Mémoire sur la Théorie des Déblais et des Remblais. Hist. de l'Acad. des Sciences de Paris 666–704 (1781)
20. S.T. Rachev, L. Rüshendorf, *Mass Transportation Problems. Vol I: Theory, Vol II: Applications*. Probability and Its Applications (Springer, New York, 1998)
21. N.S. Trudinger, A note on global regularity in optimal transportation. Bull. Math. Sci. **3**, 551–557 (2013)
22. N.S. Trudinger, X.-J. Wang, On the second boundary value problem for Monge-Ampère type equations and optimal transportation. Ann. Sc. Norm. Super. Pisa Cl. Sci. (5) **8**(1), 143–174 (2009)
23. N.S. Trudinger, X.-J. Wang, On strict convexity and continuous differentiability of potential functions in optimal transportation. Arch. Ration. Mech. Anal. **192**(3), 403–418 (2009)
24. C. Villani, *Topics in Optimal Transportation*. Graduate Studies in Mathematics, vol. 58 (American Mathematical Society, Providence, 2003)
25. C. Villani, *Optimal Transport, Old and New*. Grundlehren des mathematischen Wissenschaften [Fundamental Principles of Mathematical Sciences], vol. 338 (Springer, Berlin/New York, 2009)

MIX
Papier aus verantwortungsvollen Quellen
Paper from responsible sources
FSC® C105338

If you have any concerns about our products,
you can contact us on
ProductSafety@springernature.com

In case Publisher is established outside the EU,
the EU authorized representative is:
Springer Nature Customer Service Center GmbH
Europaplatz 3, 69115 Heidelberg, Germany

Printed by Libri Plureos GmbH
in Hamburg, Germany